Efficient Technology for the Pretreatment of Biomass II

Efficient Technology for the Pretreatment of Biomass II

Editors

Ivet Ferrer
Cigdem Eskicioglu
Georgia Antonopoulou
Audrey Battimelli

MDPI • Basel • Beijing • Wuhan • Barcelona • Belgrade • Manchester • Tokyo • Cluj • Tianjin

Editors
Ivet Ferrer
Universitat Politecnica de
Catalunya
Spain

Cigdem Eskicioglu
University of British Columbia
Canada

Georgia Antonopoulou
Institute of Chemical
Engineering Sciences
(ICE-HT/FORTH)
Greece

Audrey Battimelli
Montpellier Université
France

Editorial Office
MDPI
St. Alban-Anlage 66
4052 Basel, Switzerland

This is a reprint of articles from the Special Issue published online in the open access journal *Molecules* (ISSN 1420-3049) (available at: www.mdpi.com/journal/molecules/special_issues/pretreatment_biomass_ii).

For citation purposes, cite each article independently as indicated on the article page online and as indicated below:

LastName, A.A.; LastName, B.B.; LastName, C.C. Article Title. *Journal Name* **Year**, *Volume Number*, Page Range.

ISBN 978-3-0365-1794-0 (Hbk)
ISBN 978-3-0365-1793-3 (PDF)

Contents

About the Editors . . . . **vii**

Preface to "Efficient Technology for the Pretreatment of Biomass II" . . . . **ix**

Helen Coarita Fernandez, Diana Amaya Ramirez, Ruben Teixeira Franco, Pierre Buffière and Rémy Bayard
Methods for the Evaluation of Industrial Mechanical Pretreatments before Anaerobic Digesters
Reprinted from: *Molecules* **2020**, *25*, 860, doi:10.3390/molecules25040860 . . . . **1**

Emna Feki, Audrey Battimelli, Sami Sayadi, Abdelhafidh Dhouib and Sonia Khoufi
High-Rate Anaerobic Digestion of Waste Activated Sludge by Integration of Electro-Fenton Process
Reprinted from: *Molecules* **2020**, *25*, 626, doi:10.3390/molecules25030626 . . . . **15**

Ümmihan Günerhan, Ender Us, Lütfiye Dumlu, Vedat Yılmaz, Hélène Carrère and Altınay N. Perendeci
Impacts of Chemical-Assisted Thermal Pretreatments on Methane Production from Fruit and Vegetable Harvesting Wastes: Process Optimization
Reprinted from: *Molecules* **2020**, *25*, 500, doi:10.3390/molecules25030500 . . . . **29**

Gokce Kor-Bicakci, Timothy Abbott, Emine Ubay-Cokgor and Cigdem Eskicioglu
Occurrence of the Persistent Antimicrobial Triclosan in Microwave Pretreated and Anaerobically Digested Municipal Sludges under Various Process Conditions
Reprinted from: *Molecules* **2020**, *25*, 310, doi:10.3390/molecules25020310 . . . . **51**

Georgia Antonopoulou, Dimitrios Vayenas and Gerasimos Lyberatos
Biogas Production from Physicochemically Pretreated Grass Lawn Waste: Comparison of Different Process Schemes
Reprinted from: *Molecules* **2020**, *25*, 296, doi:10.3390/molecules25020296 . . . . **71**

Marinela Zhurka, Apostolos Spyridonidis, Ioanna A. Vasiliadou and Katerina Stamatelatou
Biogas Production from Sunflower Head and Stalk Residues: Effect of Alkaline Pretreatment
Reprinted from: *Molecules* **2019**, *25*, 164, doi:10.3390/molecules25010164 . . . . **85**

Aurélie Bichot, Mickaël Lerosty, Laureline Geirnaert, Valérie Méchin, Hélène Carrère, Nicolas Bernet, Jean-Philippe Delgenès and Diana García-Bernet
Soft Microwave Pretreatment to Extract *P*-Hydroxycinnamic Acids from Grass Stalks
Reprinted from: *Molecules* **2019**, *24*, 3885, doi:10.3390/molecules24213885 . . . . **101**

Ana Sílvia de Almeida Scarcella, Alexandre Favarin Somera, Christiane da Costa Carreira Nunes, Eleni Gomes, Ana Claudia Vici, Marcos Silveira Buckeridge and Maria de Lourdes Teixeira de Moraes Polizeli
Matrix Discriminant Analysis Evidenced Surface-Lithium as an Important Factor to Increase the Hydrolytic Saccharification of Sugarcane Bagasse
Reprinted from: *Molecules* **2019**, *24*, 3614, doi:10.3390/molecules24193614 . . . . **123**

David Steinbach, Dominik Wüst, Simon Zielonka, Johannes Krümpel, Simon Munder, Matthias Pagel and Andrea Kruse
Steam Explosion Conditions Highly Influence the Biogas Yield of Rice Straw
Reprinted from: *Molecules* **2019**, *24*, 3492, doi:10.3390/molecules24193492 . . . . **139**

Yen-Keong Cheah [1], Joan Dosta [1,2,*] and Joan Mata-Álvarez[1,2,]
Enhancement of Volatile Fatty Acids Production from Food Waste by Mature Compost Addition
Reprinted from: *Molecules* **2019**, *24*, 2986, doi:10.3390/molecules24162986 . . . . **153**

Long Lin, Ehssan Hosseini Koupaie, Armineh Azizi, Amir Abbas Bazyar Lakeh, Bipro R. Dhar, Hisham Hafez and Elsayed Elbeshbishy
Comparison of Two Process Schemes Combining Hydrothermal Treatment and Acidogenic Fermentation of Source-Separated Organics
Reprinted from: *Molecules* **2019**, *24*, 1466, doi:10.3390/molecules24081466 . . . . **171**

Chrysoula Mirtsou-Xanthopoulou, Ioannis V. Skiadas and Hariklia N. Gavala
On the Effect of Aqueous Ammonia Soaking Pre-Treatment on Continuous Anaerobic Digestion of Digested Swine Manure Fibers
Reprinted from: *Molecules* **2019**, *24*, 2469, doi:10.3390/molecules24132469 . . . . **185**

Maria Patsalou, Charis G. Samanides, Eleni Protopapa, Stella Stavrinou, Ioannis Vyrides and Michalis Koutinas
A Citrus Peel Waste Biorefinery for Ethanol and Methane Production
Reprinted from: *Molecules* **2019**, *24*, 2451, doi:10.3390/molecules24132451 . . . . **197**

Shangyuan Tang, Chunming Xu, Linh Tran Khanh Vu, Sicheng Liu, Peng Ye, Lingci Li, Yuxuan Wu, Mengyu Chen, Yao Xiao, Yue Wu, Yining Wang, Qiong Yan and Xiyu Cheng
Enhanced Enzymatic Hydrolysis of *Pennisetum alopecuroides* by Dilute Acid, Alkaline and Ferric Chloride Pretreatments
Reprinted from: *Molecules* **2019**, *24*, 1715, doi:10.3390/molecules24091715 . . . . **213**

Joana R. Bernardo, Francisco M. Gírio and Rafał M. Łukasik
The Effect of the Chemical Character of Ionic Liquids on Biomass Pre-Treatment and Posterior Enzymatic Hydrolysis
Reprinted from: *Molecules* **2019**, *24*, 808, doi:10.3390/molecules24040808 . . . . **227**

About the Editors

Ivet Ferrer is Full Professor at the Department of Civil and Environmental Engineering of the Universitat Politècnica de Catalunya·BarcelonaTech. Currently, she leads the Research Group of Environmental Engineering and Microbiology (GEMMA-UPC). Her main research topic is the optimization of biomass anaerobic digestion by applying pretreatments and co-digestion. She has also addressed the implementation and assessment of low-tech digesters for rural areas. Her research is now focused on the recovery of resources from waste streams, including high-value bioproducts and biogas.

Cigdem Eskicioglu is a Full Professor and NSERC/Metro Vancouver Senior Industrial Research Chair in the area of Resource Recovery from Wastewater in the School of Engineering at University of British Columbia (UBC)'s Okanagan Campus. She is also the Founder/Leader of the Bioreactor Technology Group (BTG), which focuses on advanced wastewater treatment processes for more efficient contaminant removal, energy conservation and production, resource recovery and mitigation of trace contaminants of emerging concern.

Georgia Antonopoulou is a Senior Research Associate in the Institute of Chemical Engineering Sciences (ICE-HT) of Foundation for Research and Technology (FORTH), in Greece. She is a Chemical Engineer and has conducted her PhD and MSc at the Department of Chemical Engineer of Patras University. Her research interests focus on waste and wastewater treatment via biological processes, anaerobic digestion and fermentation, bioreactor design, pretreatment of lignocellulosic biomass and modeling of microbial processes.

Audrey Battimelli is a research engineer at the LBE, Laboratory of Environmental Biotechnology INRAE, of the French National Research Institute for Agriculture, Food and the Environment. She is the person in charge of Bio2E Platform (https://www6.montpellier.inrae.fr/narbonne/La-plateforme-BIO2E) specialised in application bioprocesses and physico-chemical pretreatments in collaborative work with professionals.

Preface to "Efficient Technology for the Pretreatment of Biomass II"

Biomass can be used as feedstock for the production of biomaterials, chemicals, platform molecules and biofuels. It is the most reliable alternative to reduce fossil fuel consumption and greenhouse gas emissions. Within the framework of the circular economy, resource recovery from organic waste, including sewage sludge, biowaste, manure and slaughterhouse waste, is particularly useful, as it helps saving resources while reducing environmental pollution. In contrast to energy crops, lignocellulosic biomass and algae do not compete for food production; therefore they represent an important source of biomass for bioenergy and bioproducts. However, biomass may require a pretreatment step in order to enhance its conversion into valuable products in terms of process yield and/or productivity. Furthermore, a pretreatment step may be mandatory for waste management (i.e. animal by-products).

Pretreatment technologies are applied upstream of various conversion processes of biomass into biofuels or biomaterials, including bioethanol, biohydrogen, biomethane, biomolecules or bioproducts. Pretreatments may include mechanical, thermal, chemical and biological techniques, which represent a crucial, cost-intensive step for the development of biorefineries. Thus, research is needed to help identify the most effective, economic, and environmentally friendly pretreatment options for each feedstock.

This Special Issue aimed to gather recent developments of biomass pretreatments for bioproducts and biofuels production.

Ivet Ferrer, Cigdem Eskicioglu, Georgia Antonopoulou, Audrey Battimelli
Editors

Article

Methods for the Evaluation of Industrial Mechanical Pretreatments before Anaerobic Digesters

Helen Coarita Fernandez, Diana Amaya Ramirez, Ruben Teixeira Franco, Pierre Buffière * and Rémy Bayard

DEEP Laboratory, Univ. Lyon, INSA Lyon, EA7429, F-69621 Villeurbanne, CEDEX, France;
helen.coarita@insa-lyon.fr (H.C.F.); diana.amaya-ramirez@etu.univ-lyon1.fr (D.A.R.);
ruben.teixeirafranco@insa-lyon.fr (R.T.F.); remy.bayard@insa-lyon.fr (R.B.)
* Correspondence: pierre.buffiere@insa-lyon.fr

Academic Editors: Mara G. Freire and Jalel Labidi
Received: 15 January 2020; Accepted: 11 February 2020; Published: 15 February 2020

Abstract: Different methods were tested to evaluate the performance of a pretreatment before anaerobic digestion. Besides conventional biochemical parameters, such as the biochemical methane potential (BMP), the methane production rate, or the extent of solubilization of organic compounds, methods for physical characterization were also developed in the present work. Criteria, such as the particle size distribution, the water retention capacity, and the rheological properties, were thus measured. These methods were tested on samples taken in two full-scale digesters operating with cattle manure as a substrate and using hammer mills. The comparison of samples taken before and after the pretreatment unit showed no significant improvement in the methane potential. However, the methane production rate increased by 15% and 26% for the two hammer mills, respectively. A relevant improvement of the rheological properties was also observed. This feature is likely correlated with the average reduction in particle size during the pretreatment operation, but these results needs confirmation in a wider range of systems.

Keywords: anaerobic digestion; mechanical pretreatments; agricultural wastes; rheology; physical properties

---

1. Introduction

In the field of anaerobic digestion, a vast number of reviews have been published in the last twenty years about substrate conditioning and pretreatments of organic biowaste [1–3], or more specifically, on different products such as municipal wastewater sludge [4], municipal organic solid wastes [5–8], lignocellulosic biomass [9–12], and agricultural wastes [13]. Even with an extensive and diverse research literature about different substrates, pretreatments, and the biomass recovery sector, research on mechanical pretreatments is still, to our knowledge, limited [14]. However, mechanical pretreatments are the most commonly used devices at full-scale [15,16]. Many functions and objectives are attributed to mechanical (physical) pretreatments. One of the most important objectives is to upgrade the digester feeding conditions in order to avoid floating layers [1] and reduce the size of materials and, consequently, to improve mixing, heat, and mass transfer [17]. Biomass pretreatment also extends to the types of feedstocks usable in anaerobic digestion The reduction in particle size leads to an increment in the accessible surface area and may increase methane production, either by increasing the methane yield or by increasing the methane production rate [18]. Nevertheless, physical pretreatments require considerable energy, ranging from 1 to 50 kWh per ton of fresh matter [3,18,19].

Different techniques are cited in the bibliography in order to perform substrate comminution, grinding or milling. Instruments, such as ball mills, knife mills, vibratory mills, hammer mills, and extruders, are used. Kratky and Jirout [3] concluded that the adequate device will mainly depend on the moisture content of the substrate. Knife and hammer mills were cited as the most suitable techniques for dry matter comminution, while extruders are more suitable for wet matter. In addition, they point out that particle size reduction will depend on the feeding system and the equipment operation (e.g., rotational rate). However, studies on mechanical devices are principally focused on energy requirements [3,20], and generally their functions and integral performance description are not explained. Thus, the performances of industrial mechanical pretreatments are mostly found in commercial booklets. In this context, knife mills are supposed to comminute and shear the biomass. Hammer mills break up the matter through a crippling effect which is a more advanced treatment. On the other hand, discs mills fragment and compress the biomass; string mills pull fibers of biomass reducing their size and cylindric mills dilacerate and decompact the biomass. There are also grinding pumps, pulpers, presses, and extruders, and their main functions are to reduce particle size and improve the fluidity. However, some mechanical pretreatments have been evaluated at full-scale such as cross-flow grinders [14,21], ball mills [22], and knife mills [23]. In these studies, the recurrent parameters investigated were the methane production yield (BMP) and the methane production rate, before and after pretreatment. Some authors also evaluated the potential release of inhibiting compounds during pretreatments through BMP tests and continuous experiments [14,15,24–26]. Tsapekos et al. [27] proposed physicochemical analysis, such as electrical conductivity tests, soluble chemical oxygen demand tests, and enzymatic hydrolysis on lignocellulosic biomass, as part of a pretreatment evaluation in order to find a correlation with the methane yield. Other methods, such as the measurement of the accessible surface area and of the water retention capacity, were suggested to evaluate the accessibility as an indicator of pretreatment efficiency [28]. The results obtained were interesting but not conclusive. By their side, Cai et al. [29] summarized some physicochemical characterization methods on lignocellulosic biomass such as particle size, grindability, density, flowability, and moisture adsorption, among others. Those analytical methods were pointed out as relevant for the process performance. Indeed, evaluation methods of a pretreatment must be in accordance with its main objectives. In summary, kinetics, biochemical, and solubilization parameters are the most common indicators to evaluate the pretreatment performance. However, physical parameters evaluations are also important, and their evaluation could be innovative to assess the benefits of the process in terms of energy consumption (lower mixing) and maintenance cost.

Parameters, such as size particle distribution, water retention capacity (WRC) and fluidity, are modified by a physical pretreatment. Water retention capacity is defined as the permanently bound amount of water attached to substrate when an external suction force is applied [30]. It is related with some properties of the substrate's nature, such as porosity, and it can be an indicator of accessibility improvement [31]. In addition, water availability is important to make a substrate more flowable [32]. A substrate's rheological properties play an important role in reactor mixing; among other parameters, homogeneity, mixing energy, and heat and mass transfer depend on these properties [33,34]. Many studies have demonstrated that rheological properties are related to a substrate's physico-chemical properties, total solids (TS) content, particle size, and temperature [35]. Further research on the influence of mechanical pretreatments at the industrial scale on the rheological behavior of untreated and treated substrates with high TS and long-sized fiber contents is important in order to evaluate their performance. Mönch-Tegeder et al. [36] reported that smaller particles resulted in an improvement of the flow behavior. Indeed, energy efficiency in anaerobic digestion from unit operations is deeply associated to flow rheology characteristics. The pattern and the specific features, such as agitators and pumps, depend on raw and digested manure's rheological behavior [37]. As an example, mixing in the reactor is an important energy factor to be considered, because it avoids significant gradients in the concentration of nutrients, inhibitor substances formation, suspended biomass and solids, temperature or pH. Besides, it limits the sedimentation of heavy solids, the formation of floating foams and scum as well as promotes the transfer of biogas [36–38].

In short, contrary to chemical, thermal or enzymatic pretreatments, mechanical pretreatments before anaerobic digestion have received very little attention by the research community. Despite this, it is clear that mechanical treatments affect the biochemical properties, soluble fractions, particle size, and rheological behavior of a given substrate.

The main novelty of the present work is the proposal of a comprehensive method to evaluate mechanical pretreatments at full-scale. In addition to biochemical parameters, physical parameters, such as size reduction, rheological properties, and water retention capacity, were evaluated. This method was tested on two anaerobic digesters using two different hammer mills as pretreatments.

2. Results and Discussion

2.1. Characterization of Incoming Products

Incoming products of both sites were relatively similar (Table 1). The TS (%TW) was 19.6% and 23.9% for sites I and II, respectively. For versus VS (%TS), a higher amount was found for Site I in relation to Site II. The pH values were similar for both sites and were typical of these kinds of substrates [14,24]. The COD and TKN concentrations were similar for Sites I and II as well. The ammonia nitrogen concentration was found to be higher in Site I, but the difference for this parameter is not uncommon [39].

Table 1. Physical and biochemical characteristics of untreated and pretreated samples.

Parameter	Unit	Site I		Site II	
		SI-US[1]	SI-PS [2]	SII-US [3]	SII-PS [4]
TS	%(TW *)	19.6 ± 1.0	18.8 ± 0.5	23.9 ± 1.1	25.8 ± 0.4
VS	% (TS)	88.4 ± 0.9	87.1 ± 0.1	83.0 ± 3.4	75.6 ± 2
pH	-	8.23	8.04	8.28	8.27
COD	mg/g VS_{RS} **	1305	1317	1215	1219
TKN	mg/g VS_{RS} **	27.0	25.7	21.0	23.5
NH_4-N	mg/g VS_{RS} **	6.2	3.9	1.7	1.5
VFA	mg/gVS_{RS} **	1.7	24.7	4.3	2.7
BMP	mL/gVS_{R3} **	275 ± 7	269 ± 10	199 ± 15	224 ± 7
k	j^{-1}	0.066 ± 0.005	0.077 ± 0.004	0.055 ± 0.004	0.070 ± 0.004

[1] SI-US: untreated samples; [2] SI-PS: pretreated samples; [3] SII-US: untreated samples; [4] SII-PS: pretreated samples.
* TW: total weight; ** RS: raw sample.

2.2. Effect of Mechanical Pretreatments on Biochemical Characteristics

An example of the methane production is shown in Figure 1. The methane production followed a first-order kinetics for all samples. The biochemical methane potential of raw cattle manure was much higher in Site I than in Site II (275 and 199 NL/gVS, respectively). This indicates that the manure of Site II is probably stored longer before use (which is confirmed by the elevated amount of ammonia nitrogen). However, the single factor ANOVA test found no significant differences between the BMP values of untreated and pretreated samples for both sites. This shows that mechanical pretreatments had no significant impact on the intrinsic biodegradability of the substrates. Divergent results can be found in the literature. Indeed, previous works made at full-scale found an enhancement of methane production by 10% with horse manure as a substrate [21]. Likewise, an enhancement of 26% on methane production was also reported with a cross-flow grinder pretreatment for horse manure [14]. Other available data reports an enhancement between 15% and 45% using different mechanical devices with manure fibers as substrate at laboratory-scale [15]. Dahunsi [40] found an average improvement of 22% of the BMP on different types of lignocellulosic substrates. Overall, the substrate biodegradability with mechanical pretreatments application depends on different factors such as the substrate nature and mechanical devices and their specificities.

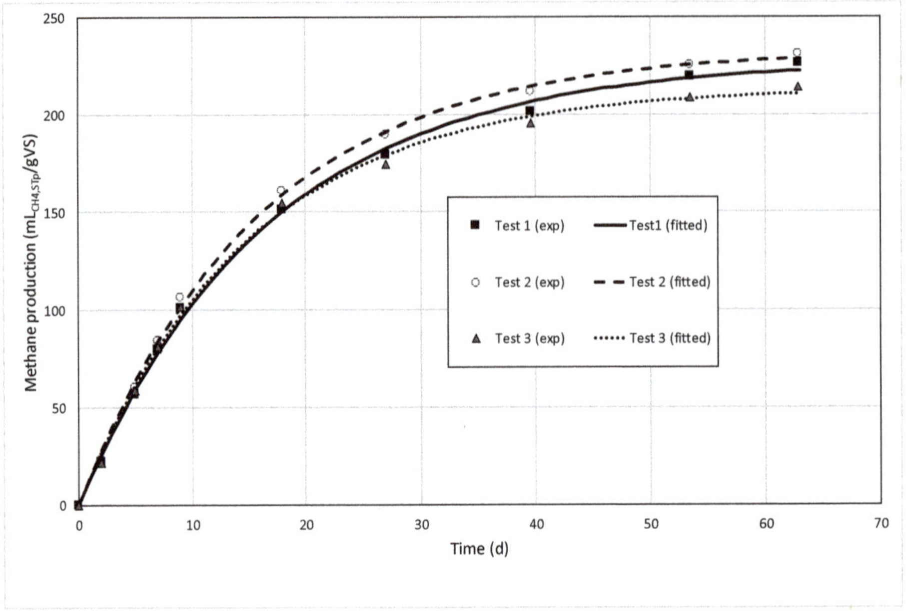

Figure 1. Methane production versus time for the biochemical methane potential (BMP) tests performed on Site II, after pretreatments (triplicate tests). The fitted curves represent the fitted first-order kinetic model.

The first-order kinetic constant k measured during the BMP tests was considered an indicator of the methane production rate. This parameter was improved by the mechanical pretreatments. Single factor ANOVA tests found significant differences in k for the untreated and treated samples for both sites. An improvement of 15% and 27% with the pretreatment application was found in Site I and Site II, respectively. Published results with manure as a substrate and mechanical pretreatments at full-scale are scarce and difficult to compare. Mönch-Tegeder et al. [21] pointed out a small effect on degradation kinetics with a cross-flow grinder device application at full-scale. These results agreed with Moset et al. [41], who found similar k values between excoriating and chopping grass pretreatment, unfortunately untreated samples were not evaluated. For their part, Herrmann et al. [25] described an enhancement of approximately 18% on the methane production rate with a chopping pretreatment of lignocellulosic feedstocks at farm-scale. In contrast, a remarkable enhancement (approximately 43%) was found by Dumas et al. [23] at the laboratory-scale with wheat straw mechanical pretreatment. In short, the methane production rate not only depends on the size reduction but also on other parameters such as the fiber structure and the chemical characteristics [42,43].

Probably, laboratory-scale devices provide higher energy levels than full-scale ones which results in a higher extent of biomass degradation and, consequently, a higher impact on both the BMP and on the methane production rate. For instance, a 500 W laboratory-scale blender treating 1 kg of manure in 5 min delivers around 42 kWh/t of fresh manure which is four times higher than the energy consumed by the hammer mills (10 kWh/t). In the work by Dahunsi, the energy required for the pretreatment comprised between 300 and 350 kWh/t (of total solids), while in our case, at real scale, the energy was lower (10 kWh/t of fresh matter at 20% TS was 50 kWh/t of TS).

Volatile fatty acids (VFAs) content increased with pretreatment for Site I, from 1.1 to 24.7 mg/gVS in contrast to Site II which presented almost no difference. However, this increase was less than 3% of the initial volatile solid content. No losses of VFA were reported through warming by mechanical pretreatments of lignocellulosic silage and manure [14,22]. An increment of VFA concentration could be attributed to a better VFA solubilization [22,26]. In addition, the substrate nature's, such as the initial proteins contents, and its fermentation may affect VFA concentration [26]. In all samples, no soluble sugars were detected.

2.3. *Effect of Mechanical Pretreatments on Physical Characteristics*

2.3.1. Size Reduction

Figure 2 shows the particle size distributions of the untreated and treated samples for Site I and Site II. For Site I, almost 50% of the total solids were composed of fibers larger than 31.5 mm. The second largest proportion was fibers shorter than 0.25 mm (approximately 15%) followed by fibers between 20 and 10 mm (approximately 11%); the other measured proportions (20–31.5 mm, 4–10 mm, 1–4 mm, 0.5–1 mm, 0.25–0.5 mm) were less than 10%. For Site II, the same proportions (approximately 25%) were found on samples higher than 31.5 mm, between 31.5 and 20 mm, and lower than 0.25 mm. Fibers between 10 and 20 mm accounted for more than 12% of the total amount of solids. Thus, initial differences may be found between the untreated Site I and Site II samples' fiber sizes. Site I presented a bigger fraction of larger fibers than Site II. Nevertheless, similarities were also found between the two evaluated samples; the biggest proportion for each sample were fibers longer than 31.5 mm and shorter than 0.25 mm.

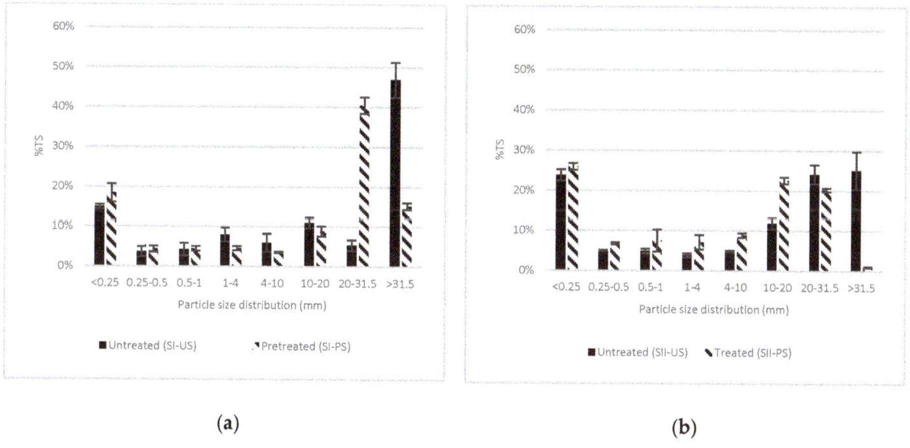

(**a**) (**b**)

Figure 2. Particle size reduction before and after hammer milling pretreatment: (**a**) Site I; (**b**) Site II.

Size reduction behavior after mechanical pretreatment was relatively similar for both sites. Site I sample fibers >31.5 mm were mainly reduced to 20–31.5 mm in size. Indeed, the decrease of longest measured fibers was approximately 22%, and the amount of fibers between 20 and 31.5 mm in length increased by approximately 35%. For Site II, the fibers >20 mm were mainly reduced to 10–20 mm in size. Initially, fibers longer than 31.5 mm were reduced by more than 20%. Then, fibers between 20 and 31.5 mm were reduced, and the amount of fibers between 10 and 20 mm increased by 10%. The proportion of fibers sized 4–10 mm, 1–4 mm, 0.5–1 mm, 0.25–0.5 mm, and lower than 0.25 mm were in all the cases reduced to a smaller size. In short, the reduction in particle size was produced mainly on longer fibers and a slight increase on the smallest particles was found with mechanical pretreatments. Sun and Cheng [44] pointed out a reduction of lignocellulosic substrate particle size between 10 and 30 mm with a chipping pretreatment and 0.2–2 mm with a grinding or milling operation. Indeed, particle size reduction depends on many factors related to the mechanical device and its operation in addition to the substrate's properties such as humidity and initial particle size [3,10,25,45].

Furthermore, size reduction is commonly reported as a factor explaining higher methane production yields and methane production rates due to the increase in the accessible surface area [46,47].

2.3.2. Water Retention Capacity

The WRC results were relatively similar for Site I and Site II (Figure 3). On the one hand, the WRC increased with the pretreatment for Site I from 5.6 ± 0.2 g water/TS to 6.7 ± 0.3 g water/TS (around +19%). On the other hand, for Site II, no significant changes were obtained (WRC of 5.6 ± 0.3 g water/TS to 5.2 ± 0.1 g water/TS). The ANOVA tests found significant differences for WRC values for Site I. These results suggest that the pretreatment improved the water retention capacity as consequence of particle size reduction [32]. In contrast, an excess of particle size reduction could lead to a decrease of WRC. Indeed, water behavior was not changed for Site II, and it could probably be due to the fact of its higher amount of particles smaller than 0.25 mm and contents with shorter fibers than Site I. Dumas et al. [23] noted that water behavior does not change considerably if the particle size becomes lower. Other authors [48,49] have suggested improvements in terms of biodegradability when the WRC increases, because it may favor nutrients and substrate dissolution and diffusion. No relation was found between biodegradability and WRC in our case.

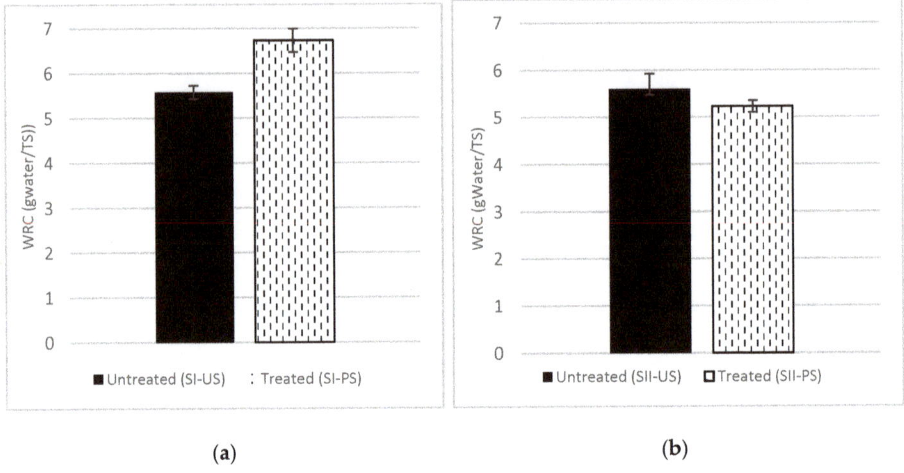

(a)   (b)

Figure 3. Water retention capacity of untreated and treated samples: (**a**) Site I; (**b**) Site II.

2.3.3. Rheological Properties

Yield stress and apparent viscosity decreased when mechanical pretreatments were applied at similar TS contents of untreated and treated samples (Table 2). A reduction in the yield stress of 63% and 18% was observed for Sites I and II, respectively. SI-US apparent viscosity in Site I could not be measured due to the equipment limitations. In Site II, a reduction of 85% of the apparent viscosity was recorded with the pretreatment. The difference among the results of each site can be explained by the fact that the rheological properties are also linked to particle size distribution [50]. Particle sizes > 31.5 mm (Figure 2) were approximately 20% higher at Site I than Site II before pretreatment. The yield stress values were three times higher for Site I than Site II at similar TS. According to Tian et al. [38], the required energy for appropriate mixing of the digester may be reduced by 9.2% with a particle size reduction from 20 to 80 mesh (with corn stover as a substrate). In addition, the mixing energy is (in first approximation) directly proportional to the apparent viscosity of the digester content [51]. Thus, it was found that higher contents of long particle size increased the yield stress and viscosity values, and mechanical device application had a positive impact on rheological properties. A similar trend was reported with tests made with agricultural wastes as a substrate by different authors [14,38,44].

In addition, flow and viscosity behavior did not necessarily correspond at similar biomass TS values according to published data [33,35].

Table 2. Rheological properties for untreated and treated samples.

Parameter	Unit	Site I		Site II	
		Untreated *	Pretreated *	Untreated **	Pretreated **
τ_y	Pa	755.3 ± 18.4	281.0 ± 4.2	250.7 ± 6.6 **	205.0 ± 16.1
η_{app}	Pa.s	ND ***	28.6 ± 3.9	38.7 ± 2.1	5.8 ± 0.2

* TS = 12%; ** TS = 11%; *** ND = non-determined.

As mentioned previously, the reduction in the yield stress and viscosity values could mean an important reduction of energy consumption in terms of pumping and mixing at full-scale sites [36,52]. Moreover, the relationship between high solids concentration and different particle sizes should be investigated more in detail for a better comprehension of rheological properties, biodegradability, hydrophilicity, and mechanical pretreatments influence.

3. Materials and Methods

3.1. Biomass Sampling and Handling

Two full-scale plants with mechanical pretreatment operations were selected for this study. In both units, cattle manure was the main feedstock.

Site I: The mechanical pretreatment was a mobile *hammer mill*. Samples were collected before and after the hammer mill device and just before the anaerobic digestion tank. Untreated and treated samples were stored at 4 °C and analyzed within a few days. At this site, a mesophilic digester is operated. Mechanical pretreatment is used to improve the feeding system and to avoid digester difficulties such as clogging.

Site II: A fixed (in-line) *hammer mill* was evaluated. Samples were collected before and after the mechanical pretreatment operation. This site operates with a mesophilic digester and produces heat and electricity from biogas. This mechanical operation is mainly used to reduce the fiber size. At each site, 50 kg of untreated and 50 kg of treated manure was sampled. The samples were transported and stored in a 4 °C cold chamber and processed less than 48 h after storage for further analysis.

3.2. Biochemical Characterization

The biochemical characterization of the samples followed the procedure suggested by Teixeira Franco et al. [53]. A fractionation of the sample was made through a leaching test with water. The leaching test was performed with 10:1 water/dry matter proportion over 2 h under a constant bottle rotation. Subsequently, the sample was centrifuged (20 min at $5000\times g$). The liquid fraction was filtrated (1.2 µm), while the particulate fraction was dried at 70 °C and ground to 1 mm length. The measurements carried out on the biomass samples are described below.

Analyses on the raw (untreated) sample: parameters as total solids (TS); volatile solids (VS); and biochemical methane potential (BMP) test assays were carried out. The samples were dried at 105 °C with a drying oven in order to measure TS; afterwards, samples were burned at 550 °C then versus were calculated.

Analyses on the liquid fraction: liquid fraction was used to measure pH, dissolved chemical organic demand (COD); ammonia nitrogen (NH_3-N); total Kjeldahl nitrogen (TKN); volatile fatty acids (VFAs); BMP; TS; and VS. The pH was measured with a consort C3020 device with SP10B. Lactic and formic acids were analyzed with a high-performance liquid chromatography (LC Module 1 plus, Waters) equipped with a Supelcogel™ C-610 H column (300 mm × 7.8 mm, Sigma–Aldrich), with both refractive index (RID) and UV detectors and operating with H_3PO_4 0.1%v/v as solvent (flow rate of 0.5 mL/min). Acetic, propionic, butyric, valeric, and caproic acids contents were analyzed by gas chromatography (Shimadzu Corp.) with an HP-FFAP fused silica capillary column (30 m × 0.25 mm, Agilent Technologies), a flame ionization detector, and H_2 as the carrier gas. The sum of lactic, formic, acetic, propionic, butyric, valeric, and caproic acids was considered as the total VFA. The TKN and NH_3-N were determined according to the NF EN 25,663 standard procedures.

Analyses on the particulate fraction: Parameters, such as TKN, COD, total solids (TS) and volatile solids (VS) were determined.

The BMP tests were performed according the guidelines given by Holliger et al. [54]. Batch assays were prepared in 1 L glass bottles in a 35 °C temperate room. The inoculum used was a digested sludge from a wastewater treatment plant (La Feyssine, Lyon, France). The total versus content was 20 g/L (sample + inoculum), and the substrate to inoculum ratio was 0.5 on VS basis. A mineral solution was added (as recommended by the ISO 117734:1995 standard). The mineral solution contained essential elements in order to give optimal conditions to the microbial growth and act as a buffer solution. Then glass bottles were purged with 80/20%v/v N_2/CO_2 and finally incubated. Control assays without substrate were performed as well in order assess the background methane production from inoculum. The gas production was followed by manometric measurements (Digitron precision manometer). The biogas was vented when the pressure went above 1200 hPa. The biogas composition was analyzed with an Agilent 300 micro gas chromatography with a thermal conductivity detector (GC-TCD). Argon and helium were used as carrier gases and Molsieve 5 A (14 m length; pore size: 5 Å) and PoraPlot (10 m length; 0.320 mm ID) columns as stationary phases. Each sample was tested in triplicate.

The background methane production (from inoculum) was retrieved, and the methane volume produced was normalized and expressed in ml standard conditions for temperature and pressure (0 °C, 1 atm) per gram of volatile solid. From the methane production, a first-order model was fitted to the data in order to get further information from the experiments:

$$V_{CH4}(t) = BMP_{max}\left(1 - e^{-kt}\right) \tag{1}$$

- the final (or ultimate) maximal biochemical methane potential, BMP_{max};
- the first-order kinetic parameter k that can be compared prior and after the mechanical pretreatment as an indicator of methane production rate improvement.

3.3. Physical Characterization

3.3.1. Size Reduction Evaluation

Particle size distribution was measured to evaluate the effects of the mechanical devices on particle size. Different procedures are proposed in the literature. Herrman et al. [25] used a quantification of ensiled feedstock by image analyses with 10 to 20 g of sample. Samples were weighted, scanned, and evaluated with an image analyzer software. In contrast, Mönch-Tegeder et al. [21] performed a wet sieving for horse manure and ensiled feedstocks at a determined size and then evaluated them with an image analyzer software. Lindner et al. [22] also used a wet sieving with a vibrating sieve shaker in order to verify the size reduction of digestate composed of cattle manure and silage. Samples were soaked in distilled water, and then samples were sieved (10 min), weighted, and dried. On the other hand, Tsapekos et al. [27] used a visual method to analyze ensiled meadow grass particle distribution. Indeed, these procedures depend on the substrate conditions and research aim for evaluating different particle sizes. Sieving is often used to measure particle size distribution due to the fact of their ample range. Nevertheless, some caution must be taken to prevent from clogging, which would result lower amounts of small particles [55].

The procedure carried out in this work was also a wet sieving method with a vibrating sieve shaker (Retsch AS 200 basic) with water recirculation. The sieve openings were 0.25 mm, 0.5 mm, 1.0 mm, 4.0 mm, 10.0 mm, 20.0 mm, and 31.5 mm. First, 50–80 g of samples was weighted, disaggregated, and carefully introduced into the upmost sieve (31.5 mm). The amount of sample was chosen in order to prevent any clogging of the sieves. Then, samples were sieved with 1 L of recirculated demineralized water for 15 min. A flow of 938 mL/min until a 40 amplitude vibration was performed. Water was retrieved, and it was reinjected for 5 min of sieving. All fractions were recovered with aluminum paper (weighted previously) and dried over 24 h at 105 °C. The size distributions were calculated in %TS, and their proportion corresponded to the total dried weight. For each sample, this operation was done in triplicate.

3.3.2. Rheological Properties

According to different authors, fermenting agricultural biomasses has been shown to have a rheofluidifying behavior characteristic of non-Newtonian fluids [33,35–37,56] where viscosity decreases non-linearly under a shear stress. Conventional methods for measuring rheological properties are mostly performed with rotational or tubular rheometers in order to build a rheogram and, consequently, to identify the behavior fluid model [56]. These measurements relate shear stress in Pascal (Pa) and shear rate (s^{-1}) depending on the force applied to the fluid [52]. In addition, other characteristics may influence the rheological properties as the temperature, the TS content, the particle size, and the presence of dispersed gas bubbles [36]. Indeed, rotary and tubular devices are not adapted for long and heterogeneous straw fibers present in cattle manure. Their size and capacity to measure only small and homogeneous quantities of sample are not representative at full-scale. By consequence, these measurements with heterogeneous substrate are difficult to investigate. Thus, information about measurement methods of rheological properties on heterogeneous samples with important fiber sizes at the industrial scale is still limited [42,44].

A few studies in this field were found in the literature, one of them by Mönch-Tegeder et al. [36], who developed an in-line process viscometer at full-scale in order to determine the rheological properties of untreated and pretreated mixed substrates with grass silage, maize silage, and solid and liquid manure. The viscosity increased with the fibrous material, and the substrate's mechanical disintegration positively influenced the rheological properties; however, an important effect of higher TS was pointed out.

Moreover, Ruys [50] proposed a dedicated rheometer of dimensions larger than commercial ones and adapted to heterogeneous substrates before anaerobic digestion at industrial scale. However, this kind of prototype is not commercially available; moreover, it is complex to operate at industrial scale. In this context, Garcia-Bernet et al. [48] developed a fast method to measure the yield stress for anaerobically digested solid waste with a slump test. It involves a cylindrical shape PVC chamber with a 10 cm diameter and 18 cm height; this chamber is filled with the sample, then it is quickly lifted vertically to allow the medium to collapse, and the difference between the initial and final heights is termed the "slump".

We tested two devices commonly used for fresh concrete characterization. These devices are able to measure the yield stress [57] and the apparent viscosity [58] using a slump test and a V funnel test, respectively. Yield stress value represents the minimum stress to cause the fluid flow, and its measurement is essential for understanding flow properties [57]. Apparent viscosity represents a fluid's property to resist forces causing it to flow; thus, a flow's velocity is controlled by these internal resistances of the fluid [52]. These internal variations become a key factor during the application of a force on the fluid for its correct handling and management [59].

The slump and V funnel tests were carried out according to the PR NF EN 12350-2 and PR NF EN 12350-9 standards used for fresh concrete, respectively, and the TS content of the samples was standardized to 11%–12% approximately. The experiments were performed at 20 °C. The slump test was carried out with an Abram cone, and the results were evaluated with a cylindrical analytical model as Pashias et al. [57] suggested, where the slump height is related to yield stress. Dimensionless height can be expressed by Equation (1), where dimensionless slump height is defined as $s' = s/H$, dimensionless yield stress is defined by $\tau'_y = \tau_y / \rho g H$, and H is the cone's height, $H = 30$ cm.

$$s' = 1 - 2\tau'_y \left[1 - Ln\left(\tau'_y \right) \right] \tag{2}$$

Apparent viscosity η_{app} was calculated with Equations (3)–(5) described by Mokéddem [58], where dP is the difference in pressure between the V funnel input and output, t_{VF} is the flow time through the V funnel in seconds (s), V_{VF} is the V funnel volume (10 L), and L_{VF}, O, z_a, and z_b are the V funnel dimensions as shown in Figure 4.

$$dP_{mean} = \frac{dP_{max}}{2} \tag{3}$$

$$dP_{max} = \rho g (z_b - z_a) \tag{4}$$

$$\eta_{app} = \frac{dP_{mean} t_{VF} O^4}{24 L_{VF} V_{VF}} \tag{5}$$

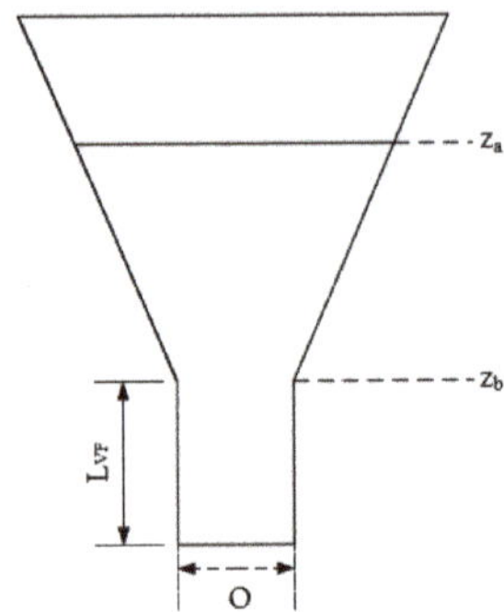

Figure 4. Schematic view of the V Funnel adapted from Mokéddem (2012) [47].

3.3.3. Water Retention Capacity (WRC)

There are various methods to measure the water retention capacity of a given sample; and most of them have been used in the food industry [30,60,61] and for lignocellulosic biomass [23,32]. A centrifugation of samples with distilled water was applied in most of the cases. Sanchez et al. [32] assembled a vacuum system with a Buchner funnel, filter paper, and a Kitasato flask. Regarding the evaluated substrate, a procedure considering a higher mass amount was privileged. Similarly, we used a column with a filter cloth system and a vacuum pump inserted into an Erlenmeyer flask, and 250–500 g of raw sample and distilled water were inserted into the column. Water was carefully added in order to have total contact with the sample at ambient temperature for 2 h. Water was pulled out via gravity by opening the tap and under slight vacuum for 120 s. The measurement was calculated as suggested by Raghavendra et al. [61]; the water quantity was retained over the TS substrate. Analyses were performed in triplicate for all samples.

4. Conclusions

The present work developed a comprehensive methodology to investigate the effect of mechanical pretreatments in the context of agricultural waste anaerobic digestion including biochemical and physical characteristics. Conventional methods included solid/liquid separation, BMP measurement, and solubilization characteristics. Physical methods included particle size distribution, water retention capacity, and rheological characteristics. Our methodology was tested on sites with two different hammer mills.

The results showed no improvement in the methane yield (BMP) before and after the pretreatments. However, the methane production rate increased significantly (+15% and +27% for Sites I and II, respectively). These results were quite in line with the increase in water retention capacity, indicating that the organic matter was somewhat more easily accessible to hydrolysis.

Rheological properties were assessed by rapid tests (the Abram cone and the V funnel) that can be easily used on site. Pretreatments improved both the yield stress (−18% and −63% for Sites I and II, respectively) and the apparent viscosity (−85% for Site II). The average particle size reduction played a significant role in these improvements. Since rheological parameters are directly linked with the energy required for digester mixing, these tests are very promising tools for further evaluation of pretreatments at full scale.

Author Contributions: H.C.F. contributed to the experimental part and to the writing of the manuscript. D.A.R. contributed to the development and to the implementation of the rheological methods. R.T.F. has developed the experimental procedure for the biochemical characterization and contributed to the analysis of the results. P.B. contributed to the design of the experiment and to the final writing of the manuscript. R.B. is the project leader, he contributed to the design of the experiments, to the experimental investigations, and to the final writing of the manuscript. All authors have read and agreed to the published version of the manuscript.

Funding: This research was funded by the French National Environmental Agency (ADEME) as part of the PAM program (Pretreatment before Anaerobic Digestion), grant number 1706C0013. This work was performed within the framework of the EUR H$_2$O'Lyon (ANR-17-EURE-0018) of Université de Lyon (UdL), within the program "Investissements d'Avenir" operated by the French National Research Agency (ANR). Helen Coarita Fernandez held a doctoral fellowship from Ministerio de Educacion del Estado Plurinacional de Bolivia.

Acknowledgments: The authors wish to thank Nathalie Dumont and David Lebouil for their help on the analytical tests and Hervé Perier-Camby for their contribution to the rheological tests.

Conflicts of Interest: The authors declare no conflict of interest.

References

1. Carrere, H.; Antonopoulou, G.; Affes, R.; Passos, F.; Battimelli, A.; Lyberatos, G.; Ferrer, I. Review of feedstock pretreatment strategies for improved anaerobic digestion: From lab-scale research to full-scale application. *Bioresour. Technol.* **2016**, *199*, 386–397. [CrossRef] [PubMed]

2. Patinvoh, R.J.; Osadolor, O.A.; Chandolias, K.; Sárvári Horváth, I.; Taherzadeh, M.J. Innovative pretreatment strategies for biogas production. *Bioresour. Technol.* **2017**, *224*, 13–24. [CrossRef] [PubMed]

3. Kratky, L.; Jirout, T. Biomass Size Reduction Machines for Enhancing Biogas Production. *Chem. Eng. Technol.* **2011**, *34*, 391–399. [CrossRef]

4. Carrère, H.; Dumas, C.; Battimelli, A.; Batstone, D.J.; Delgenès, J.P.; Steyer, J.P.; Ferrer, I. Pretreatment methods to improve sludge anaerobic degradability: A review. *J. Hazard. Mater.* **2010**, *183*, 1–15. [CrossRef]

5. Ma, J.; Duong, T.H.; Smits, M.; Verstraete, W.; Carballa, M. Enhanced biomethanation of kitchen waste by different pre-treatments. *Bioresour. Technol.* **2011**, *102*, 592–599. [CrossRef]

6. Ariunbaatar, J.; Panico, A.; Esposito, G.; Pirozzi, F.; Lens, P.N.L. Pretreatment methods to enhance anaerobic digestion of organic solid waste. *Appl. Energy* **2014**, *123*, 143–156. [CrossRef]

7. Kondusamy, D.; Kalamdhad, A.S. Pre-treatment and anaerobic digestion of food waste for high rate methane production—A review. *J. Environ. Chem. Eng.* **2014**, *2*, 1821–1830. [CrossRef]

8. Zhang, C.; Su, H.; Baeyens, J.; Tan, T. Reviewing the anaerobic digestion of food waste for biogas production. *Renew. Sustain. Energy Rev.* **2014**, *38*, 383–392. [CrossRef]

9. Mosier, N. Features of promising technologies for pretreatment of lignocellulosic biomass. *Bioresour. Technol.* **2005**, *96*, 673–686. [CrossRef]

10. Taherzadeh, M.; Karimi, K. Pretreatment of Lignocellulosic Wastes to Improve Ethanol and Biogas Production: A Review. *Int. J. Mol. Sci.* **2008**, *9*, 1621–1651. [CrossRef]

11. Barakat, A.; de Vries, H.; Rouau, X. Dry fractionation process as an important step in current and future lignocellulose biorefineries: A review. *Bioresour. Technol.* **2013**, *134*, 362–373. [CrossRef]

12. Shrestha, S.; Fonoll, X.; Khanal, S.K.; Raskin, L. Biological strategies for enhanced hydrolysis of lignocellulosic biomass during anaerobic digestion: Current status and future perspectives. *Bioresour. Technol.* **2017**, *245*, 1245–1257. [CrossRef]

13. Paudel, S.R.; Banjara, S.P.; Choi, O.K.; Park, K.Y.; Kim, Y.M.; Lee, J.W. Pretreatment of agricultural biomass for anaerobic digestion: Current state and challenges. *Bioresour. Technol.* **2017**, *245*, 1194–1205. [CrossRef] [PubMed]

14. Mönch-Tegeder, M.; Lemmer, A.; Oechsner, H. Enhancement of methane production with horse manure supplement and pretreatment in a full-scale biogas process. *Energy* **2014**, *73*, 523–530. [CrossRef]

15. Tsapekos, P.; Kougias, P.G.; Frison, A.; Raga, R.; Angelidaki, I. Improving methane production from digested manure biofibers by mechanical and thermal alkaline pretreatment. *Bioresour. Technol.* **2016**, *216*, 545–552. [CrossRef] [PubMed]

16. Lindmark, J.; Leksell, N.; Schnürer, A.; Thorin, E. Effects of mechanical pre-treatment on the biogas yield from ley crop silage. *Appl. Energy* **2012**, *97*, 498–502. [CrossRef]

17. Neshat, S.A.; Mohammadi, M.; Najafpour, G.D.; Lahijani, P. Anaerobic co-digestion of animal manures and lignocellulosic residues as a potent approach for sustainable biogas production. *Renew. Sustain. Energy Rev.* **2017**, *79*, 308–322. [CrossRef]

18. Zheng, Y.; Zhao, J.; Xu, F.; Li, Y. Pretreatment of lignocellulosic biomass for enhanced biogas production. *Prog. Energy Combust. Sci.* **2014**, *42*, 35–53. [CrossRef]

19. Hendriks, A.T.W.M.; Zeeman, G. Pretreatments to enhance the digestibility of lignocellulosic biomass. *Bioresour. Technol.* **2009**, *100*, 10–18. [CrossRef]

20. Schell, D.J.; Harwood, C. Milling of lignocellulosic biomass: Results of pilot-scale testing. *Appl. Biochem. Biotechnol.* **1994**, *45–46*, 159–168. [CrossRef]

21. Mönch-Tegeder, M.; Lemmer, A.; Jungbluth, T.; Oechsner, H. Effects of full-scale substrate pretreatment with a cross-flow grinder on biogas production. *Agric. Eng. Int. CIGR J.* **2014**, *16*, 138–147.

22. Lindner, J.; Zielonka, S.; Oechsner, H.; Lemmer, A. Effects of mechanical treatment of digestate after anaerobic digestion on the degree of degradation. *Bioresour. Technol.* **2015**, *178*, 194–200. [CrossRef] [PubMed]

23. Dumas, C.; Silva Ghizzi Damasceno, G.; Barakat, A.; Carrère, H.; Steyer, J.-P.; Rouau, X. Effects of grinding processes on anaerobic digestion of wheat straw. *Ind. Crops Prod.* **2015**, *74*, 450–456. [CrossRef]

24. Kalamaras, S.D.; Kotsopoulos, T.A. Anaerobic co-digestion of cattle manure and alternative crops for the substitution of maize in South Europe. *Bioresour. Technol.* **2014**, *172*, 68–75. [CrossRef] [PubMed]

25. Herrmann, C.; Heiermann, M.; Idler, C.; Prochnow, A. Particle Size Reduction during Harvesting of Crop Feedstock for Biogas Production I: Effects on Ensiling Process and Methane Yields. *BioEnergy Res.* **2012**, *5*, 926–936. [CrossRef]

26. De la Rubia, M.A.; Fernández-Cegrí, V.; Raposo, F.; Borja, R. Influence of particle size and chemical composition on the performance and kinetics of anaerobic digestion process of sunflower oil cake in batch mode. *Biochem. Eng. J.* **2011**, *58–59*, 162–167. [CrossRef]

27. Tsapekos, P.; Kougias, P.G.; Angelidaki, I. Biogas production from ensiled meadow grass; effect of mechanical pretreatments and rapid determination of substrate biodegradability via physicochemical methods. *Bioresour. Technol.* **2015**, *182*, 329–335. [CrossRef]

28. Karimi, K.; Taherzadeh, M.J. A critical review on analysis in pretreatment of lignocelluloses: Degree of polymerization, adsorption/desorption, and accessibility. *Bioresour. Technol.* **2016**, *203*, 348–356. [CrossRef]

29. Cai, J.; He, Y.; Yu, X.; Banks, S.W.; Yang, Y.; Zhang, X.; Yu, Y.; Liu, R.; Bridgwater, A.V. Review of physicochemical properties and analytical characterization of lignocellulosic biomass. *Renew. Sustain. Energy Rev.* **2017**, *76*, 309–322. [CrossRef]

30. Raghavendra, S.N.; Ramachandra Swamy, S.R.; Rastogi, N.K.; Raghavarao, K.S.M.S.; Kumar, S.; Tharanathan, R.N. Grinding characteristics and hydration properties of coconut residue: A source of dietary fiber. *J. Food Eng.* **2006**, *72*, 281–286. [CrossRef]

31. Williams, D.L.; Hodge, D.B. Impacts of delignification and hot water pretreatment on the water induced cell wall swelling behavior of grasses and its relation to cellulolytic enzyme hydrolysis and binding. *Cellulose* **2014**, *21*, 221–235. [CrossRef]

32. Sanchez, A.; Hernández-Sánchez, P.; Puente, R. Hydration of lignocellulosic biomass. Modelling and experimental validation. *Ind. Crops Prod.* **2019**, *131*, 70–77. [CrossRef]

33. Daniel, J. O'Neil: Rheology and Mass/Heat Transfer Aspects of Anaerobic Reactor Design. *Biomass* **1985**, *8*, 205–216. [CrossRef]

34. Miryahyaei, S.; Olinga, K.; Abdul Muthalib, F.A.; Das, T.; Ab Aziz, M.S.; Othman, M.; Baudez, J.C.; Batstone, D.; Eshtiaghi, N. Impact of rheological properties of substrate on anaerobic digestion and digestate dewaterability: New insights through rheological and physico-chemical interaction. *Water Res.* **2019**, *150*, 56–67. [CrossRef] [PubMed]

35. Björn, A.; Karlsson, A.; Ejlertsson, J.; Svensson, B.H. Rheological Characterization. *Biogas* **2012**, *1*, 63–76.

36. Mönch-Tegeder, M.; Lemmer, A.; Hinrichs, J.; Oechsner, H. Development of an in-line process viscometer for the full-scale biogas process. *Bioresour. Technol.* **2015**, *178*, 278–284. [CrossRef]
37. Hreiz, R.; Adouani, N.; Fünfschilling, D.; Marchal, P.; Pons, M.-N. Rheological characterization of raw and anaerobically digested cow slurry. *Chem. Eng. Res. Des.* **2017**, *119*, 47–57. [CrossRef]
38. Tian, L.; Shen, F.; Yuan, H.; Zou, D.; Liu, Y.; Zhu, B.; Li, X. Reducing agitation energy-consumption by improving rheological properties of corn stover substrate in anaerobic digestion. *Bioresour. Technol.* **2014**, *168*, 86–91. [CrossRef]
39. Chen, Y.; Cheng, J.J.; Creamer, K.S. Inhibition of anaerobic digestion process: A review. *Bioresour. Technol.* **2008**, *99*, 4044–4064. [CrossRef]
40. Dahunsi, S.O. Mechanical pretreatment of lignocelluloses for enhanced biogas production: Methane yield prediction from biomass structural components. *Bioresour. Technol.* **2019**, *280*, 18–26. [CrossRef]
41. Moset, V.; Fontaine, D.; Møller, H.B. Co-digestion of cattle manure and grass harvested with different technologies. Effect on methane yield, digestate composition and energy balance. *Energy* **2017**, *141*, 451–460. [CrossRef]
42. Zhang, Y.; Banks, C.J. Impact of different particle size distributions on anaerobic digestion of the organic fraction of municipal solid waste. *Waste Manag.* **2013**, *33*, 297–307. [CrossRef] [PubMed]
43. Palmowski, L.M.; Müller, J.A. Anaerobic degradation of organic materials—Significance of the substrate surface area. *Water Sci. Technol.* **2003**, *47*, 231–238. [CrossRef] [PubMed]
44. Sun, Y.; Cheng, J. Hydrolysis of lignocellulosic materials for ethanol production: A review. *Bioresour. Technol.* **2002**, *83*, 1–11. [CrossRef]
45. Shinners, K.J. *Engineering Principles of Silage Harvesting Equipment*; American Society of Agronomy Inc.: Madison, WI, USA; Crop Science Society of America: Fitchburg, WI, USA, 2003; pp. 361–403.
46. Palmowski, L.M.; Müller, J.A. Influence of the size reduction of organic waste on their anaerobic digestion. *Water Sci. Technol.* **2000**, *41*, 155–162. [CrossRef]
47. Kim, I.S.; Kim, D.H.; Hyun, S.-H. Effect of particle size and sodium ion concentration on anaerobic thermophilic food waste digestion. *Water Sci. Technol.* **2000**, *41*, 67–73. [CrossRef]
48. Garcia-Bernet, D.; Loisel, D.; Guizard, G.; Buffière, P.; Steyer, J.P.; Escudié, R. Rapid measurement of the yield stress of anaerobically-digested solid waste using slump tests. *Waste Manag.* **2011**, *31*, 631–635. [CrossRef]
49. Pommier, S.; Chenu, D.; Quintard, M.; Lefebvre, X. Modelling of moisture-dependent aerobic degradation of solid waste. *Waste Manag.* **2008**, *28*, 1188–1200. [CrossRef]
50. Ruys, V. Rhéologie des Résidus Agricoles Pour un Procédé Multi-Étapes de Méthanisation en Voie Sèche. Ph.D. Thesis, Grenoble Alpes University, Grenoble, France, 2017.
51. Wu, B. CFD simulation of mixing for high-solids anaerobic digestion. *Biotechnol. Bioeng.* **2012**, *109*, 2116–2126. [CrossRef]
52. Brambilla, M.; Romano, E.; Cutini, M.; Bisaglia, C.; Pari, L. Rheological Properties of Manure/Biomass Mixtures and Pumping Strategies to Improve Ingestate Formulation: A Review. *Trans. ASABE* **2013**, *56*, 1905–1920. [CrossRef]
53. Teixeira Franco, R.; Coarita, H.; Bayard, R.; Buffière, P. An improved procedure to assess the organic biodegradability and the biomethane potential of organic wastes for anaerobic digestion. *Waste Manag. Res.* **2019**, *37*, 746–754. [CrossRef] [PubMed]
54. Holliger, C.; Alves, M.; Andrade, D.; Angelidaki, I.; Astals, S.; Baier, U.; Bougrier, C.; Buffière, P.; Carballa, M.; de Wilde, V.; et al. Towards a standardization of biomethane potential tests. *Water Sci. Technol.* **2016**, *74*, 2515–2522. [CrossRef] [PubMed]
55. Palmowski, L.; Müller, J.; Schwedes, J. Comminution of Organic Materials to Improve Their Bioavailability. *Eng. Life Sci.* **2001**, *1*, 121–125. [CrossRef]
56. Ratkovich, N.; Horn, W.; Helmus, F.P.; Rosenberger, S.; Naessens, W.; Nopens, I.; Bentzen, T.R. Activated sludge rheology: A critical review on data collection and modelling. *Water Res.* **2013**, *47*, 463–482. [CrossRef] [PubMed]
57. Pashias, N.; Boger, D.V.; Summers, J.; Glenister, D.J. A fifty cent rheometer for yield stress measurement. *J. Rheol.* **1996**, *40*, 1179–1189. [CrossRef]
58. Mokéddem, S. Contrôle de la rhéologie d'un béton et de son évolution lors du malaxage par des mesures en ligne à l'aide de la sonde Viscoprobe. Ph.D. Thesis, Centrale Nantes University, Nantes, France, 2012.

59. Eshtiaghi, N.; Yap, S.D.; Markis, F.; Baudez, J.-C.; Slatter, P. Clear model fluids to emulate the rheological properties of thickened digested sludge. *Water Res.* **2012**, *46*, 3014–3022. [CrossRef]

60. Robertson, J.A.; de Monredon, F.D.; Dysseler, P.; Guillon, F.; Amado, R.; Thibault, J.-F. Hydration Properties of Dietary Fibre and Resistant Starch: A European Collaborative Study. *LWT Food Sci. Technol.* **2000**, *33*, 72–79. [CrossRef]

61. Raghavendra, S.N.; Rastogi, N.K.; Raghavarao, K.S.M.S.; Tharanathan, R.N. Dietary fiber from coconut residue: Effects of different treatments and particle size on the hydration properties. *Eur. Food Res. Technol.* **2004**, *218*, 563–567. [CrossRef]

Sample Availability: Samples of the untreated and pretreated manure are available from the authors. However, the storage conditions may have affected their physical or biochemical properties.

Article

High-Rate Anaerobic Digestion of Waste Activated Sludge by Integration of Electro-Fenton Process

Emna Feki [1], Audrey Battimelli [2], Sami Sayadi [3,*], Abdelhafidh Dhouib [1] and Sonia Khoufi [1,*]

[1] Laboratory of Environmental Bioprocesses, Centre of Biotechnology of Sfax, BP 1177, Sfax 3018, Tunisia; emnafeki2010@hotmail.fr (E.F.); abdelhfidh.douib@cbs.rnrt.tn (A.D.)

[2] INRAE, Université de Montpellier, Laboratoire de Biotechnologie de l'Environnement, 102 avenue des Etangs, 11100 Narbonne, France; audrey.battimelli@inrae.fr

[3] Center for Sustainable Development, College of Arts and Sciences, Qatar University, Doha 2713, Qatar

* Correspondence: sami.sayadi@gmail.com (S.S.); sonia.khoufi@cbs.rnrt.tn or soniakhoufi@gmail.com (S.K.); Tel.: +216-74-874-452 (S.K.)

Academic Editors: Ivet Ferrer, Cigdem Eskicioglu, Georgia Antonopoulou and Audrey Battimelli
Received: 24 October 2019; Accepted: 29 January 2020; Published: 31 January 2020

Abstract: Anaerobic digestion (AD), being the most effective treatment method of waste activated sludge (WAS), allows for safe disposal. The present study deals with the electro-Fenton (EF) pretreatment for enhancing the WAS biogas potential with low-cost iron electrodes. The effect of pretreatment on the physicochemical characteristics of sludge was assessed. Following EF pretreatment, the pH, conductivity, soluble chemical oxygen demand (SCOD), and volatile fatty acids (VFA) increased to 7.5, 13.72 mS/cm, 4.1 g/L, and 925 mg/L, respectively. Capillary suction time (CST) analysis highlighted the dewaterability effect of EF on WAS, as demonstrated by the decrease in CST from 429 to 180 s following 30 min of pretreatment. Batch digestion assays presented an increase in the biogas yield to 0.135 L/g volatile solids (VS) after 60 min of EF pretreatment in comparison to raw sludge (0.08 L/g VS). Production of biogas was also found to improve during semi-continuous fermentation of EF-pretreated sludge conducted in a lab-scale reactor. In comparison to raw sludge, EF-pretreated sludge produced the highest biogas yield (0.81 L biogas/g VS) with a high COD removal rate, reaching 96.6% at an organic loading rate (OLR) of 2.5 g VS/L. d. Results revealed that the EF process could be an effective WAS disintegration method with maximum recovery of bioenergy during AD.

Keywords: anaerobic digestion; biogas yield; waste activated sludge; electro-Fenton; disintegration; dewaterability

---

1. Introduction

Municipal wastewater treatment holds an important role in environmental concerns since wastewater discharges have greatly evolved in quantity and quality during recent decades. Activated sludge process is the main effective method used for treatment of municipal wastewater and thus leads to the production of large quantities of waste activated sludge (WAS). This residue, composed of cellular biomass (20–40%), refractory organic matter (40–60%), and mineral matter (10–30%), contains various minerals, organic micropollutants, and pathogenic organisms which can be potentially harmful for the environment and present genuine management problems [1]. With such environmental and economic concerns, there is growing interest focusing on the reduction in volume and stabilization of WAS [2].

Nowadays, thanks to its environmental and economic benefits, anaerobic digestion (AD) is considered to be the best option for managing WAS [3]. It is widely used as the most cost-effective way for stabilization, pathogen removal, and energy recovery. In addition, AD favors the reduction of sludge volume while producing clean energy in the form of biogas [4]. This biogas can be upgraded for direct utilization as transport fuel or injection into the gas grid. It can also be utilized in combined heat and power systems for providing electricity and heat to the wastewater treatment plant (WWTP) or for export in cases of overproduction. Any improvement in the AD efficiency should therefore lead to a further reduction in sludge volume for transport and disposal. It is likely that the biogas yield will increase, hence producing a greater amount of renewable energy and resulting in higher environmental performance and savings for the plant [5]. AD comprises a series of reactions including hydrolysis, acidogenesis, acetogenesis, and methanogenesis. Hydrolysis is the rate-limiting step for the AD of activated sludge because most organic matter present in WAS is enclosed in cell walls and membranes that protect the intracellular components [6]. The cells are shielded against osmotic lysis thanks to the semi rigid structure of the cell envelope [7]. Sludge retention time (SRT) or hydraulic retention time (HRT) are believed to be key parameters in sludge AD [8]. Various studies focus on the effect of HRT on reactor performance such as biogas production and volatile solid destruction [9]. It has often been demonstrated that a high HRT is necessary (20–30 days) to obtain a 30–50% degradation efficiency of organic solids [10]. Cell lysis has been referred as a possible method for releasing intracellular organics and increasing the rate and efficiency of the digestion process [6]. Indeed, this is possible when extracellular polymeric substances (EPS) become more bioavailable [11].

Recently, different pretreatments have been examined for improving the physicochemical characteristics of organic waste and consequently the performance of anaerobic digestion [12]. During several studies performed in half-scale and lab-scale plants, these methods, including chemical and thermal methods [12–14], as well as mechanical [15] and biological hydrolysis with enzymes [16,17], were investigated for sludge disintegration purposes. By increasing the digestion rate, biogas yields were maximized in smaller digesters while the HRTs decreased [11]. Currently, advanced oxidation processes (AOPs) are considered as valuable sludge pretreatments that might reduce hydraulic retention times and increase methane production rates [18,19]. These innovative technologies are widely used for the treatment of polluted waters. They apply the concept of producing hydroxyl radicals (HO·) which are capable of decomposing a number of organic substances via oxidation. AOPs include a series of powerful technologies: photo-catalysis, Fenton reaction, photo-Fenton, etc. Recently, researchers have focused on the disintegration of WAS by electro-Fenton (EF), although few studies have evaluated the performance of EF pretreatment on WAS anaerobic biodegradability as well as its biogas potential [19,20]. The electro-Fenton process is an advanced electrochemical oxidation process that comprises several steps. It involves electrochemical reactions that generate the reagents used for the Fenton reaction in situ. The generated reagents depend on the solution and on the nature of the electrodes. Generally, with an inert electrode, oxidation occurs via the hydroxyl radicals formed during the electrolysis of water [19].

Here, the EF process was investigated as a pretreatment step for improving the AD of activated sludge and its biogas potential. Low-cost iron electrodes were employed during EF pretreatment for in situ generation of Fe^{2+}. The effect of this pretreatment on the solubilization and dewaterability of flocs was investigated. Biochemical methane potential tests were conducted to optimize the pretreatment time in terms of maximum biogas yields. The performance of an anaerobic reactor fed with EF-pretreated sludge and operated under semi-continuous conditions was also assessed.

2. Results

2.1. Disintegration of Activated Sludge by Electro-Fenton Process

2.1.1. Physicochemical Analysis

During the EF reaction, pH, conductivity, total and soluble COD, and VFA analyses were performed (Table 1). Sludge pH increased with treatment time and reached 7.5 after 2 h of reaction. At the same time, conductivity increased to 13.72 mS/cm. The soluble COD of raw sludge is about 1.7 g/L. By applying an EF reaction, a gradual increase was observed up to a value of 4.1 g/L after 120 min of treatment time (Figure 1a). Meanwhile, a decrease in total COD occurred during the treatment, thus indicating a partial mineralization of organic matter. The volatile fatty acids concentration during the EF reaction increased as a function of the treatment time (Figure 1b). In the case of raw sludge, the total concentration of VFA is about 84.3 mg/L. This concentration increased to 1100 mg/L after 1 h then decreased with treatment time (925 mg/L at 2 h) (Figure 1b).

Table 1. Physicochemical characteristics of waste activated sludge (WAS) before and after pretreatment.

Parameters	Raw WAS	EF-Pretreated WAS
pH	6.95 ± 0.2	7.5 ± 0.8
Conductivity (mS/cm)	3.72 ± 0.3	13.72 ± 0.1
TS (g/L)	19.45 ± 1.4	14.28 ± 2
VS (g/L)	12.67 ± 1.2	10.34 ± 1.3
TSS (g/L)	15.16 ± 0.9	10.5 ± 0.5
VSS (g/L)	7.27 ± 1.3	3 ± 0.6
TCOD (g/L)	20.4 ± 4	26 ± 1.2
SCOD (g/L)	1.73 ± 2	4.1 ± 0.3
NTK (mg/L)	914 ± 10	920 ± 3
VFA (mg/L)	84.3 ± 13	925 ± 35

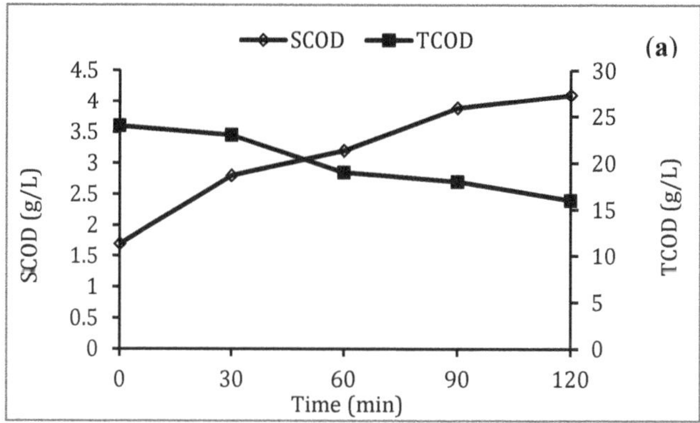

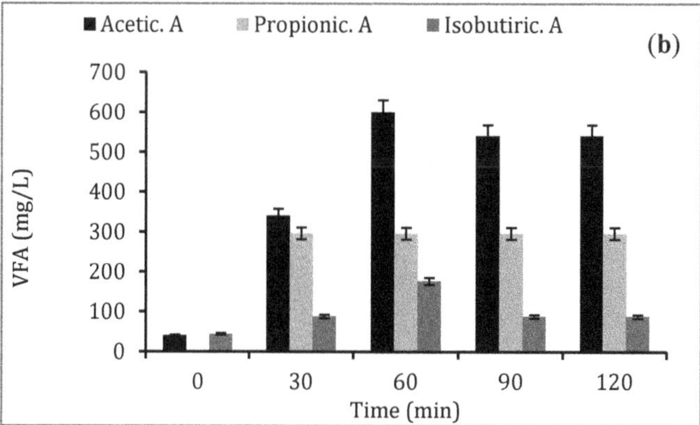

Figure 1. *Cont.*

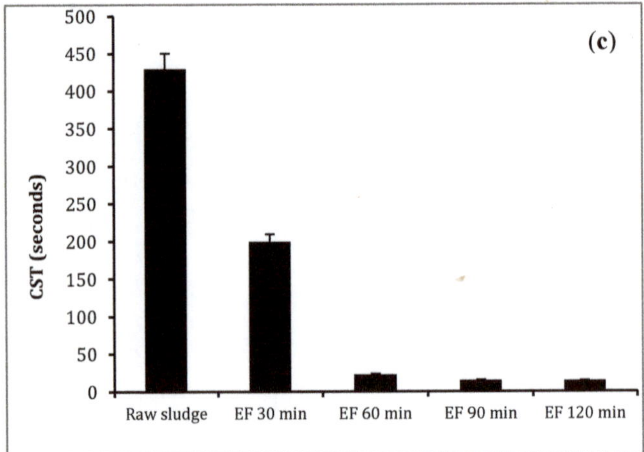

Figure 1. Evolution of soluble and total chemical oxygen demand (**a**), volatile fatty acid concentrations (**b**), and capillary suction time (**c**) during electro-Fenton (EF) pretreatment of WAS.

2.1.2. Capillary Suction Time (CST) Analysis

To identify the effect of EF treatment on sludge dewaterability, the capillary suction time (CST) of raw and pretreated sludge was determined. CST is a fast and reliable method for determining the filterability of activated sludge.

Figure 1c illustrates the decrease in elapsed time during filtration with increasing sludge pretreatment time. The CST decreased from 429 to 180 s by treating sludge by EF during 30 min. This decrease was even higher after 1 h, while dewaterability improved by about 95%.

2.1.3. Fourier Transform Infrared Spectroscopy Analysis (FTIR)

The characterization of organic material in a sludge sample was performed by FTIR. Spectrums of the raw and EF-pretreated sludge presented in Figure 2 showed absorption bands related to biomass. Both samples are characterized by a strong peak around 3100–3500 cm^{-1}. The second major band was identified between 1500 and 1700 cm^{-1} with a peak at 1634 and 1631 cm^{-1} for raw and pretreated sludge, respectively. The difference between the two spectrums is the fact that the intensity of the two major bands increased after EF pretreatment. New peaks also emerged in the pretreated sludge spectrum. The main absorption bands were thus observed in the range between 950 and 1100 cm^{-1}. Other vibrations at 2015, 2176 and 2359 cm^{-1} between 2800 and 1900 cm^{-1} were also only observed in the pretreated sludge in comparison to raw WAS.

2.1.4. Biogas Potential

To determine the effect of EF pretreatment time on anaerobic biodegradability and to optimize pretreatment time, batch anaerobic fermentations were performed. WAS samples pretreated by EF at different times (0, 30, 60, 90 and 120 min) were thus used as substrates. Figure 3 provides the kinetics of cumulative biogas yields obtained during these fermentations. In comparison to raw sludge fermentation, no time lag for biogas production was observed at the beginning of pretreated sludge fermentations. A biogas yield of about 0.080 L biogas/g vs. was achieved with raw WAS. However, an increase in biogas yields was observed for all pretreated samples in comparison to raw samples. Thus, yields calculated at the end of fermentation were 0.100, 0.113, 0.135 and 0.129 L biogas/g VS, respectively, for sludge pretreated by EF during 30, 60, 90 and 120 min.

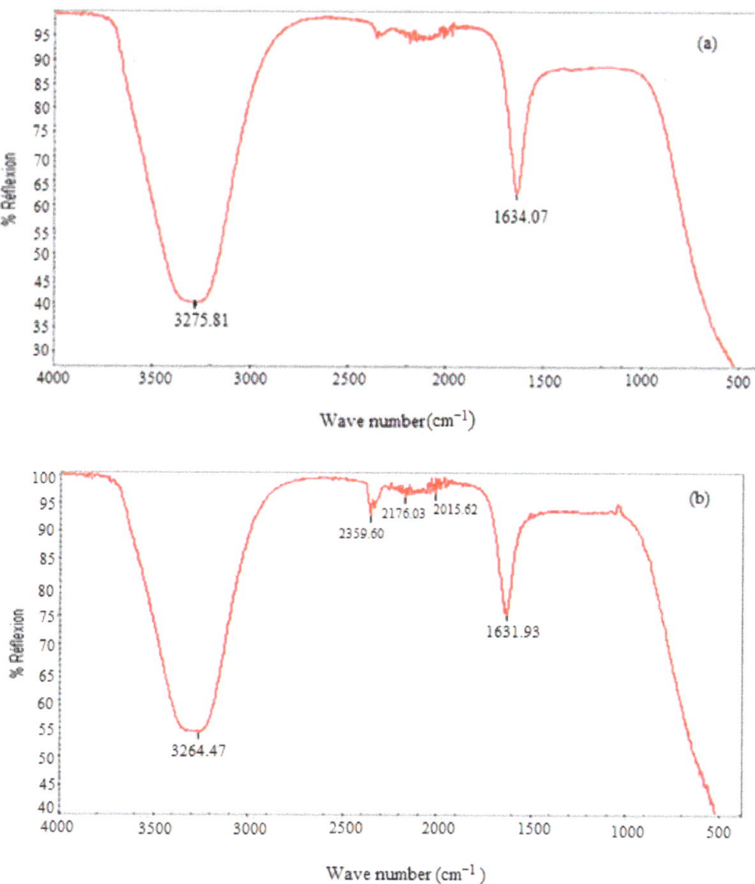

Figure 2. Spectroscopy of WAS before (**a**) and after (**b**) EF pretreatment.

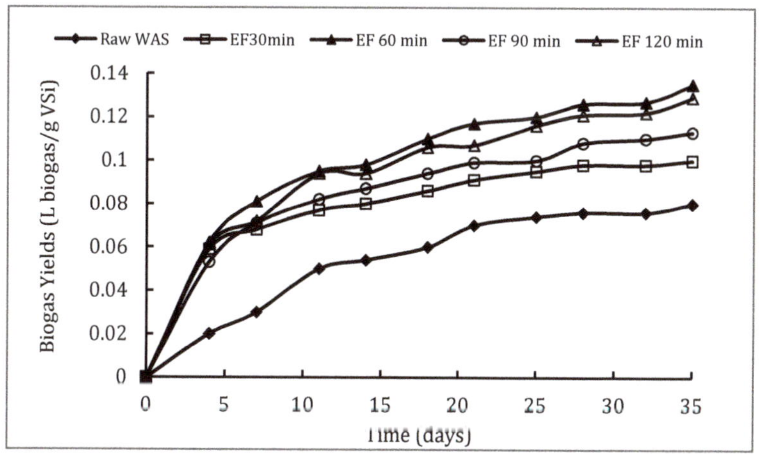

Figure 3. Biogas yields during batch anaerobic fermentation of raw and EF-pretreated (0, 30, 60, 90, and 120 min) sludge.

2.2. Semi-Continuous Fermentation of Raw and Pretreated WAS

In a first step of this study, the digester was fed with raw sludge at an organic loading rate (OLR) increasing from 0.17 g VS/L.d to 0.54 g VS/L.d (day 130). For the pretreated sludge, the first OLR applied to the reactor was 0.45 g VS/L.d. This parameter was then gradually increased to 2.5 g VS/L.d (Figure 4a). During these two fermentations, different HRTs (20, 14, 10 and 7 days) were applied which decreased with increasing OLR.

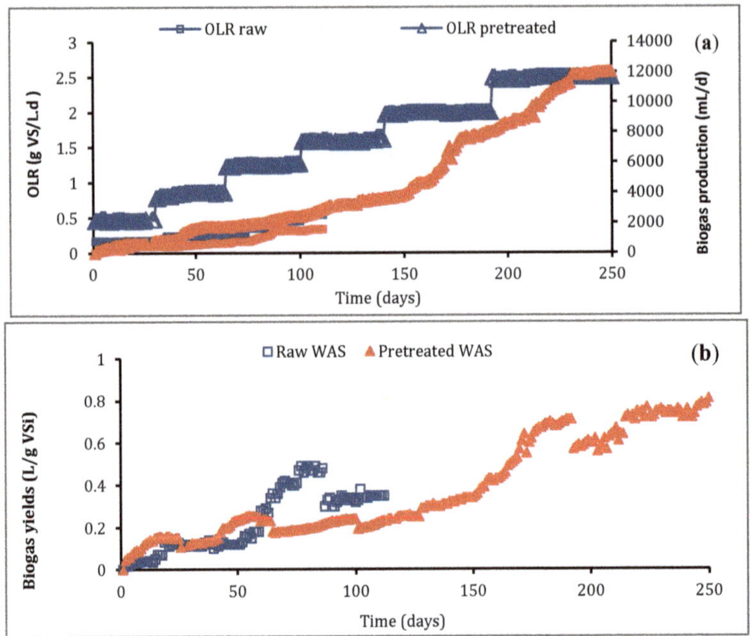

Figure 4. Evolution of organic loading rate (**a**) and biogas yield (**b**) during semi-continuous fermentation of raw and pretreated sludge in an up-flow anaerobic sludge blanket reactor.

According to data presented in Figure 4a, the daily biogas production in the two cases was low during the first days of fermentation and then increased with the increase of OLR. Figure 4b provides the biogas yields for raw and pretreated sludge. Low biogas yields were observed during the first 20 days: 0.03 and 0.14 L/g VS, respectively, in the case of raw and pretreated sludge. Improvement in the biogas yield was observed after 20 days of feeding which represent the reactor HRT. Indeed, maximum daily biogas production rates of about 1.56 and 12 L/d were recorded during the fermentation of raw and pretreated sludge, respectively, which corresponds to a biogas yield of about 0.35 and 0.81 L/g vs. (Figure 4b).

The reactor effluent COD was determined during this study. Figure 5 illustrates the evolution of this COD effluent during the fermentation of EF-pretreated sludge. When fermentation began, the COD effluent was very high. This confirms the low degradation of raw sludge during the fermentation study of raw sludge when the COD values were around 20 g/L (data not shown). The total COD of pretreated sludge was observed to be about 26 g/L. This value then decreased after passing through the bed sludge reactor. COD removal was enhanced during the fermentation step, reaching 96.7% at an OLR of 2.5 g VS/L.d.

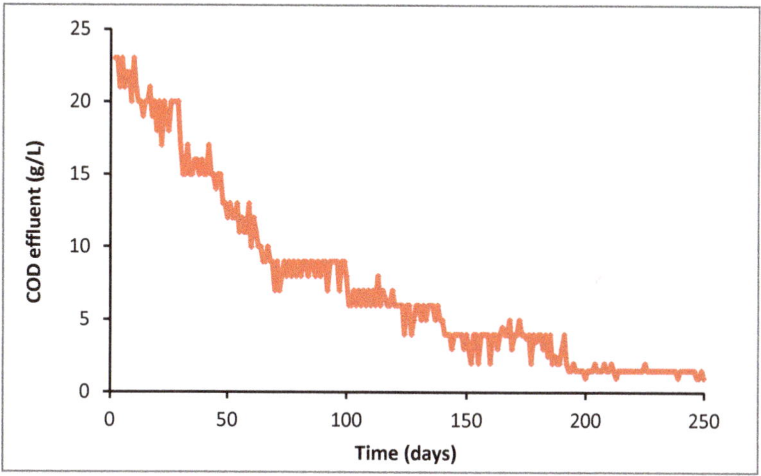

Figure 5. Evolution of total COD of effluent from UASB reactor digesting EF-pretreated sludge.

3. Discussion

The physicochemical characteristics of WAS before and after EF pretreatment, provided in Table 1, point to the improvement of sludge quality after subsequent anaerobic treatments, both in terms of pH and organic matter solubilization (COD, VFA). On one hand, the increase of pH can be explained by the generation of hydroxyl radicals by electrochemical reactions and by the decrease in H^+ production [21]. On the other hand, after 2 h of treatment, residual flocs precipitated with iron, which was continuously dissolved from the iron anode as governed by the law of Faraday. This observation has also been made for industrial effluents treated with an electro-coagulation process using a cast iron anode [22]. The increase in conductivity was probably related to the release of mineral salts during cell lysis but also to the mineralization of organic matter [23]. According to COD analyses, a solubilization reaction occurred during the pretreatment. However, the decrease of total COD during the second hour of treatment can be explained by the mineralization of organic matter due to the oxidation of simple molecules released into the solution. The release of VFA (acetic acid, isobutyric acid, propionic acid) was also observed in correlation with a soluble COD increase which probably results from the solubilization of the cellular content during the treatment [24]. This increase was higher than that obtained by Xu et al. [25] who found a total VFA concentration of 396.2 mg/L after 40 min of electrochemical sludge treatment (Ti/RuO$_2$ anode) relative to an initial concentration in raw sludge of 86.7 mg/L. Therefore, the presence of VFA in pre-treated sludge should obviously promote the subsequent anaerobic digestion process.

Examination of WAS dewaterability pointed out that EF tends to decrease CST values of pretreated samples as a function of treatment time. Theoretically, high CST values indicate a low dehydration rate, whereas low CST values indicate greater dehydration properties [26]. Consequently, EF reactions considerably reduce the required time for dewatering sludge. This implies that dissociation of sludge flocs had occurred. The decrease in the biosolid resistance to dewatering in terms of CST was also observed with Fenton pretreatment of biological sludge [27]. According to Jin et al. [28], proteins and polysaccharides, which are the main constituents of EPS, actively contribute to the water retention capacity in sludge flocs. It is therefore conceivable that material released following the disruption of the cell wall increases the available surface area of these compounds [28]. Many previous studies have also demonstrated how decantation, bioflocculation, and sludge dewatering show an excellent relationship with the EPS content and EPS spatial distribution [29,30]. Liu et al. [31] and Xu et al. [32] reported that the disruption of flocs and cells by Fenton pretreatment led to the solubilization of EPS and acceleration of solid–water separation.

The FTIR method is based on the absorption of infrared radiation by the analyzed material. It allows for both the detection of the characteristic vibrations of the chemical bonds and the analysis of the chemical functions present in the material [31]. Infrared spectrophotometer analysis in the 400–4000 cm^{-1} range is most often performed because this is where most of the frequencies of the functional groups are located [32]. Relevant peaks attributed to special functional groups according to the literature are summarized in Table 2.

Table 2. Functional groups or compounds analyzed by Fourier transform infrared spectroscopy (FTIR) analysis in raw and EF-pretreated sludge.

Wave Number (cm^{-1})	Vibration and Functional Groups
1634	C=O carboxylic acids C=C alkenes OH adsorbed water
3275	O-H hydroxyl group and water NH$_2$ amine

According to the results, the FTIR spectrum of pretreated sludge presented a change in the intensity and shift of peaks in comparison to the raw sample. The strong peak around 3100–3500 cm^{-1} observed in the two spectrums corresponded to NH and OH stretching vibrations including hydrogen bonds. The increase in intensity and the shift towards a lower frequency of this band in the spectrum of the treated sample reflect an increase in hydrogen bonds which justifies the production of proteins with amine functions (NH$_2$). The rising intensity of the main bands after EF pretreatment can be explained by the increase in reducing sugar (C=O carboxyl functions) and protein (NH$_2$ amine functions) concentrations which are released into the supernatant after cell lysis. The appearance of new peaks observed around 1100–950 cm^{-1} has been attributed to polysaccharides by Naumann et al. [33]. In the present case, these polysaccharides are released from exopolymers (EPS) dissociated during the disintegration of WAS. This can account for the improvement in dewaterability of pretreated samples. However, the apparent peak at 1030 cm^{-1} in the pretreated sludge spectrum is related to the vibration of the Si–O–Si function [32], due to the release of silicon organic compounds (siloxanes, silanols) present in the waste activated sludge. Siloxanes are generally adsorbed to EPS flocs because of their low solubility in water. These compounds have been detected in biogas during the AD of sludge, resulting from their release during organic matter degradation and the increase in temperature within the anaerobic digester [34]. Indeed, their presence in the pretreated sludge can be explained by EPS dissociation during the disintegration of sludge by EF reaction. The 2800–1900 cm^{-1} range is related to elongations of the triple bonds C≡C and C≡N (acetylene, cyanogen) and cumulative double bonds X=Y=Z [32]. Compared with the raw WAS spectrum, this range presented vibrations at 2015, 2176, and 2359 cm^{-1} in the case of pretreated sludge. This could be due to the effect of EF on the transformation of certain molecules into molecules with triple or double bonds.

Batch anaerobic fermentations of raw and pretreated samples performed in similar conditions led to an improvement in the biogas yield after EF pretreatment. The low biogas yield of raw WAS characterizes its low biodegradability as well as the rate-limiting stages of anaerobic digestion as hydrolysis due to the strong protection of EPS and cell walls. Therefore, as demonstrated in the present study with EF process application, a pretreatment step is crucial for biodegradability as well as the entire sludge AD performance to be enhanced [14]. Pretreated sample biogas yields have thus been improved in comparison with raw sludge. In addition, EF pretreatment allows for AD to be accelerated, since there is no more time lag for biogas production. Biogas improvement has been observed to correlate with treatment time. Indeed, a significant improvement in the biogas yield occurs after 60 min of pretreatment compared with the other treatment times. An increase in the biogas potential of about 68% was obtained, coinciding with the highest VFA concentration released during the treatment

(Figure 1b). These results imply that EF pretreatment during 30 or 60 min has a positive effect on the anaerobic digestion process by increasing the biodegradability and biogas yield.

In order to comprehend the effect of EF pretreatment on the AD process, semi-continuous fermentations on raw and pretreated sludge were investigated in a lab-scale reactor. The low biogas yields observed during the first 20 days of both fermentations could be explained by the absence of degradable substrates in the upper part of the digester which concerns methanogens. Comparison between results highlighted how the biogas yields improved when EF-pretreated sludge was used as feed for the reactor, thus confirming the batch test results. A 2.3-fold increase in biogas yields was achieved. Table 3 summarizes the results from both fermentations and also indicates the improvement in the methane percentage of biogas during the fermentation of EF-pretreated sludge that reached 68% against 56% in the case of raw sludge fermentation.

Table 3. Evolution of organic loading rate (OLR), biogas yield and methane percentage during anaerobic digestion of raw and pretreated sludge in UASB reactor.

Feed Sample	OLR (g VS/L.d)	Biogas Yield (mL/g VS)	CH$_4$ (%)
Raw	0.25 ± 0.02	434 ± 0.017	48 ± 4
	0.35 ± 0.01	480 ± 0.031	54 ± 2
	0.50 ± 0.02	440 ± 0.024	56 ± 1
Pretreated	0.50 ± 0.016	400 ± 0.028	58 ± 3
	1.60 ± 0.018	625 ± 0.030	66 ± 2
	2 ± 0.027	735 ± 0.032	68 ± 1
	2.50 ± 0.030	685 ± 0.029	67 ± 3

In addition, a strong COD removal rate (96.7%) was measured in the effluent even at low HRT (7 days) and at an OLR of 2.5 g VS/L.d. This improvement in COD removal could result from the increase in sludge biodegradability and from the purification performance of the Up flow anaerobic sludge blanket reactor (UASB).This performance is higher than those obtained by Xu et al. [35] and Li et al. [36] who reported a COD removal of about 49.2% and 12.5%, respectively. Yuan et al. [37] also demonstrated that a combination of electrochemical and sodium hypochlorite pretreatments significantly enhanced the biogas yields by about 1.83-fold the sludge anaerobic digestion and shortened the stabilization period. According to Li et al. [38] and Chong et al. [13], the anaerobic UASB reactor is the most robust digester for the sludge treatment. Indeed, this type of reactor contains microorganisms that form granules through which, in an upward movement, the distributed effluent passes through the base of the reactor [39]. Hence, the proposed sludge treatment system within the UASB can potentially improve treatment efficiency. It can also reduce the discharge of pollutants into the environment as well as reduce the sludge toxicity in aquatic environments [40,41].

All these results confirm the efficiency of EF sludge pretreatment for improving biogas production. This technique therefore allows for sludge to become more accessible to the anaerobic consortium, consequently enhancing the anaerobic process and biogas yield. In addition, the UASB system has proven its effectiveness by significantly improving sludge anaerobic degradation, since the different microbial communities presented in the digester are probably well balanced.

4. Materials and Methods

4.1. WAS and Anaerobic Inoculum

The WAS sample was obtained from a municipal wastewater treatment plant (WWTP) located in the Sidi Mansour region (north of Sfax, Tunisia) with a capacity of 17,900 m^3/d. Collected samples were stored at 4 °C for subsequent experiments. The characteristics of the WAS sample are provided in Table 1. For anaerobic digestion experiments (batch and semi-continuous fermentations), the inoculum was sampled from a semi-pilot anaerobic bioreactor installed in the laboratory.

4.2. Electro-Fenton Pretreatment

Electro-Fenton treatment was carried out under optimum conditions determined during a previous study [42]. The reaction took place in a 500 mL glass reactor using an ASF type 400/40.10 electric generator to apply a current density of 2.5 A/dm^2. It comprised one pair of anodic and cathodic electrodes (cast iron plates) which were positioned approximately 2 cm apart from each other and were dipped in the effluent. The total effective surface area of electrodes was 0.16 dm^2. During this treatment, the main electrochemical reactions occurred at the anode with oxidation of iron and at the cathode with the reduction of water. It was based on the principle of soluble iron anodes. Iron electrodes were used for in-situ generation of Fe^{2+}. In each run, 300 mL of raw WAS were treated and operated in batch mode. The pH of the WAS sample was adjusted to 3 by adding an HCl solution (2N). The H_2O_2 (30% v/v) was added after adjustment of pH to the desired value; the pH of the solution was not controlled again during the reaction. The aqueous solution of reactants was homogenized by magnetic agitation to avoid sedimentation of WAS particles and produced at room temperature.

4.3. Batch Anaerobic Digestion

Batch AD tests were conducted to determine the effect of EF pretreatment time on biogas production. WAS samples pretreated by EF at different times (0, 30, 60, 90 and 120 min) were used as substrates. For each treatment time, three batch reactors with a capacity of 120 mL were tested in parallel. Batch reactors treating raw sludge were also conducted for comparison. In each reactor, substrate and inoculum were introduced with a vs. substrate/VS inoculum ratio equal to 1. All batches were adjusted to 7.2 then purged with a gas mixture of 75% N_2 and 25% CO_2 for 3 min to maintain anaerobic conditions and finally incubated under mesophilic conditions (37 ± 1 °C) for 35 days. During fermentation, biogas production was measured using a gas displacement device. The Biochemical methane potential (BMP) experiments and the related analysis were performed by the Bio2E platform [43].

4.4. Semi-Continuous Anaerobic Reactor

An up-flow anaerobic sludge blanket (UASB) reactor was used to study the semi-continuous fermentation of WAS before and after 1 h of EF pretreatment. The working volume of the digester was 7 L. For maintaining a constant temperature (37 °C), it had a PVC double wall filled with heated water from a heated bath circulator. Before beginning the experiments, the bioreactor was inoculated with the anaerobic microbial consortium and fed with raw sludge (0.05 g VS/L.d) during 2 months. This preparation period ensured biomass enrichment and process stability. The hydraulic retention time (HRT) during fermentation of raw and pretreated sludge was set to 20 days during the first period. It was then reduced to 14, 10 and 7 days. Feeding and withdrawing were done once a day, using a pump. The volume of biogas was measured by liquid displacement. During the experiments, the COD of influent and effluent, biogas production, pH, and VFA were monitored.

4.5. Analytical Methods

The pH was measured with a pH meter (Metrohm). Total and soluble COD were quantified with a titration method after a total digestion with H_2SO_4 and potassium dichromate at 150 °C for 2 h [42]. BOD_5 was determined by the manometric method with a respirometer (BSB-Controller Model 620 T (WTW)). Total solids (TS) and Total suspended solids (TSS) were measured by weighing samples before and after overnight drying at 105 °C. Volatile solids (VS) and volatile suspended solids (VSS) were analyzed by loss on ignition at 600 °C for 2 h. The total Kjeldahl nitrogen content (TKN) and the ammoniacal nitrogen (N–NH_4^+) were analyzed according to the Kjeldahl-N method.

Conductivity was measured using a conductimeter (CONSORT). Total volatile fatty acids (VFA) were analyzed by centrifuging the samples for 15 min at 8000 rpm and then filtering them on 0.45 μm pore size syringe filters. The resulting filtrates were acidified (pH 3) with HCl (0.1 N) before being analyzed and quantified by high-performance liquid chromatography (HPLC: SHIMADZU 10 AVP). Fourier transform infrared spectroscopy (FTIR 380 Nicolet model) equipped with a He-Ne laser and a telluride, mercury, and cadmium detector (MCT) at a frequency of 400–4000 cm^{-1} was used to analyze the organic functional groups present in raw and pretreated WAS. Before the FTIR analysis, sludge samples were centrifuged for 20 min at 4500 rpm and then filtered through a 0.45 µm pore size membrane filter.

The capillary suction time (CST) was measured with a CST analyzer (Triton Electronics Ltd., United Kingdom) using a 7 × 9 cm^2 size CST paper. All sludge samples were used in their initial state without centrifugation and filtration. A volume of 5 mL of sample was placed into a metal tube and the ring time was recorded as CST.

5. Conclusions

The present study has highlighted the effect of EF pretreatment on the efficiency of disintegration and AD of sludge. The pretreatment has shown to improve organic matter solubilization by increasing soluble COD and volatile fatty acids. FTIR analysis also revealed the release of reducing sugars, polysaccharides, and proteins, thus confirming the dissociation of flocs. This result can account for the improvement in sludge filterability indicated by low CST results of 22.4 and 14 s after 1 and 2 h of pretreatment, respectively. Results from batch and semi-continuous anaerobic fermentations have confirmed the positive effect of the EF process in enhancing the biogas potential and stability of the anaerobic system. The EF process therefore promises to be a more reliable and robust solution for the enhancement of WAS anaerobic treatment.

Author Contributions: S.K., S.S. and A.D. conceived and designed the study experiments; E.F. performed all the experiments and wrote the original draft preparation. A.B., responsible for CST analysis, BMP experiments and data. S.K., the corresponding author, designed the overall study and interpreted the results; S.K. and A.B. revised and finalized the manuscript. All authors have read and approved the final manuscript.

Funding: This research was funded by the Tunisian Ministry of Higher Education and Scientific Research and the International Foundation for Science (Research Grant Agreement N° W/5412-1).

Acknowledgments: The authors would like to thank Nidhal Baccar for his assistance in the FTIR analyses and the Bio2E platform (DOI 10.15454/1.557234103446854E12) for experimental support.

Abbreviations

AD	anaerobic digestion
COD	chemical oxygen demand
CST	capillary suction time
EF	electro-Fenton
FTIR	Fourier transform infrared spectroscopy
OLR	organic loading rate
VFA	volatile fatty acids
WAS	waste activated sludge

References and Note

1. Ennouri, H.; Miladi, B.; Diaz, S.Z.; Gûelfo, L.A.F.; Solera, R.; Hamdi, M.; Bouallagui, H. Effect of thermal pretreatment on the biogas production and microbial communities balance during anaerobic digestion of urban and industrial waste activated sludge. *Bioresour. Technol.* **2016**, *214*, 184–191. [CrossRef] [PubMed]
2. Neczaj, E.; Grosser, A. Biogas production by thermal hydrolysis and thermophilic anaerobic digestion of waste-activated sludge. *Ind. Munic. Sludge* **2019**, 741–781.
3. Akgul, D.; Cella, M.A.; Eskicioglu, C. Influences of low-energy input microwave and ultrasonic pretreatments on single-stage and temperature-phased anaerobic digestion (TPAD) of municipal wastewater sludge. *Energy* **2017**, *123*, 271–282. [CrossRef]

4. Kim, J.; Yu, Y.; Lee, C. Thermo-alkaline pretreatment of waste activated sludge at low-temperatures: Effects on sludge disintegration, methane production, and methanogen community structure. *Bioresour. Technol.* **2013**, *144*, 194–201. [CrossRef] [PubMed]

5. Tong, H.; Tong, Y.W.; Peng, Y.H. A comparative life cycle assessment on mono- and co-digestion of food waste and sewage sludge. *Energy Procedia* **2019**, *158*, 4166–4171. [CrossRef]

6. Devlin, D.C.; Esteves, S.R.R.; Dinsdale, R.M.; Guwy, A.J. The effect of acid pretreatment on the anaerobic digestion and dewatering of waste activated sludge. *Bioresour. Technol.* **2011**, *102*, 4076–4082. [CrossRef]

7. Müller, J.; Lehne, G.; Schwedes, J.; Battenberg, S.; Naveke, R.; Kopp, J.; Dichtl, N.; Scheminski, A.; Krull, R.; Hempel, D.C. Disintegration of sewage sludges and influence on anaerobic digestion. *Water Sci. Technol.* **1998**, *38*, 425–433. [CrossRef]

8. Chen, Y.; Fu, B.; Wang, Y.; Jiang, Q.; Liu, H. Reactor performance and bacterial pathogen removal in response to sludge retention time in a mesophilic anaerobic digester treating sewage sludge. *Bioresour. Technol.* **2012**, *106*, 20–26. [CrossRef]

9. Nges, I.A.; Liu, J. Effects of solid retention time on anaerobic digestion of dewatered-sewage sludge in mesophilic and thermophilic conditions. *Renew. Energy* **2010**, *35*, 2200–2206. [CrossRef]

10. Appels, L.; Baeyens, J.; Degreve, J.; Dewil, R. Principles and potential of the anaerobic digestion of waste-activated sludge. *Prog. Energy Combust. Sci.* **2008**, *34*, 755–781. [CrossRef]

11. Gonzalez, A.; Hendriks, A.T.W.M.; Vanlier, J.B.; Kreuk, M. Pre-treatments to enhance the biodegradability of waste activated sludge: Elucidating the rate limiting step. *Biotechnol. Adv.* **2018**, *36*, 1431–1469. [CrossRef] [PubMed]

12. Tulun, S.; Bilgin, M. Enhancement of anaerobic digestion of waste activated sludge by chemical pretreatment. *Fuel* **2019**, *254*, 115671. [CrossRef]

13. Li, X.; Zhao, J.; Wang, D.; Yang, Q.; Xu, Q.; Deng, Y.; Yang, W.; Zeng, G. An efficient and green pretreatment to stimulate short chain fatty acids production from waste activated sludge anaerobic fermentation using free nitrous acid. *Chemosphere* **2016**, *144*, 160–167. [CrossRef] [PubMed]

14. González-Fernández, C.; Sialve, B.; Bernet, N.; Steyer, J.P. Thermal pretreatment to improve methane production of Scenedesmus biomass. *Biomass Bioenergy* **2012**, *40*, 105–111. [CrossRef]

15. Ariunbaatar, J.; Panico, A.; Esposito, G.; Pirozzi, F.; Lens, P.N. Pretreatment methods to enhance anaerobic digestion of organic solid waste. *Appl. Energy* **2014**, *123*, 143–156. [CrossRef]

16. Lee, K.; Chantrasakdakul, P.; Kim, D.; Kong, M.; Park, K.Y. Ultrasound pretreatment of filamentous algal biomass for enhanced biogas production. *Waste Manag.* **2014**, *34*, 1035–1040. [CrossRef]

17. Ehimen, E.A.; Hom-Nielsen, J.B.; Poulsen, M.; Boelsmand, J.E. Influence of different pre-treatment routes on the anaerobic digestion of a filamentous algae. *Renew. Energy* **2013**, *50*, 476–480. [CrossRef]

18. Luo, K.; Yang, Q.; Li, X.; Yang, G.; Liu, Y.; Wang, D.; Zheng, W.; Zeng, G. Hydrolysis kinetics in anaerobic digestion of waste activated sludge enhanced by α-amylase. *Biochem. Eng. J.* **2012**, *62*, 17–21. [CrossRef]

19. Ana, R.R.; Olga, C.N.; Manuel, F.R.P.; Adrián, M.T.S. An overview on the advanced oxidation processes applied for the treatment of water pollutants defined in the recently launched Directive 2013/39/EU. *Environ. Int.* **2015**, *75*, 33–51.

20. Anjum, M.; Al-Talhi, H.A.; Mohamed, S.A.; Kumar, R.; Barakat, M.A. Visible light photocatalytic disintegration of waste activated sludge for enhancing biogas production. *J. Environ. Manag.* **2018**, *216*, 120–127. [CrossRef]

21. Chen, Y.; Chen, H.; Li, J.; Xiao, L. Rapid and efficient activated sludge treatment by electro- Fenton oxidation. *Water Res.* **2019**, *152*, 181–190. [CrossRef] [PubMed]

22. Khoufi, S.; Feki, F.; Sayadi, S. Detoxification of olive mill wastewater by electrocoagulation an sedimentation processes. *J. Hazard. Mater.* **2007**, *142*, 58–67. [CrossRef] [PubMed]

23. Israilides, C.J.; Vlyssides, A.G.; Mourafeti, V.N.; Kavouni, G. Olive oil wastewater treatment with the use of an electrolysis system. *Bioresour. Technol.* **1997**, *61*, 163–170. [CrossRef]

24. Chen, W.; Gao, X.; Xu, H.; Cai, Y.; Cui, J. Influence of extracellular polymeric substances (EPS) treated by combined ultrasound pretreatment and chemical re-flocculation on water treatment sludge settling performance. *Chemosphere* **2017**, *170*, 196–206. [CrossRef]

25. Xu, J.; Yuan, H.; Lin, J.; Yuan, W. Evaluation of thermal, thermal-alkaline, alkaline and electrochemical pretreatments on sludge to enhance anaerobic biogas production. *J. Taiwan Inst. Chem. Eng.* **2014**, *45*, 2531–2536. [CrossRef]

26. Jiang, Y.; Su, M.; Li, D.P. Removal of sulfide and production of methane from carbon dioxide in microbial fuel cells-microbial electrolysis cell (MFCs-MEC) coupled system. *Appl. Biochem Biotechnol.* **2014**, *172*, 11. [CrossRef]

27. Gulbin, E.; Filibeli, A. Improving anaerobic biodegradability of biological sludges by Fenton pre-treatment: Effects on single stage and two-stage anaerobic digestion. *Desalination* **2010**, *251*, 58–63.

28. Jin, B.; Wilén, B.-M.; Lant, P. Impacts of morphological, physical and chemical properties of sludge flocs on dewaterability of activated sludge. *Chem. Eng. J.* **2004**, *98*, 115–126. [CrossRef]

29. Hernando, A.; Hernando, R.; Plastino, A. Space–time correlations in urban sprawl. *J. R. Soc. Interface* **2014**, *11*, 20130930. [CrossRef]

30. Peng, Z.; Fang, F.; Chen, Y.P.; Shen, Y.; Zhang, W.; Yang, J.X.; Li, C.H.; Guo, J.S.; Liu, S.Y.; Huang, Y.; et al. Composition of EPS fractions from suspended sludge and biofilm and their roles in microbial cell aggregation. *Chemosphere* **2014**, *117*, 59–65.

31. Zhang, L.; Vrieze, J.D.; Hendrickx, T.L.G.; Wei, W.; Temmink, H.; Rijnaarts, H.; Zeeman, G. Anaerobic treatment of raw domestic wastewater in a UASB-digester at 10 °C and microbial community dynamics. *Chem. Eng. J.* **2018**, *334*, 2088–2097. [CrossRef]

32. Liu, H.; Yang, J.; Zhu, N.; Zhang, H.; Li, Y.; He, S.; Yang, C.; Yao, H. A comprehensive insight into the combined effects of Fenton's reagent and skeleton builders on sludge deep dewatering performance. *J. Hazard. Mater.* **2013**, *258*, 144–150. [CrossRef] [PubMed]

33. Xu, H.; Shen, K.; Ding, T.; Cui, J.; Ding, M.; Lu, C. Dewatering of drinking water treatment sludge using the Fenton-like process induced by electro-osmosis. *Chem. Eng. J.* **2016**, *293*, 207–215. [CrossRef]

34. Wu, H.; Zhao, Y.; Long, Y.; Zhu, Y.; Wang, H.; Lu, W. Evaluation of the biological stability of waste during landfill stabilization by thermogravimetric analysis and Fourier transform infrared spectroscopy. *Bioresour. Technol.* **2011**, *102*, 9403–9408. [CrossRef] [PubMed]

35. Smidt, E.; Meissl, K. The applicability of Fourier transform infrared (FT-IR) spectroscopy in waste management. *Waste Manag.* **2007**, *27*, 268–276. [CrossRef]

36. Naumann, A.; Heine, G.; Rauber, R. Efficient discrimination of oat and pea roots by cluster analysis of Fourier transform infrared (FTIR) spectra. *Field Crop. Res.* **2010**, *119*, 78–84. [CrossRef]

37. Dewil, R.; Appels, L.; Baeyens, J.; Buczynska, A.; Vaeck, L.V. The analysis of volatile siloxanes in waste activated sludge. *Talanta* **2007**, *74*, 14–19. [CrossRef]

38. Xu, S.; Zhang, L.; Huang, S.; Zeeman, G.; Rijnaarts, H.; Liu, Y. Improving the energy efficiency of a pilot-scale UASB-digester for low temperature domestic wastewater treatment. *Biochem. Eng. J.* **2018**, *135*, 71–78. [CrossRef]

39. Chong, S.; Kanti Sen, T.; Kayaalp, A.; Ming Ang, H. The performance enhancements of upflow anaerobic sludge blanket (UASB) reactors for domestic sludge treatmentA State-of-the-art review. *Water Res.* **2012**, *46*, 3434–3470. [CrossRef]

40. Abbasi, T.; Abbasi, S.A. Formation and impact of granules in fostering clean energy production and wastewater treatment in upflow anaerobic sludge blanket (UASB) reactors. *Renew. Sustain. Energy Rev.* **2012**, *16*, 1696–1708. [CrossRef]

41. Nair, A.T.; Ahammed, M.M. The reuse of water treatment sludge as a coagulant for post-treatment of UASB reactor treating urban wastewater. *J. Clean. Prod.* **2015**, *96*, 272–281. [CrossRef]

42. Feki, E.; Khoufi, S.; Loukil, S.; Sayadi, S. Improvement of anaerobic digestion of waste activated sludge by using H_2O_2 oxidation, electrolysis, electro-oxidation and thermo-alkaline pretreatments. *Environ. Sci. Pollut. Res.* **2015**, *22*, 12. [CrossRef] [PubMed]

43. Bio2E, INRA, 2019. Environemental Biotechnology and Biorefinery Platform.

Sample Availability: Not available.

Article

Impacts of Chemical-Assisted Thermal Pretreatments on Methane Production from Fruit and Vegetable Harvesting Wastes: Process Optimization

Ümmihan Günerhan [1], Ender Us [1], Lütfiye Dumlu [2], Vedat Yılmaz [3], Hélène Carrère [4] and Altınay N. Perendeci [5,*]

[1] Istanbul Water and Sewerage Administration, Ataköy Advanced Biological Wastewater Treatment Plant, Şevketiye Street, Havaalanı Avenue, 34156 Istanbul, Turkey; ugunerhanus@iski.istanbul (U.G.); enderus@iski.istanbul (E.U.)
[2] Ministry of Environment and Urbanization, General Directorate of Environmental Management, Department of Water and Soil Management, 06520 Ankara, Turkey; lutfiye.dumlu@csb.gov.tr
[3] Environmental Engineering Department, Artvin Çoruh University, 08100 Artvin, Turkey; vedatyilmaz@artvin.edu.tr
[4] Univ Montpellier, INRAE, LBE, 102 Avenue des Etangs, 11100 Narbonne, France; helene.carrere@inra.fr
[5] Department of Environmental Engineering, Akdeniz University, 07058 Antalya, Turkey
* Correspondence: aperendeci@akdeniz.edu.tr; Tel.: +90-242-310-6334; Fax: 90-242-310-6306

Academic Editor: Ivet Ferrer
Received: 7 December 2019; Accepted: 16 January 2020; Published: 23 January 2020

Abstract: The increasing population creates excess pressure on the plantation and production of fruits and vegetables across the world. Consumption demand during the whole year has made production compulsory in the covered production system (greenhouse). Production, harvesting, processing, transporting, and distribution chains of fruit and vegetables have resulted in a huge amount of wastes as an alternative source to produce biofuels. In this study, optimization of two pretreatment processes (NaOH and HCl assisted thermal) was investigated to enhance methane production from fruit and vegetable harvesting wastes (FVHW) that originate from greenhouses. NaOH concentration (0–6.5%), HCl concentration (0–5%), reaction temperature (60–100 °C), solid content (1–5%), time of reaction (1–5 h), and mixing speed (0–500 rpm) were chosen in a wide range of levels to optimize the process in a broad design boundary and to evaluate the positive and negative impacts of independent variables along with their ranges. Increasing NaOH and HCl concentrations resulted in higher COD solubilization but decreased the concentration of soluble sugars that can be converted directly into methane. Thus, the increasing concentrations of NaOH and HCl in the pretreatments have resulted in low methane production. The most important independent variables impacting COD and sugar solubilization were found to be chemical concentration (as NaOH and HCl), solid content and reaction temperature for the optimization of pretreatment processes. The high amount of methane productions in the range of 222–365 mL CH_4 gVS^{-1} was obtained by the simple thermal application without using chemical agents as NaOH or HCl. Maximum enhancement of methane production was 47–68% compared to raw FVHW when 5% solid content, 1-hour reaction time and 60–100 °C reaction temperature were applied in pretreatments.

Keywords: biogas production; fruit and vegetable harvesting wastes; process optimization; thermo chemical pretreatment

---

1. Introduction

The acceptance of the generated wastes as a resource within the scope of the circular economy, recovery, and reuse is the major condition for sustaining life on earth. The increasing population coupled with economic development has resulted in huge agricultural production [1]. Based on the consumption requirements, the plantation and production of fruits and vegetables have increased across the world and production reached 1960 million tons in 2017 [2]. The continuation of consumption demand during the whole year has made production compulsory in the covered production system (greenhouse).

China is placed in the first rank of the world with a 27000 km^2 covered area for the production of fruits and vegetables. South Korea, Spain, and Japan followed China in the ranking. Turkey, holding approximately 772 km^2 of covered production area, occupies fifth place across the world for the massive production of fresh fruit and vegetables from greenhouses. It is expected that the global commercial greenhouse market will be 32.31 billion USD with an annual growth rate of 8.8% by 2021. Judging by this enormous production, it is predictable that huge amounts of fruits and vegetables wastes have been generated during the production, harvesting, processing, transporting and distribution chain. Fruits and vegetable wastes are the main components of municipal solid waste. The conventional disposal methods of FVHW from greenhouses are the storage in forest area, uncontrolled burning and landfilling by farmers [3]. Moreover, processing, transporting and distribution chain wastes are disposed of landfills by the municipalities. Although EU greenhouse gas emissions from landfills have already declined due to the well-practiced MSW managements [4], the degradation of FVHW during landfilling causes the production of leachates and emissions of greenhouse gases all around the world. Furthermore, uncontrolled burning has also resulted in greenhouse gases and toxic compounds [5].

Maintaining high agricultural production capacities for Turkey, as well as for other countries, has three main objectives. These are increasing agricultural income and its contribution to national income, meeting agricultural and food demand of nationals domestically, and keeping farmers on their lands to secure agricultural production [6]. About 250 km^2 of greenhouses have been located in Antalya, south of Turkey. It is around 47% of Turkey's greenhouse production hold in Antalya. Eighty percent of fruit and vegetables, such as tomato, cucumber, pepper, eggplant, courgette, banana, strawberry and any other are grown in the greenhouses. Apart from all these positive contributions, the region has been under heavy pressure in terms of FVHW to be managed. The production area urgently required the implementation of relevant policies for managing the waste disposal and recovering energy from renewable and cost-free lignocellulosic wastes.

On the other side, fossil fuels have supplied over 80% of global energy needs. However, these fuels are not renewable and lead to climate change by the production of CO_2 and other greenhouse gases [7]. Consequently, building a sustainable society requires reducing dependence on fossil fuels, minimizing the amount of waste produced, and increasing recycling. Generating renewable bioenergy from wastes makes it possible to realize these targets at the same time. The anaerobic digestion (AD) of FVHW combines these advantages, as it produces methane and simultaneously treats the wastes. AD is a trustful and approved technology with a long exercise for the degradation of organic wastes from industry, municipality, and agriculture over the decades. It is known that the predicted energy gain/input ratio for AD was 28.8 MJ/MJ. This finding has been better compared to other technologies for energy production from biomass in energy efficiency [8–10].

However, the complex structure of agricultural residues is the major limitation, which makes lignocellulosic biomass refractory to anaerobic digestion and yields in low methane production [11,12]. This restricts the hydrolysis of material in which enzymes fail to degrade complex structures. Commercial limitations of methane production from lignocellulosic residues should be solved by the pretreatment processes [12,13]. To solve this limitation, different kinds of pretreatment methods such as mechanical, chemical (acidic, alkali, oxidative), thermal, biological, and combinations have been applied to the countless number of lignocellulosic feedstock [10].

On the other hand, FVHW, primarily non-commercial parts of plants such as leaves, bodies, stems as well as unmerchantable fruits and vegetables, have a milder lignocellulosic structure than the other agricultural residues and require less severe pretreatment conditions before AD. There have been very limited studies investigating the methane production potentials of mixed FVHW combined with pretreatment methods in the literature [3,5,14–17].

In this study, it was proposed to combine the cost-effective integrated process of chemical-assisted thermal pretreatment and anaerobic digestion for the improvement of methane production from FVHW. The impacts of sodium hydroxide (NaOH) and hydrochloric acid (HCl) assisted thermal pretreatment on methane production were investigated. In the first stage of the work, impacts of variables (reaction temperature, time of reaction, amounts of chemical, solid content, mixing speed) on soluble chemical oxygen demand (sCOD) and soluble sugars (sSugar) were determined with very detailed experimental work for NaOH- and HCl-assisted thermal pretreatments. In the second part of the work, process optimizations were performed for both pretreatment processes; optimum conditions for methane production were discussed and biochemical methane potential (BMP) tests were implemented under the determined conditions. Finally, molecular bond characterization and surface properties of raw and pretreated FVHW were compared.

2. Materials and Methods

2.1. Sources of FVHW and Characterization Analysis

FVHW were obtained from Antalya region. The area of Kumluca has been well known for the extensive production of fruits and vegetables in greenhouses. Kumluca area makes a significant contribution to the fruits and vegetables supplies of Turkey and its trade partners during the four seasons. FVHW included leaves; stalks; stems; roots; and unmerchantable fruits from pepper, cucumber, tomato, courgette, and eggplant. FVHW were sun-dried and milled to 1 mm size for the characterization tests, while size-reduced (4–5 mm) fresh wastes were subjected to chemical-assisted thermal pretreatment experiments and BMP tests. The analyses of total solids (TS), volatile solids (VS) and total chemical oxygen demand (COD) were performed according to standard methods 2540C and 5220B, respectively [18]. The contents of lignin, cellulose, hemicellulose, and soluble matter were determined using procedures proposed by Van Soest (1963) [19] by Gerhard FBS6 (Gerhard, Königswinter, Germany). Carbohydrate and the soluble reducing sugar (sRedSugar) concentrations were measured via the Anthrone method [20] and the Dinitrosalicylic acid (DNS) method [21], respectively. Protein concentration was done by the Lowry method [22]. Extractive matter and lipids of the wastes were measured by soxhlet extraction using petroleum ether [23]. Elemental composition of the wastes was determined by CHNS analyzer (LECO, CHNS-932, St. Joseph, MI, USA). The Kjeldahl nitrogen content of the FVHW was quantified by an analyzer (Büchi Digest Automat K-438, Büchi Auto Kjeldahl Unit K-370 and Radiometer TitraLab 840, Büchi, Flawil, Switzerland).

2.2. Chemical-Assisted Thermal Pretreatments Experiments

Fresh FVHW were pretreated in a 0.3 L stainless steel reactor (Parr series 5500 high-pressure reactor; Parr Instrument Company, Moline, IL, USA) equipped with a temperature controller and magnetically-driven turbine agitator. NaOH and HCl were selected as chemical agents in the chemical-assisted thermal pretreatments experiments to enhance methane yield. Selected independent variables and their ranges for NaOH- and HCl-assisted thermal pretreatment are presented in Table 1. For the specified conditions, chemically-assisted thermal pretreatment runs were performed in parallel. NaOH and HCl solutions were loaded into the reactor along with the computed amount of FVHW and heated to predefined reaction temperatures. When the predefined reaction temperature was sustained, the time of reaction was initiated. Later, completing the target reaction time, the reaction vessel was soaked into water and an ice bath to cool down to 40 °C in a short time. Evaluation of the pretreatment performance was made by sCOD, sRedSugar, and BMP. Samples were separated into a liquid and solid phase by centrifugation after cooling. sCOD and sRedSugar analyses were performed from the liquid phase. The analysis of sCOD was performed by a Hach-Lange DR5000 spectrophotometer (Hach Lange GmbH, Duesseldorf, Germany) and a Lange LT200 (Keison Products, Grasshut, Germany) with

COD kits. Since the studies done in literature to determine the effectiveness of chemically-assisted pretreatment processes are generally based on the usage of the DNS-reduced sugar method instead of the Anthrone method for the sSugar analysis [24–28], sRedSugar concentrations were quantified by the DNS method [21]. Solubilization degrees for COD and Sugar (or variation in sCOD and sSugar compared to raw FVHW) were calculated according to Mottet et al. (2009) [29]. Pretreated FVHW containing liquid and solid fractions were preserved at −20 °C for the BMP test.

Table 1. Selected independent variables and their ranges for NaOH- and HCl-assisted thermal pretreatments.

NaOH-Assisted Thermal Pretreatment			
A: Reaction Temperature, (°C)	60	80	100
B: Reaction Time, (Hour)	1	2.5	4
C: Mixing Speed, (rpm)	0	250	500
D: NaOH Concentration, (%, *w/v*)	0	3.25	6.5
E: Dry Matter (DM) Content, (%)		5	

HCl-Assisted Thermal Pretreatment			
Independent Variables	**Coded Variable Levels**		
	Minimum (−1)	**Center (0)**	**Maximum (+1)**
A: HCl Concentration, (%, *w/v*)	0	2.5	5
B: Reaction Temperature, (°C)	60	80	100
C: Dry Matter (DM) Content (%)	1	2.5	5
D: Mixing Speed, (rpm)	0	250	500
E: Reaction Time, (Hour)		1	

Chemical-assisted pretreatments experiments were planned by central composite design (CCD) of Response Surface Methodology (RSM). Four independent variables with the three levels were utilized for CCD in Design-Expert®11 software trial version (Minneapolis, MN, USA) for NaOH- and HCl-assisted thermal pretreatments. Fifty-one pretreatment experiments containing three experiments at the center point along with the duplicates of each run were planned for each chemical (NaOH and HCl) assisted thermal pretreatment by CCD. Consequently, 108 values for each response (sCOD, sRedSugar) were used in the optimization of each chemical-assisted pretreatment processes.

Results of experiments were modeled and acquired regression models were assessed by the analysis of variance (ANOVA), regression coefficients along with the *p*- and *F*- values. The quality of models was concluded via the coefficient of determination (R^2) and an adjusted determination coefficient (adj-R^2). With the help of the obtained model equations, optimizations of chemical-assisted thermal pretreatment processes were performed by using maximization and minimization criteria for independent and dependent variables [3,30]. The goal settings of the variables have been adjusted by using the plus (+) symbols in the module of optimization of a Design-Expert® software program for process optimization. The significance of the goals was chosen as (+++++), the highest, for all variables in software. Design-Expert® software solved all goals for all variables to obtain the desirability function. The optimization module investigated the merging of independent variable values that meet the demands placed on each of the dependent variables. The targets were chosen as maximum sSugar and sCOD production, and optimizations were performed for economical process conditions along with the high sSugar and sCOD production.

2.3. Biochemical Methane Potential (BMP)

The impact of chemical-assisted thermal pretreatment was also evaluated by a BMP test. The BMP test according to Carrère et al. (2009) and Us and Perendeci (2012) [3,31] was accomplished. Raw and pretreated FVHW, micro and macro nutrients, anaerobic seed sludge along with the buffer solution were incubated at mesophilic temperature (35 °C). Seed sludge was obtained from a wastewater treatment plant's anaerobic digester unit in Antalya, Turkey. The substrate-to-seed ratio in every reactor was 0.5 (gVS/gVS for samples). After adding the sample, seed sludge, necessary nutrients, and buffer solution to the reactors, volume was set to 400 mL by adding deionized water. pH was set to neutral, and reactors were flushed with N_2/CO_2 (70%/30%) gas mixture to provide an anaerobic environment. Triplicate BMP tests were done for all pretreated samples, raw waste, and seed sludge control. BMP tests ended after the 30 days. During fermentation, CH_4 and CO_2 components formed in the BMP reactors were detected by Varian CP-4900 Micro-Gas Chromatography (Micro-GC) (Agilent Tech., Santa Clara, USA). Varian CP-4900 was supplied with a thermal conductivity detector (online-TCD) and a 10 m PPQ column. Injector and column temperatures in Micro-GC were 110 °C and 70 °C, respectively. High purity grade helium gas was used as a mobile phase with a flow rate of 25 mL/min. For measuring total gas volume, a gas-liquid displacement device was used filled with a saturated pH 1 acidic salt solution prepared according to Standart Method 2720 [18]. The amount of methane produced from seed sludge used for BMP tests was calculated as $mLCH_4$/gVS and then this value was subtracted from the methane amount produced in reactors to normalize the produced methane amount.

2.4. Surface Properties and Bond Structure

Scanning electron microscopy (SEM) was utilized to evaluate microstructural changes on the surface of wastes after the application of chemical-assisted thermal pretreatments. Raw and pretreated FVHW were lyophilized and coated with gold-palladium for 120 seconds under 18 mA vacuum and samples were examined by SEM with a Zeiss Leo 1430 electron microscope (ZEISS International, Oberkochen, Germany) at a voltage of 15 kV.

To investigate the impacts of thermochemical pretreatment on waste structure molecular-bond characterization, samples were examined by an FTIR-Varian 1000 Scimitar series FTIR spectrometer (Agilent Tech., Santa Clara, USA). Samples were lyophilized and pelletized with KBr before analyses. Measurements were performed in 500 cm^{-1} and 4000 cm^{-1} wavelength range with a spectral resolution of 4 cm^{-1} and with 16 scans per sample.

3. Results and Discussion

3.1. Impacts of Chemical Assisted Thermal Pretreatments

High carbohydrate content as cellulose and hemicelluloses was obtained as 56.86% from the FVHW (Table 2). Carbon content (C%) of FVHW from the elemental analysis was found to be 34.16% (49.1% VS basis). The analysis results suggested that FVHW is a suitable waste substrate for methane production.

Table 2. Characterization analysis results of FVHW.

Parameters	Value
Total Solid, TS, g TS/kg dry sample	913.93
Volatile Solid, VS, g VS/kg TS	694.09
Total Kjeldahl Nitrogen, TKN, g N/kg VS	24.80
Total Organic Carbon, g TOC/kg VS	501.77
Soluble Protein (g SProtein/kg VS)	94.56
Soluble Sugar (g SGlucose/kg VS)	279.50
Extractable Material and Lipid (g/kgVS)	42.17
Van Soest Fractionation (TS basis)	
Celulose, %	33.12
Hemicellulose, %	23.74
Lignin, %	9.32
Soluble matter, %	33.82
Elemental Analysis (TS basis)	
Carbon, C, %	34.16
Hydrogen, H, %	5.03
Nitrogen, N, %	2.39
Sulphur, S, %	0.82
C/N	14.29

The main purpose of NaOH- and HCl-assisted thermal pretreatment of FVHW is to make more cellulose and hemicellulose available for the anaerobic microorganism and to enhance methane production. Since the enhancement of methane production was related to sCOD and sSugar [3,32], impacts of NaOH- and HCl-assisted thermal pretreatment on sCOD and sSugar were evaluated for the FVHW. The results of the experiments provided a clear indication that NaOH- and HCl-assisted thermal pretreatment boosted a remarkable increase in sCOD concentration, as illustrated in Figure 1a,c. As shown in Figure 1a, an average sCOD concentration of 564 mgsCOD gVS^{-1} was obtained with no NaOH addition, almost at the same concentration level, which varied in the narrow range of 486–629 mgsCOD gVS^{-1}, regardless of the changes of other independent variables. On the other hand, increasing NaOH concentration from 0% to 3.3% and 6.5% boosted the COD solubilization. The average sCOD values of 1112 mgsCOD gVS^{-1} and 1243 mgsCOD gVS^{-1} were measured for 3.3% and 6.5% NaOH concentration, respectively, with a moderate fluctuation between 832–1248 mgsCOD gVS^{-1} and 1034–1356 mgsCOD gVS^{-1}. Furthermore, no remarkable changes with other independent variables were found. The most significant increase in sCOD of 1356 mgsCOD gVS^{-1} was observed in the pretreatment experiment performed at 6.5% NaOH concentration, 100 °C reaction temperature, 1 h reaction time, and 500 rpm mixing speed, yielding more than a 240% increment compared to raw sCOD of the FVHW. Conclusively, the addition of NaOH resulted in more dissolved organic material, measured as sCOD. Thus, the concentration of sCOD increased with increasing NaOH concentration.

Figure 1c presents that HCl-assisted thermal treatment also raised the sCOD release, but it produced a lower impact on sCOD than by the application of NaOH-assisted thermal pretreatment. When no HCl was applied in the pretreatment experiments, the level of measured sCOD was in the range of 504–613 mgsCOD gVS^{-1} with the average sCOD of 554 mgsCOD gVS^{-1}. A similar trend observed in sCOD by increasing the amount of NaOH was observed by increasing the amount of HCl. However, the measured amounts were at the level of 613–1149 mgsCOD gVS^{-1} (average 800 mgsCOD gVS^{-1}) and 590–1232 mgsCOD gVS^{-1} (average 908 mgsCOD gVS^{-1}), for 2.5% and 5% HCl, respectively, used in the pretreatments.

Basically, Figure 1a,c indicated that (i) the increase in sCOD obtained by increasing the NaOH concentration was higher than the increase in sCOD measured by increasing the HCl concentration. (ii) The sCOD increment consistently stayed higher when the NaOH was increased to 3.3–6.5% while it remained at the same level when other independent variables were kept in their range. (iii) The amount of sCOD was also increased when the HCl was elevated to 2.5–5%, but apart from the increase in HCl, the increment of sCOD was affected by the dry matter content, reaction temperature and mixing speed. This observation is particularly essential so that the selected chemical and its concentration should be adjusted to solubilize the lignocellulosic structure, but not to oxidize them to produce further metabolites in the process optimization. Furthermore, apart from the selected chemical and the amount applied, each of the other independent variables should also be evaluated in process optimization.

Teghammar et al. (2010) [33] investigated the steam explosion and nonexplosive hydrothermal pretreatment with NaOH for paper tube residuals to improve biogas production. The effect of the reaction temperature, reaction time and NaOH concentration were statistically evaluated on sCOD. sCOD of the untreated paper tube was measured as 262 mg L^{-1}, whereas sCOD of the nonexplosively pretreated paper tube at 190 °C with 2% NaOH was found at 2280 mg L^{-1}. The results from Teghammar et al. (2010) [33] revealed that increasing the NaOH concentration yielded high sCOD production, which is somehow coupled with high phenolic compounds solubilization. Wang et al. (2009) [34] applied hydrothermal alkaline pretreatment to sorted municipal solid waste to enhance biogas production. sCOD concentration increased from 5931 mg L^{-1} at 1 g NaOH/100 g solid to 12007 mg L^{-1} at 4 g NaOH/100 gr solid. They indicated that the concentration of sCOD increased with the addition of NaOH. Jard et al., 2013 [35] investigated the effect of different thermochemical pretreatments on the solubilization and anaerobic degradability of the red macroalgae *Palmaria palmata*. They also indicated that thermo-chemical pretreatment induced a change in the structure of macroalgae, increasing the cell-wall porosity and thus facilitating the solubilization of COD. The results from our study were found to be consistent with Wang et al. (2009), Teghammar et al. (2010) and Jard et al. (2013) [33–35].

The NaOH- and HCl-assisted thermal pretreatment experiments also pursued to observe the alteration of soluble reducing sugars (Figure 1b,d). The soluble reducing sugar profile displayed in Figure 1b clearly explains the impact of NaOH-assisted thermal pretreatment; with the caustic addition, produced sugar from cellulose and hemicellulose was transformed to further metabolites. The highest sRedSugar concentration was measured as 335 mg glucose gVS^{-1} from the FVHW pretreated at 0% NaOH concentration, 100 °C reaction temperature, 4 h reaction time, and 0 rpm mixing speed, where the maximum increase in reducing sugar was calculated as 67.2%. On the other hand, soluble sugar concentration increased by increasing the HCl concentration from 0% to 2.5%, but the sugars produced by increasing HCl dosage to 5% were transformed. Figure 1d apparently shows the impact of HCl-assisted thermal pretreatment: With HCl addition, thermal treatment at a temperature of 100 °C and 3% dry matter content greatly intensified its solubility to 831 mg glucose gVS^{-1} and resulted in a sharp decrease to the average level of 68 mg glucose gVS^{-1} when the HCl concentration was raised to 5%. It should be remembered that the decrease in sugar concentration necessitates the optimization of the amount of chemical usage in the pretreatment process.

NaOH-Assisted Thermal Pretreatment

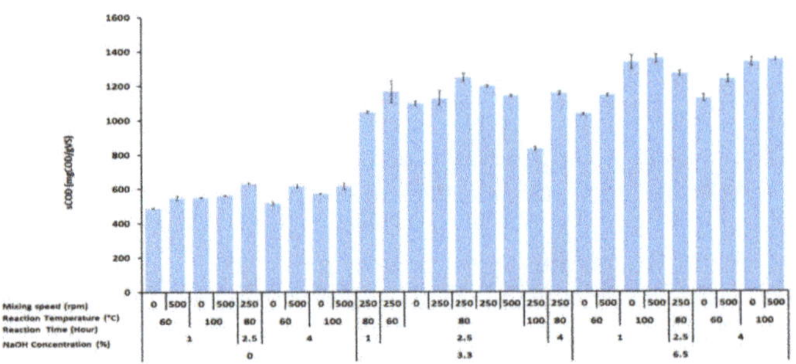

(**a**)

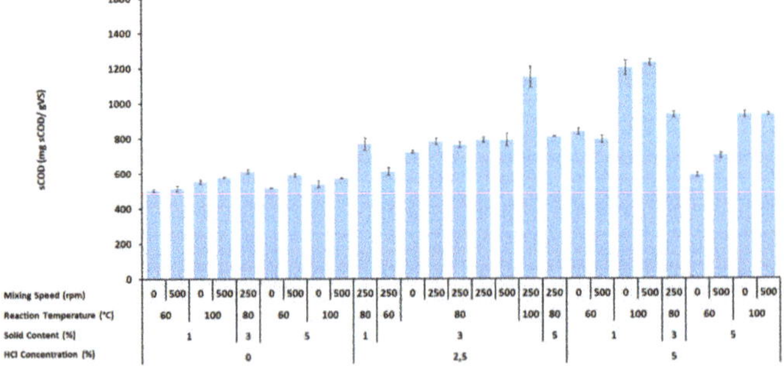

(**b**)

HCl-Assisted Thermal Pretreatment

(**c**)

Figure 1. *Cont.*

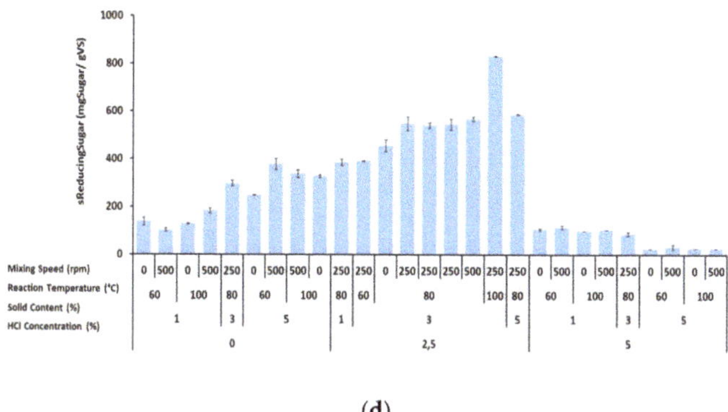

(d)

Figure 1. Effects of NaOH-assisted thermal pretreatment on sCOD (**a**) and sRedSugars (**b**); and effects of HCl-assisted thermal pretreatment on sCOD (**c**) and sRedSugars (**d**).

3.2. Optimization of Chemical Assisted Thermal Pretreatment Process for sCOD and sSugar

RSM and CCD were applied to derive optimum conditions effective on FVHW for NaOH and HCl-assisted thermal pretreatment. Statistical evaluation was made and the models for sCOD and sSugar were constructed for both pretreatments. Analysis of variance (ANOVA) was employed to investigate the quality of model equations. Quadratic regression models for sCOD and sSugar were significant, as apparent from the Fisher's *F*-test with the low value (P model > *F* = 0.0001) for NaOH- and HCl-assisted thermal pretreatments. To check the significance of the fit of equations for sCOD and sSugar of NaOH- and HCl-assisted thermal pretreatments, ANOVA was made and results are presented in Tables 3 and 4, respectively.

Table 3. ANOVA results for chemical-assisted thermal pretreatments for sCOD.

NaOH-Assisted Thermal Pretreatment					
Source	Sum of Square	Degrees of Freedom	Mean Square	F-Value	Prob $> F$
sCOD					
Model	4.907×10^6	14	3.505×10^5	53.31	<0.0001
A—Reaction Temperature	43898.63	1	43898.63	6.68	0.0136
B—Reaction Time	23782.27	1	23782.27	3.62	0.0646
C—Mixing Speed	29589.73	1	29589.73	4.50	0.0403
D—NaOH Concentration	4.147×10^6	1	4.147×10^6	630.71	<0.0001
AB	5767.11	1	5767.11	0.8772	0.3547
AC	10885.13	1	10885.13	1.66	0.2058
AD	62377.77	1	62377.77	9.49	0.0038
BC	553.20	1	553.20	0.0841	0.7733
BD	3.86	1	3.86	0.0006	0.9808
CD	245.92	1	245.92	0.0374	0.8476
A^2	54067.88	1	54067.88	8.22	0.0066
B^2	0.3052	1	0.3052	0.0000	0.9946
C^2	1573.40	1	1573.40	0.2393	0.6274
D^2	1.185×10^5	1	1.185×10^5	18.02	0.0001
$R^2 = 0.9503$, $Adj\text{-}R^2 = 0.9325$, $Adeq\ Precision = 20.0622$ $C.V.\ \% = 8.33$					

HCl-Assisted Thermal Pretreatment					
Source	Sum of Square	Degrees of Freedom	Mean Square	F-Value	Prob $> F$
sCOD					
Model	2.104×10^6	14	1.503×10^5	36.35	<0.0001
A—Reaction Temperature	4.591×10^5	1	4.591×10^{05}	111.09	<0.0001
B—HCl Concentration	1.132×10^6	1	1.132×10^6	273.80	<0.0001
C—Solid Content	69230.38	1	69230.38	16.75	0.0002
D—Mixing Speed	10722.26	1	10722.26	2.59	0.1153
AB	2.004×10^5	1	2.004×10^5	48.49	<0.0001
AC	13407.44	1	13407.44	3.24	0.0794
AD	604.22	1	604.22	0.1462	0.7043
BC	1.182×10^5	1	1.182×10^5	28.60	<0.0001
BD	216.27	1	216.27	0.0523	0.8203
CD	5219.10	1	5219.10	1.26	0.2680
A^2	18583.84	1	18583.84	4.50	0.0404
B^2	11063.97	1	11063.97	2.68	0.1099
C^2	4855.75	1	4855.75	1.17	0.2851
D^2	20711.11	1	20711.11	5.01	0.0310
$R^2 = 0.9288$, $Adj\text{-}R^2 = 0.9033$, $Adeq\ Precision = 22.3369$, $C.V.\ \% = 8.53$					

Table 4. ANOVA results for chemical-assisted thermal pretreatments for sSugar.

NaOH-Assisted Thermal Pretreatment					
Source	Sum of Square	Degrees of Freedom	Mean Square	F-Value	Prob > F
sSugar					
Model	8.671×10^5	14	61936.12	854.31	<0.0001
A—Reaction Temperature	2058.59	1	2058.59	28.40	<0.0001
B—Reaction Time	360.68	1	360.68	4.98	0.0315
C—Mixing Speed	83.78	1	83.78	1.16	0.2890
D—NaOH Concentration	6.434×10^5	1	6.434×10^5	8874.32	<0.0001
AB	42.60	1	42.60	0.5876	0.4480
AC	117.12	1	117.12	1.62	0.2112
AD	3549.03	1	3549.03	48.95	<0.0001
BC	8.04	1	8.04	0.1109	0.7409
BD	1264.29	1	1264.29	17.44	0.0002
CD	50.10	1	50.10	0.6911	0.4109
A^2	207.77	1	207.77	2.87	0.0985
B^2	67.99	1	67.99	0.9378	0.3388
C^2	12.02	1	12.02	0.1657	0.6862
D^2	82123.05	1	82123.05	1132.76	<0.0001

$R^2 = 0.9967$, *Adj-R^2* $= 0.9956$, *Adeq Precision* $= 69.1844$, *C.V.* $\% = 7.24$

HCl-Assisted Thermal Pretreatment					
Source	Sum of Square	Degrees of Freedom	Mean Square	F-Value	Prob > F
sSugar					
Model	2.318×10^6	14	1.656×10^5	30.43	<0.0001
A—Reaction Temperature	30752.88	1	30752.88	5.65	0.0224
B—HCl Concentration	2.588×10^5	1	2.588×10^5	47.56	<0.0001
C—Solid Content	43307.69	1	43307.69	7.96	0.0075
D—Mixing Speed	8252.21	1	8252.21	1.52	0.2255
AB	2294.35	1	2294.35	0.4217	0.5199
AC	45.89	1	45.89	0.0084	0.9273
AD	512.64	1	512.64	0.0942	0.7605
BC	1.395×10^5	1	1.395×10^5	25.64	<0.0001
BD	1547.35	1	1547.35	0.2844	0.5969
CD	1193.16	1	1193.16	0.2193	0.6422
A^2	19378.84	1	19378.84	3.56	0.0666
B^2	6.606×10^5	1	6.606×10^5	121.41	<0.0001
C^2	21405.59	1	21405.59	3.93	0.0544
D^2	7628.98	1	7628.98	1.40	0.2435

$R^2 = 0.9161$, *Adj-R^2* $= 0.8860$, *Adeq Precision* $= 16.5818$, *C.V.* $\% = 26.19$

The quality of the fit of models was expressed by the coefficient of determination R^2 and adjusted determination of coefficient, *Adj-R^2*. R^2 and *Adj-R^2* were found to be 0.9503 and 0.9325, and 0.9288 and 0.9033 for sCOD, indicating a high degree of correlation between the response and the independent variables in the quadratic models for NaOH- and HCl-assisted thermal pretreatments, respectively. Signal-to-noise ratio is defined as adequate precision and a ratio bigger than four is favorable. Therefore, the ratios 20.06 and 22.33 have shown adequate signals for the models. Values of Prob > F less than 0.05 indicated that model terms are significant [36]. While the linear effect of reaction temperature (A), mixing speed (C) and NaOH concentration (D), interactive terms of AD, and square terms of reaction temperature (A^2) and NaOH concentration (D^2) were significant for the sCOD model of the NaOH-assisted thermal pretreatment, the linear effect of reaction temperature (A), HCl concentration (B) and solid content (C), interactive terms of AB and BC, and square terms of reaction temperature (A^2) and mixing speed (D^2), were significant for the sCOD model of HCl-assisted thermal pretreatment.

Similarly, the values of R^2 and *Adj-R^2* were computed as 0.9967 and 0.9956, and 0.9161 and 0.8860 for quadratic models of sSugar of NaOH- and HCl-assisted thermal pretreatments, respectively. Adequate precision with the values of 69.18 and 16.58 was obtained, indicating models can be applied in the design space. Furthermore, the linear effect of reaction temperature (A), reaction time (B), and NaOH concentration (D); interactive terms of AD and BD; and square terms of NaOH concentration (D^2) were significant for the sSugar model of NaOH-assisted thermal pretreatment, while the linear effect of reaction temperature (A), HCl concentration (B) and solid content (C); interactive terms of BC; and square terms of HCl concentration (B^2) were significant for the sSugar model of HCl-assisted thermal pretreatment.

On the other hand, other variables shown in ANOVA analysis presented in Tables 3 and 4 were not significant ($p > 0.005$).

sCOD and sSugar models were operated for the optimization of NaOH- and HCl-assisted thermal pretreatment conditions by Design-Expert®software. During the optimizations of both pretreatment applications, a cost-driven approach was applied. Reaction temperature, reaction time and NaOH concentration were diminished, while sCOD and sSugar were maximized for NaOH-assisted thermal pretreatment. Mixing speed was kept in the ranges. Optimum conditions were obtained with a desirability of 0.681 at 0.67% NaOH concentration, 65 °C reaction temperature, 1 h reaction time, and 500 rpm mixing speed. Under these conditions, 688 mgsCOD mgVS^{-1} and 222 mgsSugar mgVS^{-1} were predicted for sCOD and sSugar, respectively. To validate the optimization, pretreatment experiments were executed under these proposed conditions. In these runs, 712 mgsCOD mgVS^{-1} and 236 mgsSugar mgVS^{-1} were realized for sCOD and sSugar models, respectively. Solubilization of sCOD and sSugar compared to the raw FVHW were found to be 80% and 18%, respectively, for cost optimization conditions.

To optimize the HCl-assisted thermal pretreatment, initial solid loading, sCOD and sSugar were maximized, whereas reaction temperature and HCl concentration were minimized from the point of environmentally sustainable procedures and process costs. Since mixing speed was found to be insignificant for both models, it was set within the range. Optimum conditions were obtained with the desirability of 0.647 at 1.44% HCl concentration, 60 °C reaction temperature, 5% solid content, and 324 rpm mixing speed. Under these conditions, sCOD and sSugar values were estimated as 682 mgsCOD mgVS^{-1} and 553 mgsSugar mgVS^{-1}, respectively. To validate the optimization, specific batch runs were conducted. sCOD and sSugar were determined as 698 mgsCOD mgVS^{-1} and 438 mgsSugar mgVS^{-1}, verifying the acceptable performance of the models for HCl-assisted thermal pretreatment. The increments of sCOD and sSugar compared to the raw wastes were computed as 76% and 119%, respectively for cost optimization conditions.

The impacts of independent variables of NaOH-assisted thermal pretreatment on sCOD and sSugar solubilization are demonstrated in Figure 2a,b. Figure 2a explains the individual impact of each independent variable on sCOD. Approximately, 700 mgsCOD mgVS^{-1} can be produced when 1-hour reaction time, 0.67% NaOH concentration, 500 rpm mixing speed, and 65 °C reaction temperature conditions were applied in the pretreatment. On the other hand, increasing NaOH concentration resulted in higher sCOD solubilization of approximately 1200–1300 mgsCOD mgVS^{-1}. However, as it is apparently observed from Figure 2b, increasing the NaOH concentration yielded extremely decreasing sSugar solubilization that can be converted directly into methane. Approximately, 200–250 mgsSugar mgVS^{-1} should be produced under the same pretreatment conditions.

The independent variables' impact on sCOD and sSugar solubilization after the application of HCl-assisted thermal pretreatment is also presented in Figure 2c,d. The individual impact of each independent variable on sCOD presented in Figure 2c shows that 600–700 mgsCOD mgVS^{-1} should be produced by the application of 5% solid content, 1.44% HCl concentration, 324 rpm mixing speed, and 60 °C reaction temperature conditions in the pretreatment. As it is clearly seen from Figure 2d, with the increasing HCl concentration, sSugar concentration was reduced to further compounds by the transformation and obtained very low sugar concentration in the pretreatment medium.

The most important difference between the two pretreatments was the use of different initial solid content in the pretreatment. Enhancement of methane production from FVHW can be achieved by the usage of 5% solids content along with the 60–80 °C reaction temperature and no or very low chemical agent in the pretreatment.

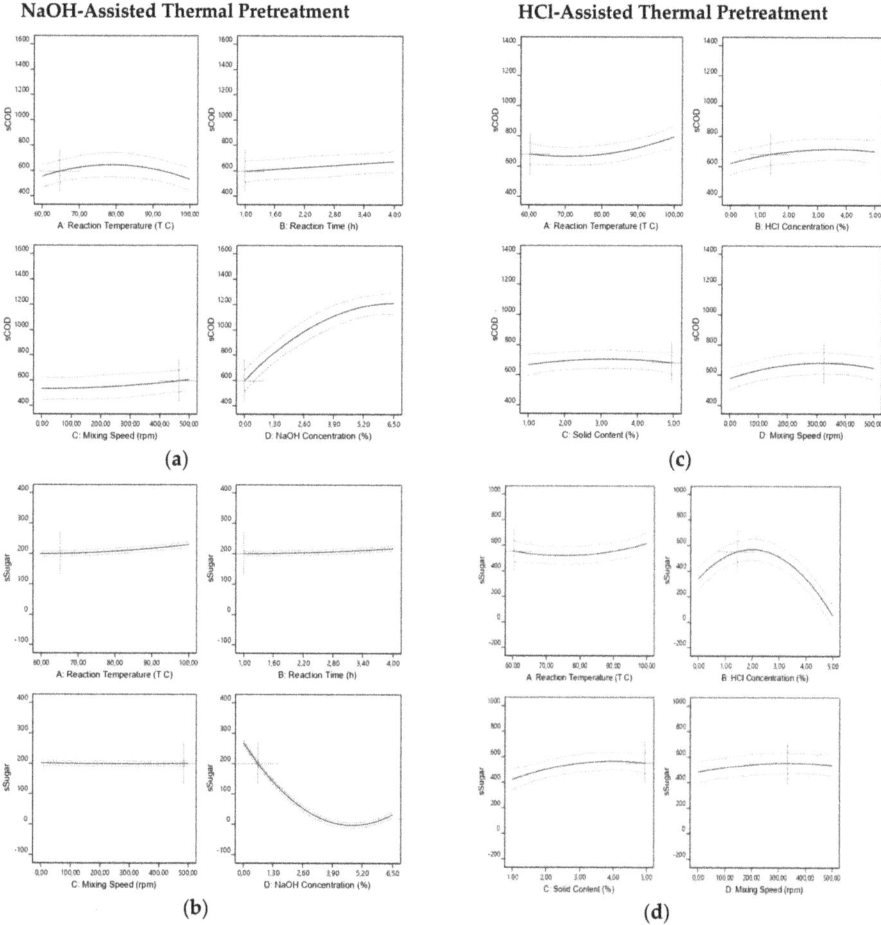

Figure 2. Individual impacts of independent variables (Reaction temperature, reaction time, mixing speed and NaOH concentration) on sCOD (**a**) and sSugar (**b**) of NaOH-assisted thermal pretreatment; individual impacts of independent variables (Reaction temperature, HCl concentration, solid content and mixing speed) on sCOD (**c**) and sSugar (**d**) of HCl-assisted thermal pretreatment.

3.3. Methane Production

Since increasing the NaOH and HCl concentrations from 0% to 6.6% and 0% to 5%, respectively, yielded with the lowly-soluble sugar concentration due to degradation of the produced sugar and possibly the conversion to refractory compounds or inhibitors, the methane productions were investigated with a reduced number of pretreatment experiments. A BMP value of raw FVHW without pretreatment, serving as a comparison, was found as 217 mLCH$_4$ gVS^{-1}. The observed methane production profiles for thermal and NaOH- and HCl-assisted thermal pretreatments are presented in Figure 3a,b.

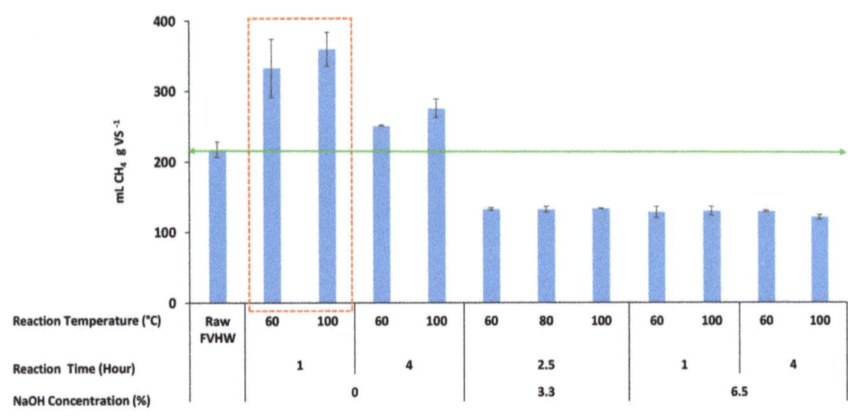

(a)

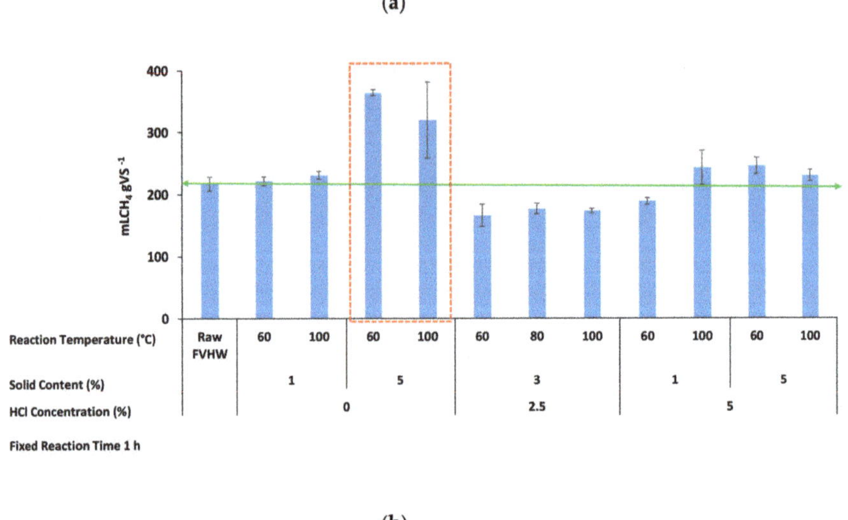

(b)

Figure 3. Methane production profiles after the NaOH-assisted thermal pretreatments (**a**); HCl-assisted thermal pretreatments (**b**).

Figure 3a,b shows that thermal pretreatments with no addition of chemicals led to the highest methane productions. Methane production was attained in the range of 222–365 mLCH$_4$ gVS^{-1} (av. 295 mLCH$_4$ gVS^{-1}) at a range of 60–100 °C reaction temperature and 1–5% solid content. The highest BMP values were 365 and 360 mLCH$_4$ gVS^{-1} acquired at 60 °C and 100 °C reaction temperature, respectively, along with 5% solid content and 1 h reaction time, as indicated in the red dash line zone.

The lowest BMP value (121 mLCH$_4$ gVS^{-1}) was measured from the FVHW pretreated at a NaOH concentration of 6.5%, a reaction temperature of 100 °C, and a reaction time of 4 h. By adding NaOH as 3.3% and 6.5% concentrations, the production of methane decreased from 251–360 mLCH$_4$ gVS^{-1} (av. 305 mLCH$_4$ gVS^{-1}) without NaOH to 132–133 mLCH$_4$ gVS^{-1} (av. 133 mLCH$_4$ gVS^{-1}) at 3.3% and further dropped to 121–129 mLCH$_4$ gVS^{-1} (av. 127 mLCH$_4$ gVS^{-1}) at 6.5%. As it is easily captured from Figure 3a, pretreatment experiments without NaOH showed an average of 41% enhancement in methane production; on the other hand, an average 40% decrease was observed in methane production with the increased NaOH concentration. The increased concentration of NaOH resulted in low methane production due to degradation of the produced sugar and possible conversion to refractory compounds or inhibitors [37–39] or increased concentration of Na$^+$ itself or combined effects of both. In fact, for the sake of clarity, Na$^+$ concentration was approximately 1.8 g Na$^+$/L in the BMP reactor, it reached 2.5 g Na+/L with the supplied NaHCO$_3$ as a buffer in the BMP reactor for the lowest BMP obtained as 121 mLCH$_4$ gVS^{-1}. Despite the fact that slight inhibition concentrations for Na$^+$ are in the range of 3.5–5.5 g/L [40], it may be concluded that two effects (degradation of sugar to refractory compounds or inhibitors and Na$^+$ concentration) caused inhibition.

As for HCl-assisted thermal pretreatment, minimum methane of 166 mLCH$_4$ gVS^{-1} was produced from the wastes pretreated at 2.5% HCl concentration, 60 °C reaction temperature, and 3% solid content. As it is clearly seen from Figure 3b, methane production decreased when the HCl concentration was increased. However, the decrease in methane production was not at the same level as the decrease in NaOH application. Methane production decreased to an average value of 172 mLCH$_4$ gVS^{-1} when the amount of HCl was increased to 2.5% and the number of solids was decreased to 3%. On the contrary to this, methane production increased to an average value of 240 mLCH$_4$ gVS^{-1} with increasing solids content to 5% and in the range of 60–100 °C reaction temperature. While an average of 31% enhancement was observed in methane production in the pretreatment experiments where HCl was not used, an average of 21% decrease occurred in methane production when 2.5% HCl was added.

Conclusively, (i) a higher amount of methane production was obtained by the thermal application without using the chemical agent of NaOH or HCl. (ii) Although the chemical usage contributes to the hydrolysis rate of the lignocellulosic structure, the selected high dose for NaOH (3.3–6.5%) and HCl (2.5–5%) should be the probable cause of low methane production. (iii) Maximum enhancement of methane production was 47–68% compared to raw FVHW when 5% solid content, 1 hour reaction time and 60–100 °C reaction temperature were used in pretreatments.

Song et al. (2014) [41] examined the impacts of seven chemical pretreatments on methane production from corn straw. Four acid reagents (H$_2$SO$_4$, HCl, H$_2$O$_2$, and CH$_3$COOH) at concentrations of 1, 2, 3, and 4% and three alkali reagents (NaOH, Ca(OH)$_2$ and NH$_3$.H$_2$O) at concentrations of 4, 6, 8, and 10% were applied in the pretreatments at 25 °C during the 7 days. Results revealed that acid and alkaline treatments increased methane production by approximately 10.3% to 115.4% higher yield compared to raw. This phenomenon can be explained by the fact that alkaline and acid pretreatments promote organic solubilization and increase the surface area available for enzymatic accessibility. Monlau et al., (2012) [42] applied five thermo-chemical pretreatments by using NaOH, H$_2$O$_2$, Ca(OH)$_2$, HCl, FeCl$_3$ at 4–10 gr/100gr TS concentration along with the 30–170 °C reaction time and 1–24 h reaction time to sunflowers stalks. They found that methane potentials of pretreated sunflower with different chemicals increased to 3–35% compared to the raw sunflower. Although corn straw and sunflower stalks do not seem to be similar to the FVHW at first glance, the cellulose and hemicellulose contents of the sunflower are close to FVHW used in this study to make a comparison. Furthermore, the enhancement of methane production results from this study was found to be coherent with the literature and revealed that pretreatment conditions necessitate optimization and depend on the structure and composition of lignocellulosic wastes. As a result, the selection of a wide of range levels of the independent variables has riveting characteristics to optimize the process in a broad design boundary and to evaluate the positive and negative impacts of independent variables along with their ranges. The results obtained from optimization in the design space have resulted in the suggestion of a low amount of NaOH and HCl usage due to the inhibition effect of the high amount of chemicals on methane production. The selection of a wide range of levels of independent variables has provided an opportunity for evaluating extreme points in optimization.

3.4. Surface Modification and Molecular Bond Changes in FVHW by Chemical Assisted Thermal Pretreatment

SEM images and FTIR spectra of FVHW pretreated at optimized conditions for sCOD and sSugar by NaOH-assisted thermal pretreatments (0.67% NaOH, 65 °C, 1 h, 500 rpm) and HCl-assisted thermal pretreatments (1.44% HCl, 60 °C, 5% DM, 324 rpm), and maximum CH_4 production conditions (100 °C, 0% NaOH, 1 h, 0 rpm and 60 °C, 0% HCl, 5% DM, 0 rpm) compared to those of the raw FVHW are demonstrated in Figure 4a,b, respectively.

SEM images presented in Figure 4a,b depicted that both NaOH- and HCl-assisted thermal pretreatments caused modification on the surfaces of FVHW. The images of raw wastes revealed that lignocellulosic material had a smooth and fluent fibrous structure on its surface before any pretreatment. Despite the robust structure of lignocellulosic material, both NaOH- and HCl-assisted thermal pretreatments hydrolyzed lignocellulosic material and destructed the smooth outer surface by opening it up and damaged the inner structure of tissues. SEM images remarked consistent results with methane production from FVHW.

In the dashed line zone of the FTIR spectra, both NaOH- and HCl-assisted thermal pretreated wastes presented a different spectrum than from raw lignocellulosic wastes. Especially, modifications in spectra at 400–500, 780, 875, 1000–1100, 1150, 1245, 1310, 1380, 1500–1600, 1750, 2300–2400 and 400–500, 775, 875, 1000–1100, 1150, 1280–1310, 1380, 1500–1600, 1750, and 2300–2400 cm^{-1} wavelengths can be apparently observed from the pretreated samples of FVHW by the application of NaOH- and HCl-assisted thermal pretreatments, respectively. Most of the observed modifications in the molecular bonds were found to be similar. While FTIR band intensities were closer to each other at 60–100 °C and 0–0.67% NaOH pretreatment conditions and gave lower density than the raw sample, FTIR band intensities of samples pretreated at 1.44% HCl condition were found to be closer to the raw sample.

At 470 and 775 cm^{-1}, FTIR spectra show vibrations of Si–O–Si and NH_2 bonds, respectively [43,44]. Si may come from small amounts of soil adhering to the roots of FVHW. Absorbance at 875 cm^{-1} wavelength represents the vibration of glucosidic linkage in hemicellulose [45], and it can be concluded that all processed wastes had the same absorbance value, indicating a transformation in the chemical bond structure caused by pretreatment. Bond characteristic changes in samples belonged to C–O stretching vibrations in cellulose/hemicellulose and the aryl-OH group in lignin at 1050 cm^{-1} [46], C–O–C vibrations at β-glucosidic linkages in cellulose and hemicellulose at 1150 cm^{-1} [47], and C–O adsorption at 1245 cm^{-1} [48]. Observed changes at 1310 cm^{-1} and 1380 cm^{-1} of FTIR spectra indicated CH_2 wagging in cellulose and hemicellulose and planar C–H bending in cellulose, hemicellulose, and lignin, respectively [45]). Since wavelengths between 1500–1600 cm^{-1} are attributed to aromatic C=C bonds, changes at 1510 cm^{-1} were found to belong to aromatic ring vibrations of lignin [49]. As changes at 1734 cm^{-1} pointed to C=O stretching vibration in acetyl groups of hemicellulose [47], FTIR spectra observed at 1750 cm^{-1} may belong to this group. Out of the dashed line zone, pretreated wastes showed a difference only at one wavelength (2400 cm^{-1}) with low densities, compared to raw FVHW. The band at 2400 cm^{-1} was attributed to C=O stretching within the carbon dioxide [50].

In summary, FTIR analysis revealed that pretreatment caused changes in molecular bond characterization especially in the region of 800–1500 cm^{-1}.

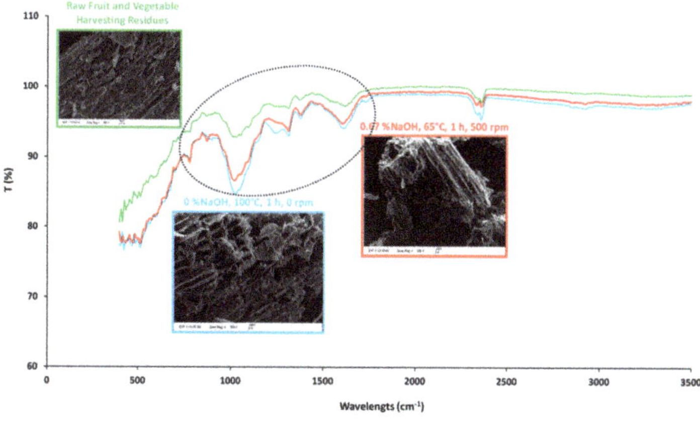

(**a**)

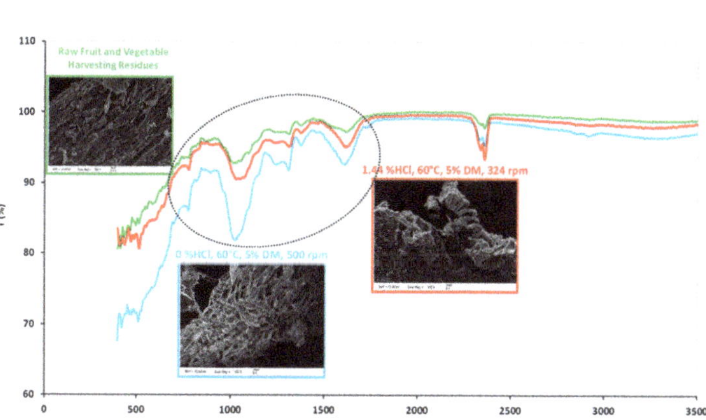

(**b**)

Figure 4. SEM images and FTIR spectra of raw and pretreated FVHW at 0% NaOH concentration, 100 °C reaction temperature, 1 h reaction time and no mixing, and 0.67% NaOH concentration, 65 °C reaction temperature, 1 h reaction time and 500 rpm mixing speed conditions (**a**); SEM images and FTIR spectra of raw and pretreated FVHW at 0% HCl concentration, 60 °C reaction temperature, 5% DM content and 500 rpm mixing speed and 1.44 %HCl concentration, 60 °C reaction temperature, 5% DM content and 324 rpm mixing speed conditions (**b**).

4. Conclusions

FVHW collected from the extensive production greenhouses area of Antalya city was thermochemically pretreated to enhance methane production. The most significant increase in sCOD of 1356 mgsCOD gVS^{-1} was observed at 6.5% NaOH concentration, 100 °C reaction temperature, 1 h reaction time, and 500 rpm mixing speed, yielding more than a 240% increment compared to raw sCOD of the FVHW. The highest sSugar concentration was measured as 335 mg glucose gVS^{-1} from the FVHW pretreated with no chemicals at a 100 °C reaction temperature, 4 h reaction time, and no mixing, where the maximum increase in reducing sugar was calculated as 67.2% compared to raw

FVHW. Increasing NaOH and HCl concentrations resulted in higher sCOD but extremely lower sSugar solubilization. Results of sCOD and sSugar from the NaOH- and HCl-assisted thermal pretreatment experiments were modeled, and statistically significant regression models were developed. The most important independent variables were found to be chemical concentration (as NaOH and HCl), solid content and reaction temperature for the pretreatment processes.

Two-hundred-and-twenty-two to 365 mLCH$_4$ gVS^{-1} methane productions were obtained by the simple thermal application without using NaOH or HCl. Maximum enhancement of methane production was 47–68% compared to raw FVHW at 5% solid content, 1-hour reaction time and 60–100 °C reaction temperature.

The optimization of chemically-assisted thermal pretreatment processes should be done with the methane production results obtained from the BMP test. On the other hand, increasing the NaOH and HCl concentrations from 0% to 6.6% and 0% to 5%, respectively, resulted in low soluble sugar concentrations. Furthermore, optimization in the design space suggested the use of low amounts of NaOH and HCl due to the inhibition risk of the high amount of chemicals. In addition, BMP tests yielded to low methane production for the samples pretreated with high a amount of chemicals (NaOH and HCl). As a result, it was found that sSugar and methane production results were correlated and sSugar data can be used in optimization in a reliable way for this substrate.

Results from this comprehensive study have been encouraging that it serves information to build the potential application and to reduce the waste-related environmental pressure in the production-dense area. Gained optimization solution in the design space resulted in the suggestion of no chemical usage due to the inhibition effect of the high amount of chemicals on methane production. The selection of a wide range of levels of independent variables provided an opportunity for evaluating extreme points in optimization.

Author Contributions: Conceptualization, A.N.P. and H.C.; Methodology, Ü.G., L.D. and E.U.; Software, A.N.P. and E.U.; Formal Analysis, Ü.G., L.D. and E.U.; Investigation, A.N.P., H.C. Writing-Original Draft Preparation, A.N.P., and V.Y.; Writing-Review & Editing, A.N.P. and H.C. All authors have read and agreed to the published version of the manuscript.

Funding: This research was funded by the Scientific and Technological Research Council of Turkey, (107Y145) under the Pia Bosphorus bilateral agreements signed by France and Turkey Governments. This study research has also been financed (Grant no: 2008.02.0121.024) by the Project Management Unit of Akdeniz University from Turkey.

Conflicts of Interest: The authors declare no conflict of interest.

References

1. Shanthi, M.; Rajesh, B.J.; Sivashanmugam, P. Synergistic effect of combined pretreatment in solubilizing fruits and vegetable residue for biogas production: Hydrolysis, energy assessment. *Fuel* **2019a**, *250*, 194–202. [CrossRef]

2. Available online: http://www.fao.org/faostat/en/#data/QC (accessed on 2 November 2019).

3. Us, E.; Perendeci, A. Improvement of methane production from greenhouse residues: Optimization of thermal and H$_2$SO$_4$ pretreatment process by experimental design. *Chem. Eng. J.* **2012**, *181*, 120–131. [CrossRef]

4. Calabro, P.S.; Gori, M.; Lubello, C. European trends in greenhouses gases emissions from integrated solid waste management. *Environ. Technol.* **2015**, *36*, 2125–2137. [CrossRef] [PubMed]

5. Soltan, M.; Elsamadony, M.; Mostafa, A.; Awad, H.; Tawfik, A. Harvesting zero waste from co-digested fruit and vegetable peels via integrated fermentation and pyrolysis processes. *Environ. Sci. Pollut. Res.* **2019**, *26*, 10429–10438. [CrossRef]

6. Ceylan, R.F.; Sayin, C.; Yelboğa, M.N.M.; Özalp, M.; İlbasmiş, E.; Sav, O. Land ownership and profitability of greenhouse production: Antalya Case. *Turk. J. Agric. Food Sci. Technol.* **2009**, *6*, 930–935. [CrossRef]

7. Dahunsi, S.O. Liquefaction of pineapple peel: Pretreatment and process optimization. *Energy* **2019**, *185*, 1017–1031. [CrossRef]

8. Chandra, R.; Takeuchi, H.; Hasegaw, T. Methane production from lignocellulosic agricultural crop wastes: A review in context to second generation of biofuel production. *Renew. Sustain. Energy Rev.* **2012**, *16*, 1462–1476. [CrossRef]

9. Chandra, R.; Takeuchi, H.; Hasegawa, T.; Kumar, R. Improving biodegradability and biogas production of wheat straw substrates using sodium hydroxide and hydrothermal pretreatments. *Energy* **2012**, *43*, 273–282. [CrossRef]

10. Paudel, S.R.; Banjara, S.P.; Choi, O.K.; Park, K.Y.; Kim, Y.M.; Lee, J.W. Pretreatment of agricultural biomass for anaerobic digestion: Current state and challenges. *Bioresour. Technol.* **2017**, *245*, 1194–1205. [CrossRef]

11. Sawatdeenarunat, C.; Nguyen, D.; Surendra, K.C.; Shrestha, S.; Rajendran, K.; Oechsner, H.; Xie, L.; Khanal, S.K. Anaerobic biorefinery: Current status, challenges, and opportunities. *Bioresour. Technol.* **2016**, *215*, 304–313. [CrossRef]

12. Xu, N.; Liu, S.; Xin, F.; Zhou, J.; Jia, H.; Xu, J.; Jiang, M.; Dong, W. Biomethane Production from Lignocellulose: Biomass Recalcitrance and Its Impacts on Anaerobic Digestion. *Front. Bioeng. Biotechnol.* **2019**, *7*, 191. [CrossRef] [PubMed]

13. Himmel, M.E.; Ding, S.Y.; Johnson, D.K.; Adney, W.S.; Nimlos, M.R.; Brady, J.W.; Foust, T.D. Biomass recalcitrance: Engineering plants and enzymes for biofuels production. *Science* **2007**, *315*, 804–807. [CrossRef] [PubMed]

14. Shanthi, M.; Banu, J.R.; Sivashanmugam, P. Solubilization of fruits and vegetable dregs through surfactant mediated sonic disintegration: Impact on biomethane potential and energy ratio. *Environ. Technol.* **2019**, *20*, 1–12. [CrossRef]

15. Perendeci, N.A.; Gökgöl, S.; Orhon, D. Impact of alkaline H2O2 pretreatment on methane generation potential of greenhouse crop waste under anaerobic conditions. *Molecules* **2018**, *23*, 1794. [CrossRef] [PubMed]

16. Zeynali, R.; Khojastehpour, M.; Ebrahimi, -N.M. Effect of ultrasonic pre-treatment on biogas yield and specific energy in anaerobic digestion of fruit and vegetable wholesale market wastes. *Sustain. Environ. Res.* **2017**, *27*, 259–264. [CrossRef]

17. Li, Y.; Jin, Y.; Li, J.; Li, H.; Yu, Z. Effects of thermal pretreatment on the biomethane yield and hydrolysis rate of kitchen waste. *Appl. Energy* **2016**, *172*, 47–58. [CrossRef]

18. APHA *Standard Methods for the Examination of Water and Wastewater*; American Public Health Association, American Water Works Association, Water Environment Federation: Washington, DC, USA, 2005.

19. Van Soest, P.J. Use of detergent in the analysis of fibrous feeds. A rapid method for the determination of fibre and lignin. *J. Assoc. Off. Anal. Chem.* **1963**, *46*, 829–835.

20. Dreywood, R. Qualitative test for carbohydrate material. *Ind. Eng. Chem. Res.* **1946**, *18*, 199. [CrossRef]

21. Miller, G.L. Use of dinitrosalicylic acid reagent for determination of reducing sugar. *Anal. Chem.* **1959**, *31*, 426–428. [CrossRef]

22. Lowry, O.H.; Rosebrough, N.J.; Fau, A.L.; Randall, R.J. Protein measurement with the Folin reagent. *J. Biol. Chem.* **1951**, *193*, 265–275.

23. Bridoux, G.; Dhulster, P.; Manem, J. Grease analysis on municipal wastewater treatment. *Plants Tech. Sci. Methodes* **1994**, *5*, 257–262.

24. Wang, Z.; Keshwani, D.R.; Redding, A.P.; Cheng, J.J. Sodium hydroxide pretreatment and enzymatic hydrolysis of coastal Bermuda grass. *Bioresour. Technol.* **2010**, *101*, 3583–3585. [CrossRef] [PubMed]

25. Hernández, S.J.M.; Villa, R.M.S.; Veloz, R.J.S.; Rivera, H.K.N.; González, C.R.A.; Plascencia, E.M.A.; Trejo, E.S.R. Comparative hydrolysis and fermentation of sugarcane and agave bagasse. *Bioresour. Technol.* **2009**, *100*, 1238–1245.

26. Sukumaran, R.K.; Singhania, R.R.; Mathew, G.M.; Pandey, A. Cellulase production using biomass feed stock and its application in lignocellulose saccharification for bio-ethanol production. *Renew. Energy* **2009**, *34*, 421–424. [CrossRef]

27. Zhao, X.; Peng, F.; Cheng, K.; Liua, D. Enhancement of the enzymatic digestibility of sugarcane bagasse by alkali–peracetic acid pretreatment. *Enzym. Microb. Technol.* **2009**, *44*, 17–23. [CrossRef]

28. Darani, K.K.; Zoghi, A. Comparison of pretreatment strategies of sugarcane baggase: Experimental design for citric acid production. *Bioresour. Technol.* **2008**, *99*, 9686–9993.

29. Mottet, A.; Steyer, J.P.; Deleris, S.; Vedrenne, F.; Chauzy, J.; Carrère, H. Kinetics of thermophilic batch anaerobic digestion of thermal hydrolysed waste activated sludge. *Biochem. Eng. J.* **2009**, *46*, 169–175. [CrossRef]

30. Ozay, Y.; Kokdemir, U.E.; İsik, Z.; Yılmaz, F.; Dizge, N.; Perendeci, N.A.; Mazmanci, M.A.; Yalvac, M. Optimization of electrocoagulation process and combination of anaerobic digestion for the treatment of pistachio processing wastewater. *J. Clean. Prod.* **2018**, *196*, 42–50. [CrossRef]

31. Carrere, H.; Sialve, B.; Bernet, N. Improving pig manure conversion into biogas by thermal and thermo-chemical pretreatments. *Bioresour. Technol.* **2009**, *100*, 3690–3694. [CrossRef]

32. Bougrier, C.; Delgenes, J.P.; Carrere, H. Effects of thermal treatments on five different waste activated sludge samples solubilization, physical properties and anaerobic digestion. *Chem. Eng. J.* **2008**, *139*, 236–244. [CrossRef]

33. Teghammar, A.; Yngvesson, J.; Lundin, M.; Taherzadeh, M.J.; Horvath, I.S. Pretreatment of paper tube residues for improved biogas production. *Bioresour. Technol.* **2010**, *101*, 1206–1212. [CrossRef] [PubMed]

34. Wang, H.; Wang, H.; Lu, W.; Zhao, Y. Digestibility improvement of sorted waste with alkaline hydrothermal pretreatment. *Tsinghua Sci. Technol.* **2009**, *14*, 378–382. [CrossRef]

35. Jard, G.; Dumas, C.; Delgenes, J.P.; Marfaing, H.; Sialvec, B.; Steyer, J.P.; Carrèrea, H. Effect of thermochemical pretreatment on the solubilization and anaerobic biodegradability of the red macroalga Palmaria palmata. *Biochem. Eng. J.* **2013**, *79*, 253–258. [CrossRef]

36. Yılmaz, F.; Kökdemir, U.E.; Perendeci, N.A. Enhancement of methane production from banana harvesting residues: Optimization of thermal-alkaline hydrogen peroxide pretreatment process by experimental design. *Waste Biomass Valorization* **2019**, *10*, 3071–3087.

37. Chen, H.; Zhang, C.; Rao, Y.; Jing, Y.; Luo, G.; Zhang, S. Methane potentials of wastewater generated from hydrothermal liquefaction of rice straw: Focusing on the wastewater characteristics and microbial community compositions. *Biotechnol. Biofuels* **2017**, *10*, 140. [CrossRef]

38. Pandey, A.; Larroche, C.; Dussap, C.G.; Gnansounou, E.; Khanal, S.K.; Ricke, S. *Biofuels: Alternative Feedstocks and Conversion Processes for the Production of Liquid and Gaseous Biofuels*, 2nd ed.; Elsevier: Amsterdam, The Netherlands, 2019.

39. Song, X.; Shi, Z.; Li, X.; Wang, X.; Ren, Y. Fate of proteins of waste activated sludge during thermal alkali pretreatment in terms of sludge protein recovery. *Front. Environ. Sci. Eng.* **2019**, *13*, 25. [CrossRef]

40. Speece, R.E. *Anaerobic Biotechnology for Industrial Wastewaters*; Archae Press: Nashville, TN, USA, 1996.

41. Song, Z.; Yang, G.; Liu, X.; Yan, Z.; Yuan, Y.; Liao, Y. Comparison of seven chemical pretreatments of corn straw for improving methane yield by anaerobic digestion. *PLoS ONE* **2014**, *9*, e93801. [CrossRef]

42. Monlau, F.; Barakat, A.; Steyer, J.P.; Carrere, H. Comparison of seven types of thermos-chemical pretreatments on the structural features and anaerobic digestion of sunflower stalks. *Bioresour. Technol.* **2012**, *120*, 241–247. [CrossRef]

43. Saikia, B.J.; Parthasarathy, G. Fourier Transform Infrared Spectroscopic Characterization of Kaolinite from Assam and Meghalaya, Northeastern India. *J. Mod. Phys.* **2010**, *01*, 206–210. [CrossRef]

44. Sumayya, A.; Panicker, C.Y.; Varghese, H.T.; Harikumar, B. Vibrational spectroscopic studies and AB inition calculations of l-glutamic acid 5-amide. *Rjc Rasayanj. Chem.* **2008**, *1*, 548–555.

45. Xu, F.; Yu, J.; Tesso, T.; Dowell, F.; Wang, D. Qualitative and quantitative analysis of lignocellulosic biomass using infrared techniques: A mini review. *Appl. Energy* **2013**, *104*, 801–809. [CrossRef]

46. Ang, N.; Ngoh, G.C.; Seak, A.; Chua, M.; Lee, M.G. Elucidation of the effect of ionic liquid pretreatment on rice husk via structural analyses. *Biotechnol. Biofuels* **2012**, *5*, 1–10.

47. Xu, X.J.-K.; Sun, Y.-C.; Sun, R.-C. Structural and hydrolysis characteristics of cypress pretreated by ionic liquids in a microwave irradiation environment. *Bioenergy Res.* **2014**, *7*, 1305–1316. [CrossRef]

48. Fu, S.F.; Wang, F.; Yuan, X.Z.; Yanf, Z.M.; Luo, S.J.; Wang, C.S.; Guo, R.B. The thermophilic (55 °C) microaerobic pretreatment of corn straw for anaerobic digestion. *Bioresour. Technol.* **2015**, *175*, 203–208. [CrossRef]

49. Sun, C.; Liu, R.; Cao, W.; Yin, R.; Mei, Y.; Zhang, L. Impacts of alkaline hydrogen peroxide pretreatment on chemical composition and biochemical methane potential of agricultural crop stalks. *Energy Fuels* **2015**, *29*, 4966–4975. [CrossRef]

50. Gerakines, P.; Schutte, W.A.; Greenberg, J.M.; van Dishoeck, E.F. The infrared band strengths of H_2O, CO and CO_2 in laboratory simulations of astrophysical ice mixtures. *Astr. Astrophys.* **1995**, *2996*, 810–818.

Sample availability: Samples of the compounds are not available from the authors.

 *molecules*

Article

Occurrence of the Persistent Antimicrobial Triclosan in Microwave Pretreated and Anaerobically Digested Municipal Sludges under Various Process Conditions

Gokce Kor-Bicakci [1,2], Timothy Abbott [1], Emine Ubay-Cokgor [2] and Cigdem Eskicioglu [1,*]

[1] UBC Bioreactor Technology Group, School of Engineering, University of British Columbia Okanagan Campus, Kelowna, BC V1V 1V7, Canada; gokce.kor@gmail.com (G.K.-B.); tim.abbott@alumni.ubc.ca (T.A.)
[2] Civil Engineering Faculty, Environmental Engineering Department, Istanbul Technical University, Maslak, 34469 Istanbul, Turkey; ubay@itu.edu.tr
* Correspondence: cigdem.eskicioglu@ubc.ca; Tel.: +1-250-807-8544

Academic Editor: Derek J. McPhee
Received: 25 November 2019; Accepted: 10 January 2020; Published: 12 January 2020

Abstract: Treatment of emerging contaminants, such as antimicrobials, has become a priority topic for environmental protection. As a persistent, toxic, and bioaccumulative antimicrobial, the accumulation of triclosan (TCS) in wastewater sludge is creating a potential risk to human and ecosystem health via the agricultural use of biosolids. The impact of microwave (MW) pretreatment on TCS levels in municipal sludge is unknown. This study, for the first time, evaluated how MW pretreatment (80 and 160 °C) itself and together with anaerobic digestion (AD) under various sludge retention times (SRTs: 20, 12, and 6 days) and temperatures (35 and 55 °C) can affect the levels of TCS in municipal sludge. TCS and its potential transformation products were analyzed with ultra-high-performance liquid chromatography and tandem mass spectrometry. Significantly higher TCS concentrations were detected in sludge sampled from the plant in colder compared to those in warmer temperatures. MW temperature did not have a discernible impact on TCS reduction from undigested sludge. However, AD studies indicated that compared to controls (no pretreatment), MW irradiation could make TCS more amenable to biodegradation (up to 46%), especially at the elevated pretreatment and digester temperatures. At different SRTs studied, TCS levels in the thermophilic digesters were considerably lower than that of in the mesophilic digesters.

Keywords: municipal sludge; anaerobic digestion; thermal pretreatment; microwave; contaminants of emerging concern; personal care products; antimicrobial disinfectants; triclosan; ultra-high performance liquid chromatography; tandem mass spectrometry

---

1. Introduction

The land application of biosolids is reported as the most sustainable, least expensive, and environmentally accepted disposal option for wastewater sludge management [1]. Wastewater sludge must be stabilized before recycling to soil; for reducing environmental health risks, meeting the minimum biosolids quality standards, and convincing public on its benefits and safety [2]. As a commonly used sludge stabilization method, anaerobic digestion (AD) can generate stable organic residues (e.g., biosolids) from high-strength organic waste (e.g., municipal wastewater sludge) while providing economical and environment-friendly renewable energy resources [3]. However, the release of emerging contaminants (e.g., antimicrobials) into the environment via biosolids land application is creating potential ecological and human health risks. For this reason, the detection and reduction of these problematic chemicals in sludge/biosolids has become a priority matter around the world. Although there are several studies and surveys available in the literature regarding the occurrence, fate, and endocrine disruptive effects of emerging contaminants in biosolids [4–6]; regulations for their occurrence in biosolids have not been established yet [7,8]. The legal framework needs to be prepared and/or updated for limiting these compounds in biosolids/agricultural soils and preventing their transfer to the groundwater and then drinking water bodies.

Triclosan (TCS) is a broad-spectrum antibacterial and antifungal agent that is effective against both Gram-positive and Gram-negative bacteria as well as yeasts and molds [9]. TCS is a polychlorinated aromatic antimicrobial which is commonly used as an active ingredient in many personal care products (e.g., soaps, shower gels, shampoos, toothpastes, and cosmetics) as well as household and industrial products (e.g., detergents, plastics, carpets, textile products, toys, and furniture) [10]. Key physical and chemical properties of antimicrobial TCS are listed in Table 1.

Table 1. Key physical and chemical properties of triclosan.

Parameters	Triclosan (TCS)
Chemical structure	(structure of 2,4,4′-trichloro-2′-hydroxydiphenyl ether)
Molecular formula	$C_{12}H_7Cl_3O_2$
Molecular weight (g/mol)	289.536
CAS registration number	3380-34-5
IUPAC name	2,4,4′-trichloro-2′-hydroxydiphenyl ether or 5-chloro-2-(2,4-dichlorophenoxy)phenol
Trade name	Irgasan DP300 or CH 3565
Use	Antiseptic and disinfectant
Chlorine content (weight %)	33.7 [a]
Log K_{ow} (at 25 °C, pH 7)	4.8 [a]
Log K_{oc} (at 25 °C, pH 7)	4.1 [b]
pK_a (at 20 °C)	8.14 [c]
Melting point (°C)	54 – 57 [c]
Boiling point (°C)	280–290 [d]
Water solubility (mg/L at 25 °C)	1.97–4.6 [a]
Vapour pressure (mm Hg at 25 °C)	4.65×10^{-6} [e]
Some of potential transformation products [f,g h,i,j]	Methyl-triclosan 2,4-dichlorophenol 2,4,6-trichlorophenol 2,3,4-trichlorophenol 2,8-dichlorodibenzo-p-dioxin Triclosan O-β-D-glucuronide Triclosan-O-Sulfate Tetra-III Penta Chloroform

Log K_{ow}: The logarithmic octanol-water partition coefficient, Log K_{oc}: Organic carbon partition coefficient, pK_a: Dissociation constant. [a] Halden and Paull [11], [b] Heidler and Halden [12], [c] Reiss et al. [13], [d] Dann and Hontela [14], [e] Ying et al. [15], [f] Tohidi and Cai [16], [g] Chen et al. [17], [h] Armstrong et al. [18], [i] Dann and Hontela [14], [j] Canosa et al. [19].

Municipal wastewater treatment plants (WWTPs) are generally designed to control easily/moderately biodegradable carbonaceous substances, nutrients and pathogens [20]; thus, emerging contaminants such as TCS may only be partially eliminated and/or transformed into its transformation products during treatment processes [21,22]. TCS has been widely detected in high levels (>1000 ng/g dry weight [ng/g-dry]) in municipal wastewater sludge (undigested), and has been shown to accumulate in digested biosolids and biosolids-amended agricultural soils [5,23–26] due to its physico-chemical characteristics, such as high lipophilicity and poor biodegradability. Among the 231 emerging contaminants analyzed in U.S. biosolids, TCS was identified as a priority chemical in terms of both its high occurrence and high bioaccumulation potential [5]. TCS was one of the most abundant contaminants with the mean concentration of 12,600 ± 3800 ng/g-dry, within the 72 pharmaceuticals and personal care products investigated in 110 U.S. biosolids samples collected from 94 WWTPs [8]. TCS is also commonly found in treated WWTP effluent, which then becomes another source of environmental contamination when discharged to the environment. According to the U.S. Geological Survey, TCS was identified as one of the top 10 water contaminants among the 95 compounds tested in 139 streams across 30 states [27]. Furthermore, TCS is an endocrine disrupting chemical, which has been reported to bioaccumulate in wide range of aquatic and terrestrial organisms [14,28,29]. TCS has also been detected in human samples such as urine [30], blood serum [31], breast milk [32], and maternal urine and cord blood plasma [33]. Moreover, TCS is the source of different types of chlorinated derivatives/transformation products such as dioxins/furans, chlorophenols and chloroform, which are toxic and carcinogenic compounds [34–36].

Until now, much attention has been paid to the identification and quantification of TCS and its transformation products in aqueous matrices (e.g., wastewater, drinking water, surface water) [9,19,37]; however, very few studies have focused on their transformation in solid matrices (e.g., municipal sludge, biosolids-amended soils, sediments) [22,35,38–40]. This is possibly caused by the highly-complex nature of solid matrices and the difficulties encountered during their quantitative analysis [41]. For this reason, it is important to develop easy, reliable, and cost-effective analytical methods for the determination of antimicrobials in solid matrices to help reduce the potential risks of their release and adverse effects on the environment. Despite the prevalence of TCS in wastewater sludge, a few studies have explored its fate during conventional AD. According to these studies, it can be concluded that very limited (between 0–25%) TCS removal has been observed during the AD process [22,42–44].

As AD is the most widely used sludge digestion technology, especially for medium- and large-scale WWTPs, it is logical to focus on solutions that can enhance the removal of TCS during anaerobic sludge digestion. In this context, different sludge pretreatment technologies applied prior to AD (i.e., advanced AD) have been used for boosting sludge disintegration and improving digestion with several advantages over the conventional AD [2]. Thermal hydrolysis, which is a highly efficient and commercially available technology, has been successfully adapted to full scale AD operations for pretreatment [45]. As one of the most studied thermal pretreatment alternative, microwave (MW) technology has received interest from the early 2000s due to its ability to speed up the rate of reaction, heat rapidly, and provide easier operation as well as to enhance bioenergy production, sludge dewaterability, and pathogen destruction when combined with an AD process [46]. Although an extensive number of studies have been conducted to investigate the effectiveness of MW pretreatment on sludge disintegration and AD operational performance parameters (such as biogas generation, solids reduction, and sludge rheology), until now, to the best of our knowledge, there has been no study reporting on how MW pretreatment affects the level of TCS in municipal waste sludge.

The main objective of this study was to investigate the occurrence and behavior of the persistent antimicrobial TCS and its potential transformation products in undigested and anaerobically digested municipal sludges, with and without MW pretreatment. In order to achieve these goals, the following tasks were applied: (i) An ultra-high performance liquid chromatography connected to triple quadrupole mass spectrometry (UHPLC-MS/MS) based analytical method was developed for the simultaneous detection and quantification of TCS and its five potential transformation products in sludge samples, and (ii) six bench-scale, semi-continuous flow anaerobic digesters were operated with municipal mixed sludge in order to determine how different MW pretreatment temperatures (80 or 160 °C), digester sludge retention times (SRTs: 20, 12, or 6 days), and digester operating temperatures (35 or 55 °C) influence the levels of selected target compounds in anaerobically digested sludge. This study also assessed how MW pretreatment itself can affect these compounds in municipal wastewater sludge by batch sludge heating experiments.

2. Results and Discussion

2.1. Environmental Occurrence of Triclosan in Municipal Sludge

The environmental (without spiking) concentration of TCS (parent compound) in municipal wastewater sludge (as a mixture of fermented primary sludge and thickened waste activated sludge (WAS)) was measured as 4450 ± 2750 ng/g-dry. However, none of the target transformation products (i.e., triclosan O-β-D-glucuronide, triclosan-O-Sulfate, tetra-III and penta, and 2,3,4-trichlorophenol) considered in this study could be detected in municipal sludge samples (below the instrument detection limits given in E-Supplementary Data). Throughout the different sampling periods, significant seasonal fluctuation in TCS concentrations of sludge samples was observed from winter to spring season (p-value = 0.011 < 0.05). As seen in Figure 1, TCS concentrations in the samples, which were collected from the Westside Regional WWTP during cold months (average weather and wastewater temperatures: −3 and 14 °C, respectively) of the year ranged from 6760 to 8240 ng/g-dry, whereas TCS concentrations in the samples during warmer months (average weather and wastewater temperatures: 8 and >16 °C, respectively) was in the range between 1700 and 3530 ng/g-dry. The sampling months shown in Figure 1 for undigested mixed sludge (digester feed) correspond to times when bench-scale anaerobic digesters were at the steady–state. These variations in the occurrence of TCS in sludge samples collected during the different seasonal temperatures might be related to the wastewater treatment processes applied at the facility. Specifically, the lower TCS concentrations detected in the sludge at warmer temperatures may have been caused by the higher removal of TCS (via biotransformation) in the biological nutrient removal process, utilized by the Westside Regional WWTP. As the microbial activity increased at higher temperatures compared to lower temperatures, TCS accumulation in the sludge (via sorption) may have decreased during biological process. This trend was also observed in previous studies [47–49]. A study by Guerra et al. [48] compared seasonal differences of TCS concentrations in WAS samples (n = 9 with TCS detected in 100% samples) taken from Canadian WWTPs. The TCS concentrations in the sludge were considerably higher in the samples collected in colder temperatures (4.2 ± 1.2 °C) in comparison to those collected in warmer temperatures (23.4 ± 2.4 °C). The levels of TCS were found between 630–3700 ng/g-dry in the cold season whereas between 250–910 ng/g-dry in the warm season. Tohidi and Cai [22] reported that TCS elimination efficiency in a conventional activated sludge process was significantly lower in cold sampling collection periods (34.8%) than warm periods (57%). Another study conducted by Trinh et al. [49] revealed that higher TCS removal was observed through a full-scale membrane bioreactor plant during the summer sampling period (24 ± 1 °C) compared to the winter sampling (15 ± 1 °C). Lozano et al. [38] also revealed the nitrification and denitrification processes as the most effective treatments for TCS removal (22.6%) compared to the conventional activated sludge process (10.4%).

Levels of TCS obtained from municipal sludge in this study were similar to previously reported TCS concentrations in Canadian sludge [48,50,51]. For example, TCS was found with median concentration of 4090 ng/g-dry (between 3540–4150 ng/g-dry) and 9130 ng/g-dry (between 5340–36,800 ng/g-dry) in mixed sludge samples taken from the Saskatoon and the Red Deer's WWTPs, respectively [50]. TCS concentrations were reported from 5500 to 17,900 ng/g-dry (median: 11,550 ± 4265 ng/g-dry) in 12 municipal raw sludge collected from cities across Canada [51]. According to a study by Guerra et al. [48], the concentration ranges for TCS (n = 6 with TCS detected in 100% samples taken from Canadian WWTPs) were between 6000–14,000 and 430–11,000 ng/g-dry in primary sludge and thickened WAS, respectively.

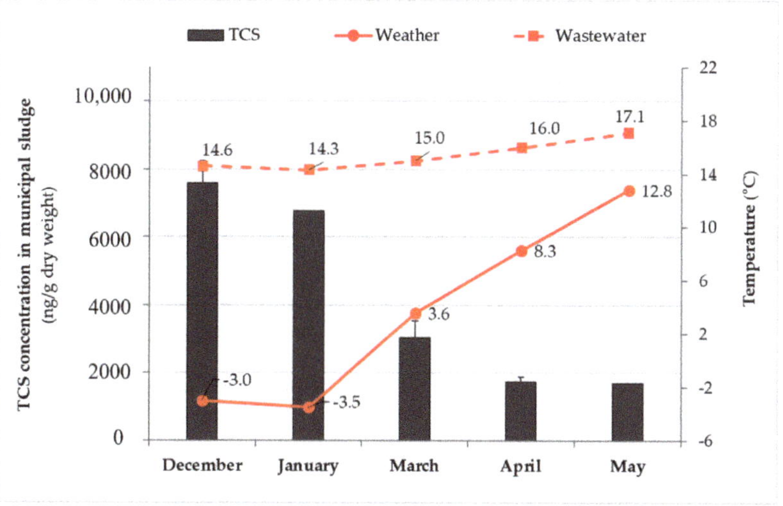

Figure 1. Seasonal variation in the occurrence of triclosan (TCS) in undigested mixed sludge during different sample collection periods coupled with average wastewater and weather temperatures.

2.2. Impact of Microwave Pretreatment on the Behavior of Triclosan in Municipal Sludge

The initial studies focused on assessing MW pretreatment impact on TCS in undigested municipal sludge by applying batch heating experiments. Two independent batches of samples collected from the Westside Regional WWTP in September were analyzed in un-pretreated and MW-pretreated mixed sludges in order to evaluate the effectiveness of MW pretreatment on the reduction of TCS sludge concentrations (Figure 2). Initially, TCS was measured at an average concentration of 4320 ± 225 ng/g-dry in un-pretreated (raw) mixed sludge. Then, MW pretreatment was applied at different temperatures (80 or 160 °C); however, no discernible changes were observed in TCS sludge concentrations. After MW pretreatment, the levels of TCS in MW-pretreated mixed sludge samples were not statistically different from un-pretreated sludge according to a One-Way ANOVA (p-value = 0.178 > 0.05). This result might be explained by the quantum energy of MW irradiation operated at a frequency of 2450 MHz. As this activation energy (1.0×10^{-5} eV) is too low, MW irradiation does not provide sufficient energy to break typical chemical bonds (such as C–C: 3.61 eV, C–O: 3.74 eV, C–H: 4.28, or OH: 4.80 eV) commonly found in organic compounds [52], although it has been proven to achieve disintegration of extracellular polymeric network and pathogen destruction [46]. Based on energies of the chemical bonds, MW may not able to disrupt the specific bonds of TCS as it is a halogenated aromatic hydrocarbon. This finding was in agreement with the results of Armstrong et al. [18] who reported that TCS and some of its transformation products (methyl-TCS, and 2,4-dichlorophenol) in mixed sludge were not impacted by *Cambi Hydrolysis Process*™ at temperatures of 150–180 °C for 30 min and at pressures of 0.37–0.95 MPa. In their study, the average concentrations of TCS, methyl-TCS, and 2,4-dichlorophenol before and after thermal hydrolysis ranged from 7489 to 6884 ng/g-dry, 280 to 248 ng/g-dry, and 283 to 204 ng/g-dry, respectively. On the other hand, Ross et al. [53] indicated that TCS was removed (<100 ng/g-dry) from biosolids (a mixture of WAS and anaerobically digested primary solids) when pyrolysis was performed at a temperature of 300 °C during the batch pyrolysis experiments. The authors also indicated that the pyrolysis reaction time required in order to eliminate TCS from biosolids was less than 5 min at 500 °C. Considering the results from this MW study, as well as other studies using conventional heating, it seems that higher temperatures (i.e., >300 °C, higher than TCS's boiling point given in Table 1) are needed to observe statistically significant changes on TCS after heating.

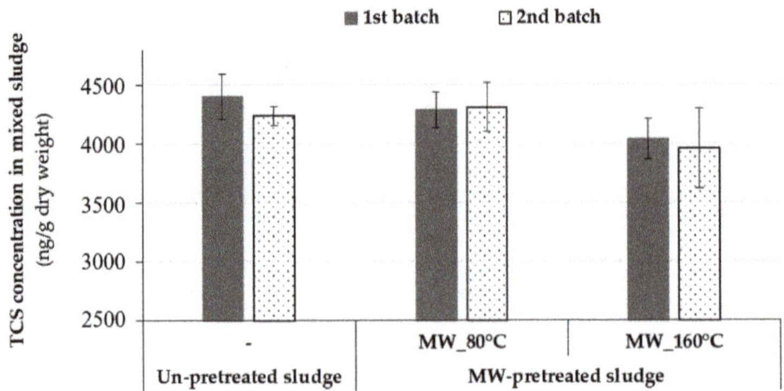

Figure 2. Levels of triclosan (TCS) in undigested mixed sludge before and after microwave (MW) pretreatment.

2.3. Operation and Performance of Bench-Scale Conventional and Advanced Anaerobic Digesters

The semi-continuous flow anaerobic digesters achieved steady-state conditions (where solids concentration, biogas production, and pH varied by 10% or less for each digester) before data was collected for each SRT. Steady-state operations were maintained for 67, 49, and 32 days, respectively, at SRTs of 20, 12, and 6 days. All digesters were stable at each operational phase and the average organic loading rates (OLRs) of digesters were 1.46 ± 0.17, 2.57 ± 0.26, and 5.25 ± 0.42 g volatile solids (VS)/L/d, at SRTs of 20, 12, and 6 days, respectively.

The evolution of thermophilic and mesophilic anaerobic digesters' biogas production through the operating periods (during steady-state) is presented in Figure 3. The average daily specific biogas yields of digesters were in the range of 485–552 and 501–590 mL/g VS_{fed}, respectively, under thermophilic and mesophilic temperatures. The biogas production was slightly higher in mesophilic digesters compared to thermophilic ones; however, neither operating temperature nor SRT had a significant effect on the biogas composition (p-value > 0.05). Throughout the AD operation, the average CH_4 and CO_2 contents in all digesters' headspace were 68 ± 2 and 30 ± 2%, respectively. At SRTs of 20 and 12 days, MW-pretreated advanced digesters achieved higher biogas yields (~15%) under both operating temperatures compared to the respective controls. As the SRT was decreased to 6 days, the highest biogas generation was obtained from the advanced digester of "M_160 °C" (590 ± 23 mL/g VS_{fed}; p-value = 0.000 < 0.05) compared to the digester of "M_Control" (532 ± 13 mL/g VS_{fed}), as shown in the Figure 3b. On the other hand, a slight decrease in biogas production to 488 ± 17 mL/g VS_{fed} was observed in the thermophilic digesters at the 6-day SRT (Figure 3a), which is most likely due to the accumulation of volatile fatty acids (VFAs: up to 1341 ± 297 mg/L) under thermophilic conditions at high OLRs. Increased VFAs concentrations within the digesters can be explained by the fact that there was enough time only for the completion of hydrolysis, acidogenesis, and acetogenesis steps, but not adequate time for the completion of methanogenesis. The inhibition behavior observed from thermophilic systems are in agreement with literature reporting on reduced microbial diversity and therefore less stable AD operation under high OLRs and/or when utilizing by-products of high intensity pretreatment at elevated digester temperatures [54,55].

Organic matter destruction efficiency across the digester is used as one of the key operational parameters while determining the performance of digesters. During the operation of three SRTs, the average removal efficiencies of total chemical oxygen demand (COD) varied between 39%–55% and 45%–56% in the thermophilic and mesophilic digesters, respectively (Figure 4). As expected, higher total COD removal efficiencies (up to 56%) were achieved at longer SRTs under both operating temperatures. Advanced digesters that were fed with MW-pretreated sludge accomplished higher total COD removals compared to the respective controls, except for the thermophilic digesters operated at the shortest SRT (due to mild VFA inhibition). Specifically, the digester of "T_160 °C" had 13% reduction in total COD removal efficiency compared to the respective control (Figure 4). Based on the improvement results in biogas production and organics removal, it can be concluded that advanced AD coupled with MW pretreatment was able to tolerate ~15% higher loading rates compared to the control digesters for the lowest SRT.

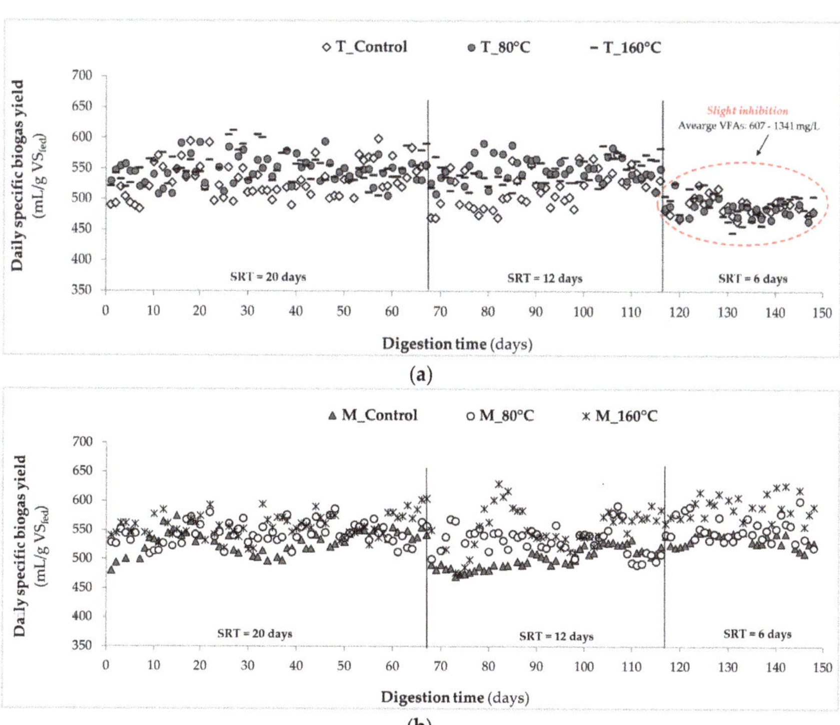

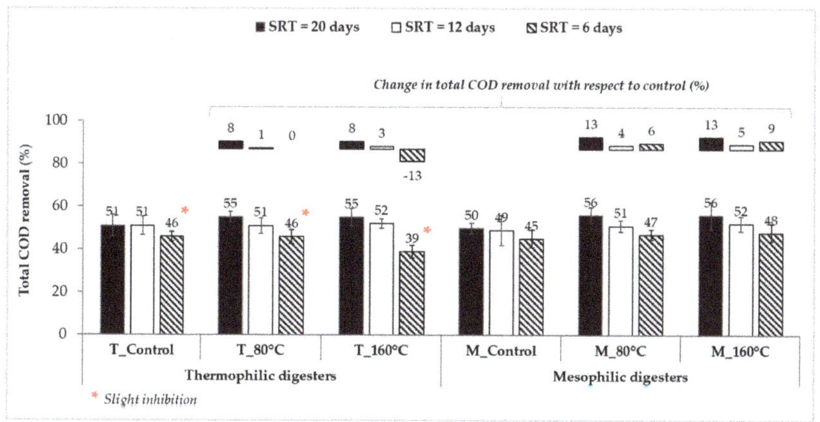

Figure 3. The evolution of anaerobic digesters' specific biogas yields (0 °C, 1 atm) under (a) thermophilic, and (b) mesophilic conditions at sludge retention times (SRTs) of 20, 12, and 6 days during steady-state.

Figure 4. Average total chemical oxygen demand (COD) removal efficiencies of thermophilic and mesophilic anaerobic digesters at SRTs of 20, 12, and 6 days.

Other conventional parameters such as pH, alkalinity, and ammonia were also analyzed during each operational phase to monitor each digester's performance. The daily pH values of digesters were stable between 7.7 and 8.1 under thermophilic temperature and between 7.3 and 7.5 under mesophilic temperature. Average alkalinity concentrations of the thermophilic and mesophilic digesters were in the range of 3900–5600 and 3000–4400 mg/L as $CaCO_3$, respectively. Average ammonia–nitrogen concentrations ranged from 1150 to 1700 and from 800 to 1400 mg N/L for thermophilic and mesophilic digesters, respectively. These results also confirmed stable operation of all digesters during different SRTs without process failure.

2.4. Triclosan Occurrence after Conventional and Advanced Anaerobic Sludge Digestion

Average concentrations of TCS in digested sludge taken from conventional (without pretreatment) and advanced anaerobic digesters that utilized MW irradiation at steady-state under thermophilic and mesophilic conditions are presented in Figure 5. At different SRTs, the levels of TCS in thermophilic digestates were lower than that of in mesophilic digestates (p-value = 0.01 < 0.05 for the 20-day and p-value = 0.03 < 0.05 for the 12-day). For instance, the digester of "T_Control" had an average TCS concentration of 10,990 ± 440 ng/g-dry in its effluent at an SRT of 20 days whereas the digester of "M_Control" had 12,790 ± 965 ng/g-dry (p-value = 0.057 > 0.05). This indicates that the long SRT combined with the higher digester temperature (20-day/55 °C) in "T_Control" created more favorable conditions for TCS reduction compared to "M_Control". As a combined effect (MW pretreatment coupled with AD), the use of MW pretreatment prior to AD process contributed to some extent on the reduction of TCS levels in digested sludge compared to conventional AD at each operational phase. At the 20-day SRT, the average TCS concentrations decreased to 9680 ± 1130 and 9280 ± 660 ng/g-dry in the thermophilic MW-pretreated digesters of "T_80 °C" and "T_160 °C", respectively (Figure 5). The lowest TCS concentration (6145 ± 450 ng/g-dry) was achieved from the digester of "T_160 °C" fed with pretreated sludge at 160 °C, under thermophilic temperature at the 12-day SRT. At the final 6-day SRT, no significant differences were observed between the average TCS concentrations in digested sludge samples from conventional and MW-pretreated digesters operated at thermophilic temperature (p-value = 0.925 > 0.05). This could be due to the reduced biological treatment process performance observed from the advanced digesters pretreated at high temperatures at the shortest SRT. Similar to the aforementioned AD performance results that reported slight inhibition caused by VFAs accumulation (in Section 2.3), mesophilic advanced digesters operated at the 6-day SRT also had better performance on the elimination of TCS from digested sludge compared to thermophilic advanced digesters (Figure 5). It is also worthy to note that although the effect of MW irradiation itself was not statistically significant on TCS concentration of mixed sludge (Figure 2), when the effect was combined with effect of AD (SRT and temperatures), the advanced digesters were more successful in TCS reduction compared to their control counterparts (Figure 5). This suggest that although MW cannot break apart the bonds of TCS during pretreatment at the temperature range studied (80 and 160 °C), it can make micropollutants (originally part of the polymeric network) more amenable (accessible) to biological degradation by disintegrating/solubilizing extracellular polymeric network and releasing them into soluble phase. This behavior was also reported for steroidal hormones in MW pretreated digesters utilizing municipal sludge cake [56].

The concentrations of TCS measured in anaerobically digested sludge in this study were similar to results of earlier studies [50,57,58]. According to study by Guerra et al. [57], TCS was found in all biosolids samples (n = 24) collected from six WWTPs at levels from 2000 to 11,000 ng/g-dry (median: 6800 ng/g-dry). TCS was measured (with 97% occurrence) in treated sludge and biosolids (n = 31) taken from the different regions of Canada with the median concentration of 6085 ng/g-dry [50]. The concentration levels of TCS in anaerobically digested biosolids collected from mesophilic digesters in the Red Deer's WWTP were between 11,700 and 13,900 ng/g-dry, whereas TCS concentrations were between 5590 and 6270 ng/g-dry in the Saskatoon WWTP, located in Canada [50]. TCS concentrations in treated biosolids obtained from four WWTPs in Ontario were also reported as in the range of 680–11,550 ng/g-dry [58].

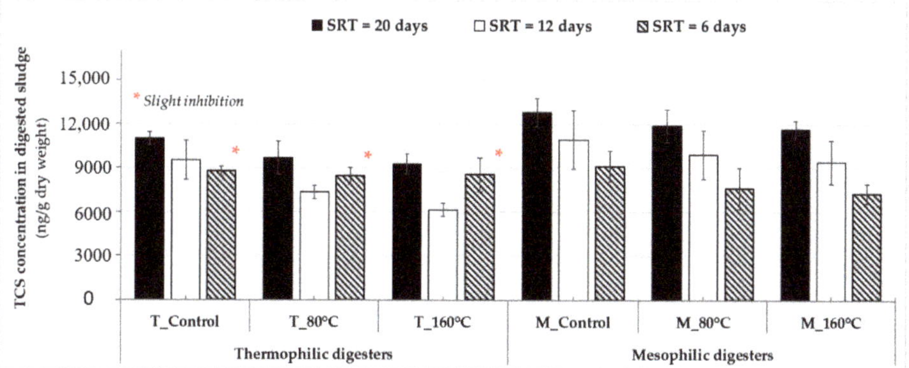

Figure 5. Average triclosan (TCS) concentrations in anaerobically digested sludge obtained from thermophilic and mesophilic digesters at sludge retention times (SRTs) of 20, 12, and 6 days during steady-state.

Although high levels of TCS were detected in digestate samples in this study, its transformation products monitored were below the instrument detection limits (*E*-Supplementary Data) in sludge samples. These results highlight the need for further investigation on the formation of TCS's other transformation products (such as methyl-TCS or 2,4-dichlorophenol) during AD.

When it comes to the effect of digester SRT, the levels of TCS in the thermophilic and mesophilic anaerobic digesters' effluents (except for the thermophilic MW-pretreated digesters at the 6-day SRT) decreased as the SRT was reduced (Figure 5), which was unexpected. Decreasing TCS concentrations in digested sludge may have been caused by the seasonal temperature variation, which followed a similar trend with the fluctuation of TCS levels monitored in the municipal sludge at the different sampling periods (in Section 2.1). In this study, the waste sludge samples used for digester feeding were collected from the WWTP between the months of "September–December", "January–March", and "March–April" during SRTs of 20, 12, and 6 days, respectively. For this reason, the concentrations of TCS in digestates might be higher at colder (winter) temperatures compared to that of at warmer (spring) temperatures due to higher levels of TCS entering AD systems. These results are consistent with the findings of the previous study conducted by Guerra et al. [48]. The authors found higher levels of TCS in aerobically digested biosolids collected in colder temperatures (190–1600 ng/g-dry) compared to those collected in warmer temperatures (69–930 ng/g-dry). Making the comparison based on rate of TCS mass loading represented as ng/d (calculated from Equation (1), *Materials and Methods* section), rather than concentration (ng/g-dry) facilitates comparison among various SRTs.

The rate of mass loading determined for TCS in the conventional and advanced AD systems at different SRTs is represented in Figure 6a. The daily average TCS mass loadings in digested sludge samples were between 8230–14,935 ng/d at the 20-day SRT, between 9385–22,145 ng/d at the 12-day SRT, and between 23,295–37,230 ng/d at the 6-day SRT. When it comes to understanding the mechanisms of changes occurred in TCS levels during AD, photodegradation and volatilization can be assumed negligible, because the nature of TCS is relatively non-volatile. This observation could only be explained by the microbial activity during AD, which may have resulted in TCS degradation via biotransformation and/or mineralization.

Similar to the trend reported in Figure 5, the mass loading comparison also confirmed that TCS removal from digested sludge was increased in the advanced digesters compared to conventional digesters (Figure 6a). The reduction percentage of TCS mass loading exiting from MW-pretreated digesters compared to the respective control digesters (calculated from Equation (2) in *Materials and*

Methods section) is shown in Figure 6b. Digesters utilizing low-temperature microwaved sludge (80 °C) improved TCS reduction in digested sludge by 18%–33%, and 21%–29%, respectively under thermophilic and mesophilic operating temperatures at different SRTs, compared to controls. The levels of TCS reduction in the advanced digesters operated at thermophilic temperatures (except at the 6-day SRT) were higher compared to that of the digesters at mesophilic temperatures. Under thermophilic conditions, the lowest rate of TCS mass exiting from the digester utilizing microwaved sludge at 160 °C was 9385 ± 745 ng/d at the 12-day SRT, and resulted in the highest TCS reduction (46%) compared to the respective control (17,500 ± 2310 ng/d). However, the reduction of TCS from the mesophilic digester of "M_160 °C" accounted only for 33% compared to the respective control (Figure 6b). On the other hand, the same mesophilic digester displayed moderate reduction of TCS (37%) over the respective control at the 6-day SRT. In general, these results confirm that compared to controls, MW irradiation makes TCS more available to microbial assisted transformation within AD and the effect is most pronounced at the highest MW temperatures (160 °C) combined with thermophilic AD under an OLR of 2.57 ± 0.26 g VS/L/d.

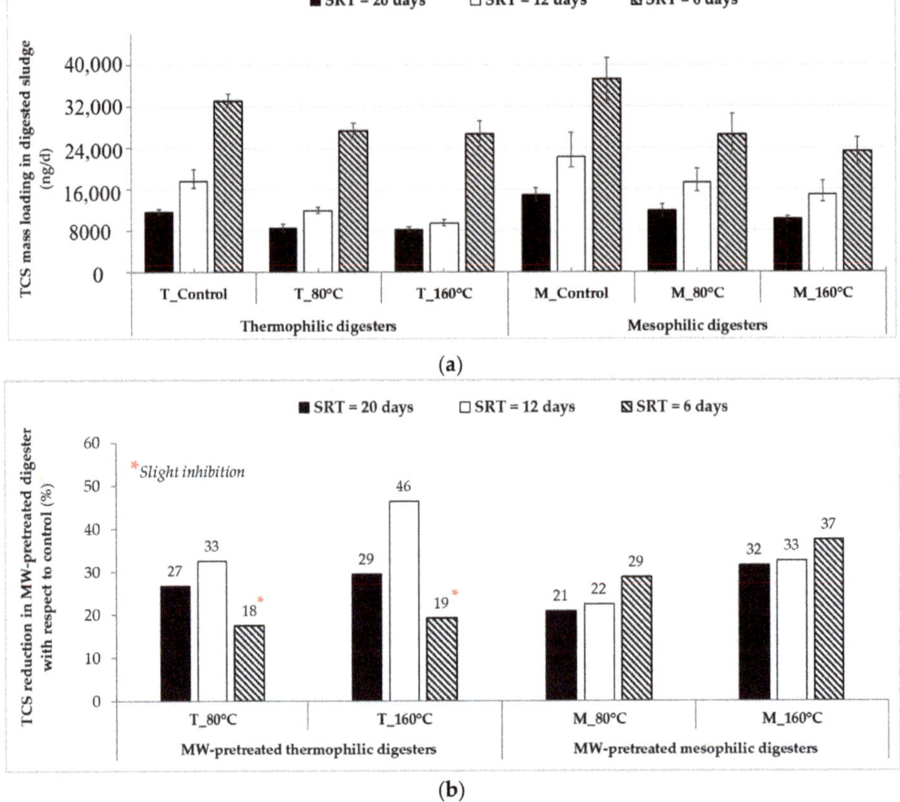

Figure 6. (**a**) Daily triclosan (TCS) mass loadings in anaerobically digested sludge obtained from thermophilic and mesophilic digesters at sludge retention times (SRTs) of 20, 12, and 6 days during steady-state, and (**b**) the reduction percentage of TCS mass loadings exiting from MW-pretreated digesters compared to their respective control digesters.

Initially it was expected that improvements (%) in TCS removals in hybrid AD systems (MW + AD) compared to controls (AD) would increase as SRT was reduced from 20 days to 12 and 6 days as controls were expected to be challenged at shorter SRTs. However, in this study, when the SRT was reduced, TCS reduction obtained from the digesters increased at both mesophilic and thermophilic (except for "T_80 °C" and "T_160 °C" that experienced slight inhibition at the 6-day SRT) temperatures (Figure 6b). Aside from the MW irradiation affect, this can be explained by the higher accessibility of micropollutants in digested sludge matrix compared to digester feed sludge during extraction step of analytical analysis. In other words, micropollutants can be more accessible to quantification in digested sludge compared to that of in undigested sludge due to the alteration of organic matter quality and quantity during AD process (both in control and pretreated ADs). Consequently, as the extent (i.e., SRT) of digestion increased, concentrations of micropollutants (such as TCS levels in this study) in effluents from AD would increase due to higher detectability/quantification compared to that in effluents at shorter SRTs (i.e., 12 or 6 days), and resulting in reduced micropollutant reduction percentages for long SRTs, as previously reported for pharmaceuticals in sludge digesters [59].

2.5. Feasibility of Technology at Full-Scale

Although the results from this study indicate that MW pretreatment was effective in decreasing TCS levels in anaerobically digested sludge when combined with AD, microwave systems at 2450 MHz would not be feasible for full-scale implementation because of inefficient heating of wastewater sludge leading to higher electricity requirements [54]. However, recent studies [60] indicate that optimizing heating frequency and pretreatment equipment design for sludge dielectric properties will increase net energy production significantly at full-scale, while retaining process benefits of sludge pretreatment in terms of digester operation and digestate quality.

3. Materials and Method

3.1. Municipal Wastewater Sludge

Fermented primary sludge and thickened WAS were taken every two weeks from the Westside Regional WWTP (West Kelowna, B.C., Canada) which has a capacity of 16,800 m^3/day. This facility employs preliminary, and primary treatment processes followed by secondary (biological nutrient removal through a three-phase modified Bardenpho process) and tertiary (filtration and UV disinfection) treatments.

3.2. Experimental Methodology

MW pretreatment of thickened WAS samples was carried out in a *Milestone 2.45 GHz MW Lab Station* (ETHOS-EZ: maximum pressure of 35 bar, maximum temperature of 300 °C, and maximum power of 1200 W). At a constant ramp rate of 2.25 °C/min, thickened WAS samples were irradiated to desired temperatures of 80 and 160 °C which represented the low- and high-temperature pretreatment conditions as below and above boiling point temperatures, respectively. Samples were held at these temperatures for 30 min, and then cooled until ambient room temperature. Microwaved thickened WAS samples were mixed with raw fermented primary sludge with a volume % ratio of 67:33, respectively, to provide MW-pretreated mixed sludge samples. Three different mixed sludge were prepared for feeding the digesters: one un-pretreated and two pretreated samples. The basic characteristics of the different sludge streams and mixed sludge samples are shown in Table 2.

Six side-armed Erlenmeyer flasks were used to construct bench-scale anaerobic digesters (fed once a day, 7 days/week). Total and liquid volumes of the digesters were 2 and 1 L, respectively. The start-up of the digesters was done by taking mesophilic and thermophilic inocula from existing bench-scale digesters which had been operating using similar mixed sludge for over one year. Digesters were operated at SRTs of 20, 12, and 6 days under thermophilic (55 ± 1 °C) and mesophilic (35 ± 1 °C) temperatures. The summary of experimental design used for MW pretreatment and anaerobic digesters is provided in Table 3. Further information about the experimental methodology of this study can be found in the previous publication of the authors [54].

Table 2. Basic characteristics of waste sludge streams and mixed sludge samples.

Waste Sludge Sample			Parameter			
			pH (-)	TS (% *w/w*)	VS (% *w/w*)	VS/TS (%)
Thickened waste activated sludge			6.15 (0.20; 5) [a]	3.93 (0.48; 12)	3.18 (0.40; 12)	80.9
Fermented primary sludge			5.27 (0.15; 5)	7.10 (0.96; 12)	6.47 (0.89; 12)	91.1
Mixed sludge	*Un-pretreated*		5.63 (0.13; 5)	3.94 (0.24; 16)	3.38 (0.20; 16)	86.0
	Pretreated	MW_80 °C	5.64 (0.05; 5)	3.40 (0.35; 16)	2.90 (0.33; 16)	85.3
		MW_160 °C	5.52 (0.20; 5)	3.37 (0.25; 16)	2.87 (0.24; 16)	85.2

TS: total solids, VS: volatile solids, MW: microwave. [a] Data displays the arithmetic mean of measurements (standard deviation; number of data points).

Table 3. Summary of the experimental methodology used.

Microwave Pretreatment		*Mixed Sludge Sample Name*	*Anaerobic Digestion*		
Temperature [a] (°C)	Exposure Duration (min)		Digester Name	Digester Type	SRT (days)
-	-	Un-pretreated	T_Control M_Control	Thermophilic Mesophilic	20, 12, and 6
80	30	MW_80 °C	T_80 °C M_80 °C	Thermophilic Mesophilic	20, 12, and 6
160	30	MW_160 °C	T_160 °C M_160 °C	Thermophilic Mesophilic	20, 12, and 6

SRT: sludge retention time, T: thermophilic, M: mesophilic, MW: microwave. [a] Samples were pretreated at a constant heating ramp rate of 2.25 °C/min.

3.3. Analytical Methodology

Once steady-state conditions were reached during each operational phase (after a period corresponding to minimum three SRTs), anaerobically digested sludge samples (n = 3 or 4) were taken from the digesters' effluent streams for the simultaneous detection and quantification of target compounds (analytes: parent compound [*TCS*] and its five transformation products [*triclosan O-β-D-glucuronide*, *triclosan-O-Sulfate*, chlorophenoxyphenol derivatives (*tetra-III* and *penta*), and chlorophenol derivative (*2,3,4-trichlorophenol*)]), and characterization of digesters performance via conventional parameters. For quantification of the target compounds in control digester feed (mixed sludge), a total of 15 independent undigested mixed sludge samples were analyzed (between the months of December to May). For sample preparation, after sampling, thickened WAS and fermented primary sludge samples were first mixed in the laboratory at a volume % ratio of 67:33, respectively (based on the plant's operation), then the mixed sample was processed for TCS quantification (Figure 1). Additionally, two independent batches of samples were taken from the plant during September to assess the effectiveness of MW pretreatment itself on these target compounds' concentrations. MW treated mixed sludge samples (4 samples in total) were prepared by mixing pretreated thickened WAS and raw fermented primary sludge samples at a volume % ratio of 67:33. Then TCS quantification was done for the pretreated mixed sludge samples along with the unpretreated mixed sludge (prepared the same way described above) (Figure 2). Although other TCS transformation products have been detected in literature (i.e., methyl-triclosan, 2,8-dichlorodibenzo-*p*-dioxin), the compounds selected for this study were based on the availability of native standards and their suitability for UHPLC analysis.

3.3.1. Analysis of Triclosan and Its Transformation Products

Chemicals

TCS (CAS# 3380-34-5; 99.7 ± 0.2% purity) was acquired by Sigma Aldrich (Oakville, ON, Canada). TCS transformation products including triclosan O-β-D-glucuronide (CAS# 63156-12-7), and triclosan-O-Sulfate (CAS# 68508-18-9) were purchased from Toronto Research Chemicals (Toronto, ON, Canada) while tetra-III (CAS# 63709-57-9) and penta (CAS# 53555-01-4) were obtained from Wellington Laboratories (Guelph, ON, Canada). 2,3,4-trichlorophenol (CAS# 15950-66-0, ≥98% purity) was acquired through Sigma Aldrich (Oakville, ON, Canada). Isotopically labelled triclocarban-d_4 (TCC-d_4, CAS# 1219799-29-7, >99% atom deuterated) was purchased from C/D/N Isotopes Inc. (Pointe-Claire, QC, Canada). Additional information about target analytes are given in Table S1 (*E*-Supplementary Data in online version). Stock solutions of native compounds were prepared in UHPLC grade methanol (A456, Fisher Scientific, Ottawa, ON, Canada).

Sample Preparation and Extraction

Sludge samples were prepared and extracted according to the acid fraction procedure of the *U.S. EPA Method 1694* with some modifications [61]. A flow chart that summarizes procedures for sample preparation, cleanup, concentration, and analysis steps of target analytes in the sludge samples, is given in Figure 7. Further detailed information regarding each step of procedure is provided in Kor-Bicakci [62] and Kor-Bicakci et al. [63].

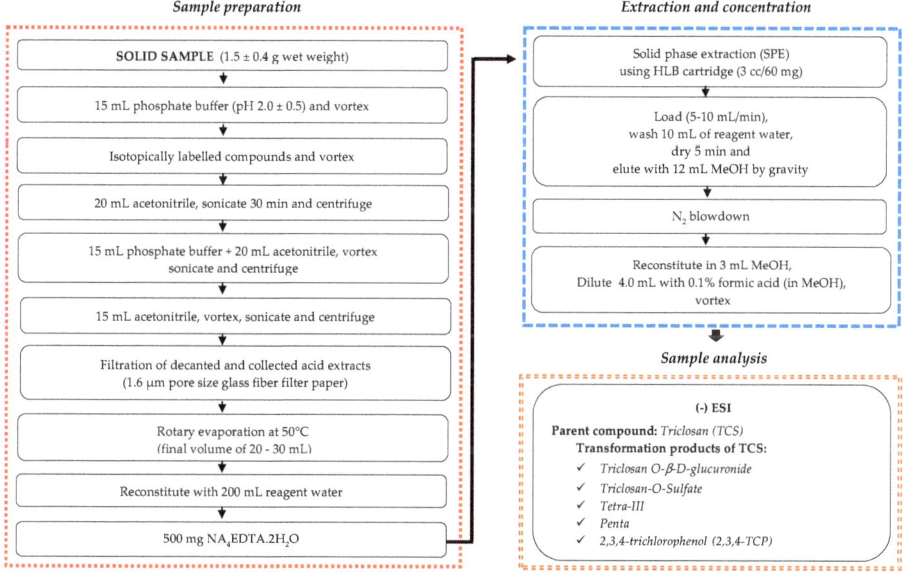

Figure 7. Flow chart for simultaneous detection and quantification of triclosan and its transformation products in sludge (adapted from *U.S. EPA Method 1694* [61]).

Instrumental Analysis

A Waters™ ACQUITY UHPLC coupled with a Waters™ Xevo TQD triple quadrupole mass spectrometer (MS) equipped with an electrospray ionization (ESI) probe was utilized for the simultaneous detection and quantification of TCS and its transformation products in sludge samples. Various methods for the extraction and detection of TCS and several TCS metabolites have been reported in literature. U.S. EPA 1694 served as the basis for sample extraction and provided a starting point for UHPLC and

MS/MS methods. Numerous trials of various aqueous phase additives, UHPLC gradients, mobile phase modifiers, injection volumes, MS conditions etc. were performed to optimize the chromatography (peak shape and compound separation) and to maximize the sensitivity of the MS, while minimizing noise. However, environmental levels of TCS (non-spiked) and the limited biodegradability of TCS during AD resulted in even lower TCS metabolite levels that were challenging to quantify. Furthermore, the low mass to charge ratio (m/z) of TCS fragment ions as well as the fragment ions of several TCS metabolites made ultra-low detection and quantification limits difficult in such a complex sludge matrix.

After multiple trials, the following analysis conditions were selected. The UHPLC was run using an aqueous mobile phase of 5 mM of UHPLC grade ammonium acetate (AX1222-5, VWR, Mississauga, ON, Canada) solution and an organic phase of UHPLC grade methanol. Analytes were separated using a Waters™ Ethylene Bridged Hybrid C18 (2.1 × 50 mm, 1.7 μm) column and matching guard column. MS acquisition was performed in multiple reaction monitoring (MRM) using the negative ion mode (ESI-) for each analyte. Retention times of analytes were within ±15 s of a native compound as per U.S. EPA Method 1694 [61]. Further details regarding optimized MS/MS conditions used for each analyte quantification are provided in the Supplemental Information in Table S2 (*E*-Supplementary File).

Method Validation

A multi-point calibration curve was prepared from native compounds to quantify each analyte (minimum of 8 and maximum of 12 calibration points). Quantification was performed by an isotope dilution technique using isotopically labelled TCC-d$_4$ as an internal standard. Good linearity was yielded with a correlation coefficients (R^2) ≥ 0.98 for all analytes. The instrument limit of detection (LOD) and limit of quantification (LOQ) were calculated using the signal-to-noise ratio of 3 and 10 or greater in native standards or 9 in samples, respectively. Recoveries for most analytes were within the EPA recommended range of 70%–130% with some exceptions [61]. Compound-specific LODs, LOQs, reporting limits, and recoveries are provided in Table S3 (*E*-Supplementary Data).

Mass Loading Calculation for Triclosan

The daily rate of TCS mass exiting from the conventional and advanced AD system (as anaerobically digested sludge) was calculated with Equation (1):

$$M_{eff} = Q_{eff} \times TS_{eff} \times C_{eff} \tag{1}$$

M_{eff} = Daily TCS mass loading in the effluent of the digester (ng/d),
Q_{eff} = Flow rate of anaerobically digested sludge (digestate) (mL/d),
TS_{eff} = Total solids concentration of anaerobically digested sludge (% by weight), and
C_{eff} = TCS concentration in anaerobically digested sludge (ng/g-dry).

The reduction percentage of TCS mass loading exiting from MW-pretreated digesters compared to the respective control digesters was then calculated from Equation (2):

$$TCS\ reduction\ (\%) = \frac{M_{control\ digester} - M_{microwave\ pretreated\ digester}}{M_{control\ digester}} \times 100 \tag{2}$$

$M_{control\ digester}$ = Daily TCS mass loading in the effluent of the control digester (ng/d), and
$M_{microwave\ pretreated\ digester}$ = Daily TCS mass loading in the effluent of MW-pretreated digester (ng/d).

3.3.2. Conventional Parameters

Total solids, VS, COD, pH, alkalinity, and ammonia were analyzed according to Standard Methods [64] procedures 2540 B, 2540 E, 5250 D, 4500-H$^+$B, 2320 B, and 4500-NH$_3$D, respectively. Digester biogas was collected in Tedlar® bags and was measured daily by a U-Tube type manometer, while its composition was analyzed by an Agilent 7820A Gas Chromatograph equipped with a thermal conductivity detector [65]. Total VFAs (sum of acetic, propionic and butyric acids) were measured by an Agilent 7890A Gas Chromatograph with a flame ionization detector and Agilent 19091F-112 capillary column [66].

3.4. Statistical Analysis

The experimental data were analyzed with an analysis of variance (ANOVA) considering a 95% confidence interval ($\alpha = 0.05$), using Minitab™ 17 statistical software.

4. Conclusions

This study assessed the behavior of the antimicrobial TCS in MW-pretreated and anaerobically digested municipal sludges under various process conditions by batch MW pretreatment and semi-continuous flow AD experiments, respectively. The combined effect of AD coupled with MW pretreatment on the occurrence of TCS in wastewater sludge was studied for the first time. All digesters maintained steady-state operation during each of the three SRTs studied. Slight inhibition was observed in the thermophilic digesters operated at the 6-day SRT; nevertheless, these digesters continued producing daily biogas with a typical methane content of 65%. Initial batch pretreatment studies indicated that TCS levels in mixed sludge after MW pretreatment applied at 80 or 160 °C did not significantly change in comparison to un-pretreated sludge. Compared to conventional AD, the combination of anaerobic sludge digestion with MW pretreatment moderately favored the reduction of TCS levels in digested sludge. The advanced thermophilic digesters achieved up to 46% TCS reduction from digested sludge over the respective controls. Compared to the thermophilic temperature, the mesophilic digester temperature was less effective on decreasing TCS levels in digested sludge (up to 37%). Higher TCS concentrations were observed in undigested and anaerobically digested municipal sludges during colder sludge sampling periods. This study confirmed the impacts of seasonal temperature variations on the levels of TCS in municipal wastewater sludge. Although TCS concentrations varied seasonally, MW pretreatment was effective in decreasing TCS levels in anaerobically digested sludge when combined with AD.

Supplementary Materials: The following are available online, Table S1: Further information about triclosan and five of its transformation products, Table S2: Optimized MS/MS conditions used for each analyte quantification, and Table S3: Validation of developed method for each analyte.

Author Contributions: Conceptualization, G.K.-B., E.U.-C. and C.E.; methodology, G.K.-B., T.A., E.U.-C. and C.E.; formal analysis, G.K.-B. and T.A.; investigation and writing—original draft preparation, G.K.-B.; writing—review and editing, T.A., E.U.-C., and C.E.; supervision, E.U.-C. and C.E.; funding acquisition, G.K.-B. and C.E. All authors have read and agreed to the published version of the manuscript.

Funding: This research was funded by "The Scientific and Technological Research Council of Turkey (TUBITAK)—International Research Fellowship Program (2214/A)" and "The Natural Sciences and Engineering Research Council (NSERC)—Collaborative Research and Development Grant (No: J462765-13)".

Acknowledgments: The authors would like to thank TUBITAK and NSERC for their financial support.

Conflicts of Interest: The authors declare no conflict of interest. The funders had no role in the design of the study; in the collection, analyses, or interpretation of data; in the writing of the manuscript, or in the decision to publish the results.

References

1. *Ex-Post Evaluation of Certain Waste Stream Directives—Final Report*; EC—DG Environment European Commission: Neuilly-sur-Seine, France, 18 April 2014.
2. Kor-Bicakci, G.; Eskicioglu, C. Recent developments on thermal municipal sludge pretreatment technologies for enhanced anaerobic digestion. *Renew. Sustain. Energy Rev.* **2019**, *110*, 423–443. [CrossRef]
3. Appels, L.; Baeyens, J.; Degrève, J.; Dewil, R. Principles and potential of the anaerobic digestion of waste-activated sludge. *Prog. Energy Combust. Sci.* **2008**, *34*, 755–781. [CrossRef]

4. Venkatesan, A.K.; Done, H.Y.; Halden, R.U. United States National Sewage Sludge Repository at Arizona State University-a new resource and research tool for environmental scientists, engineers, and epidemiologists. *Environ. Sci. Pollut. Res.* **2015**, *22*, 1577–1586. [CrossRef] [PubMed]

5. Venkatesan, A.K.; Halden, R.U. Wastewater treatment plants as chemical observatories to forecast ecological and human health risks of manmade chemicals. *Sci. Rep.* **2014**, *4*, 3731. [CrossRef] [PubMed]

6. Clarke, B.O.; Smith, S.R. Review of 'emerging' organic contaminants in biosolids and assessment of international research priorities for the agricultural use of biosolids. *Environ. Int.* **2011**, *37*, 226–247. [CrossRef] [PubMed]

7. Verlicchi, P.; Zambello, E. Pharmaceuticals and personal care products in untreated and treated sewage sludge: Occurrence and environmental risk in the case of application on soil—A critical review. *Sci. Total Environ.* **2015**, *538*, 750–767. [CrossRef]

8. McClellan, K.; Halden, R.U. Pharmaceuticals and personal care products in archived US biosolids from the 2001 EPA national sewage sludge survey. *Water Res.* **2010**, *44*, 658–668. [CrossRef]

9. McAvoy, D.C.; Schatowitz, B.; Jacob, M.; Hauk, A.; Eckhoff, W.S. Measurement of triclosan in wastewater treatment systems. *Environ. Toxicol. Chem.* **2002**, *21*, 1323–1329. [CrossRef]

10. Halden, R.U. On the need and speed of regulating triclosan and triclocarban in the United States. *Environ. Sci. Technol.* **2014**, *48*, 3603–3611. [CrossRef]

11. Halden, R.U.; Paull, D.H. Co-occurrence of triclocarban and triclosan in U.S. water resources. *Environ. Sci. Technol.* **2005**, *39*, 1420–1426. [CrossRef]

12. Heidler, J.; Halden, R.U. Meta-analysis of mass balances examining chemical fate during wastewater treatment. *Environ. Sci. Technol.* **2008**, *42*, 6324–6332. [CrossRef] [PubMed]

13. Reiss, R.; Mackay, N.; Habig, C.; Griffin, J. An ecological risk assessment for triclosan in lotic systems following discharge from wastewater treatment plants in the United States. *Environ. Toxicol. Chem.* **2002**, *21*, 2483–2492. [CrossRef] [PubMed]

14. Dann, A.B.; Hontela, A. Triclosan: Environmental exposure, toxicity and mechanisms of action. *J. Appl. Toxicol.* **2011**, *31*, 285–311. [CrossRef] [PubMed]

15. Ying, G.-G.; Yu, X.-Y.; Kookana, R.S. Biological degradation of triclocarban and triclosan in a soil under aerobic and anaerobic conditions and comparison with environmental fate modelling. *Environ. Pollut.* **2007**, *150*, 300–305. [CrossRef]

16. Tohidi, F.; Cai, Z. GC/MS analysis of triclosan and its degradation by-products in wastewater and sludge samples from different treatments. *Environ. Sci. Pollut. Res.* **2015**, *22*, 11387–11400. [CrossRef]

17. Chen, X.; Casas, M.E.; Nielsen, J.L.; Wimmer, R.; Bester, K. Identification of Triclosan-O-Sulfate and other transformation products of Triclosan formed by activated sludge. *Sci. Total Environ.* **2015**, *505*, 39–46. [CrossRef]

18. Armstrong, D.L.; Rice, C.P.; Ramirez, M.; Torrents, A. Influence of thermal hydrolysis-anaerobic digestion treatment of wastewater solids on concentrations of triclosan, triclocarban, and their transformation products in biosolids. *Chemosphere* **2017**, *171*, 609–616. [CrossRef]

19. Canosa, P.; Morales, S.; Rodriguez, I.; Rubi, E.; Cela, R.; Gomez, M. Aquatic degradation of triclosan and formation of toxic chlorophenols in presence of low concentrations of free chlorine. *Anal. Bioanal. Chem.* **2005**, *383*, 1119–1126. [CrossRef]

20. Verlicchi, P.; Al Aukidy, M.; Zambello, E. Occurrence of pharmaceutical compounds in urban wastewater: Removal, mass load and environmental risk after a secondary treatment-A review. *Sci. Total Environ.* **2012**, *429*, 123–155. [CrossRef]

21. Pycke, B.F.G.; Roll, I.B.; Brownawell, B.J.; Kinney, C.A.; Furlong, E.T.; Kolpin, D.W.; Halden, R.U. Transformation Products and Human Metabolites of Triclocarban and Triclosan in Sewage Sludge Across the United States. *Environ. Sci. Technol.* **2014**, *48*, 7881–7890. [CrossRef]

22. Tohidi, F.; Cai, Z. Fate and mass balance of triclosan and its degradation products: Comparison of three different types of wastewater treatments and aerobic/anaerobic sludge digestion. *J. Hazard. Mater.* **2017**, *323 Pt A*, 329–340. [CrossRef]

23. *Targeted National Sewage Sludge Survey Sampling and Analysis Technical Report*; EPA, U.S. Environmental Protection Agency: Washington, DC, USA, 2009.

24. Heidler, J.; Halden, R.U. Mass balance assessment of triclosan removal during conventional sewage treatment. *Chemosphere* **2007**, *66*, 362–369. [CrossRef] [PubMed]

25. Lozano, N.; Rice, C.P.; Ramirez, M.; Torrents, A. Fate of triclosan and methyltriclosan in soil from biosolids application. *Environ. Pollut.* **2012**, *160*, 103–108. [CrossRef] [PubMed]

26. Heidler, J.; Halden, R.U. Fate of organohalogens in US wastewater treatment plants and estimated chemical releases to soils nationwide from biosolids recycling. *J. Environ. Monitor.* **2009**, *11*, 2207–2215. [CrossRef] [PubMed]

27. Kolpin, D.W.; Furlong, E.T.; Meyer, M.T.; Thurman, E.M.; Zaugg, S.D.; Barber, L.B.; Buxton, H.T. Pharmaceuticals, hormones, and other organic wastewater contaminants in U.S. streams, 1999–2000: A National Reconnaissance. *Environ. Sci. Technol.* **2002**, *36*, 1202–1211. [CrossRef] [PubMed]

28. Dhillon, G.S.; Kaur, S.; Pulicharla, R.; Brar, S.K.; Cledon, M.; Verma, M.; Surampalli, R.Y. Triclosan: Current status, occurrence, environmental risks and bioaccumulation potential. *Int. J. Environ. Res. Public Health* **2015**, *12*, 5657–5684. [CrossRef]

29. Sherburne, J.J.; Anaya, A.M.; Fernie, K.J.; Forbey, J.S.; Furlong, E.T.; Kolpin, D.W.; Dufty, A.M.; Kinney, C.A. Occurrence of triclocarban and triclosan in an agro-ecosystem following application of biosolids. *Environ. Sci. Technol.* **2016**, *50*, 13206–13214. [CrossRef]

30. Calafat, A.M.; Ye, X.; Wong, L.Y.; Reidy, J.A.; Needham, L.L. Urinary concentrations of triclosan in the U.S. population: 2003–2004. *Environ. Health Perspect.* **2008**, *116*, 303–307. [CrossRef]

31. Allmyr, M.; Harden, F.; Toms, L.M.; Mueller, J.F.; McLachlan, M.S.; Adolfsson-Erici, M.; Sandborgh-Englund, G. The influence of age and gender on triclosan concentrations in Australian human blood serum. *Sci. Total Environ.* **2008**, *393*, 162–167. [CrossRef]

32. Dayan, A.D. Risk assessment of triclosan [Irgasan] in human breast milk. *Food Chem. Toxicol.* **2007**, *45*, 125–129. [CrossRef]

33. Pycke, B.F.; Geer, L.A.; Dalloul, M.; Abulafia, O.; Jenck, A.M.; Halden, R.U. Human fetal exposure to triclosan and triclocarban in an urban population from Brooklyn, New York. *Environ. Sci. Technol.* **2014**, *48*, 8831–8838. [CrossRef] [PubMed]

34. Halden, R.U.; Lindeman, A.E.; Aiello, A.E.; Andrews, D.; Arnold, W.A.; Fair, P.; Fuoco, R.E.; Geer, L.A.; Johnson, P.I.; Lohmann, R.; et al. The Florence Statement on Triclosan and Triclocarban. *Environ. Health Perspect.* **2017**, *125*, 064501. [CrossRef] [PubMed]

35. Buth, J.M.; Steen, P.O.; Sueper, C.; Blumentritt, D.; Vikesland, P.J.; Arnold, W.A.; McNeill, K. Dioxin photoproducts of triclosan and its chlorinated derivatives in sediment cores. *Environ. Sci. Technol.* **2010**, *44*, 4545–4551. [CrossRef] [PubMed]

36. Fiss, E.M.; Rule, K.L.; Vikesland, P.J. Formation of chloroform and other chlorinated byproducts by chlorination of triclosan-containing antibacterial products. *Environ. Sci. Technol.* **2007**, *41*, 2387–2394. [CrossRef] [PubMed]

37. Buth, J.M.; Grandbois, M.; Vikesland, P.J.; McNeill, K.; Arnold, W.A. Aquatic photochemistry of chlorinated triclosan derivatives: Potential source of polychlorodibenzo-p-dioxins. *Environ. Toxicol. Chem.* **2009**, *28*, 2555–2563. [CrossRef] [PubMed]

38. Lozano, N.; Rice, C.P.; Ramirez, M.; Torrents, A. Fate of triclocarban, triclosan and methyltriclosan during wastewater and biosolids treatment processes. *Water Res.* **2013**, *47*, 4519–4527. [CrossRef]

39. Yu, B.; Zheng, G.; Wang, X.; Wang, M.; Chen, T. Biodegradation of triclosan and triclocarban in sewage sludge during composting under three ventilation strategies. *Front. Environ. Sci. Eng.* **2019**, *13*, 41. [CrossRef]

40. Zheng, G.; Yu, B.; Wang, Y.; Ma, C.; Chen, T. Removal of triclosan during wastewater treatment process and sewage sludge composting—A case study in the middle reaches of the Yellow River. *Environ. Int.* **2020**, *134*, 105300. [CrossRef]

41. Zuloaga, O.; Navarro, P.; Bizkarguenaga, E.; Iparraguirre, A.; Vallejo, A.; Olivares, M.; Prieto, A. Overview of extraction, clean-up and detection techniques for the determination of organic pollutants in sewage sludge: A review. *Anal. Chim. Acta* **2012**, *736*, 7–29. [CrossRef]

42. Gonzalez-Gil, L.; Papa, M.; Feretti, D.; Ceretti, E.; Mazzoleni, G.; Steimberg, N.; Pedrazzani, R.; Bertanza, G.; Lema, J.M.; Carballa, M. Is anaerobic digestion effective for the removal of organic micropollutants and biological activities from sewage sludge? *Water Res.* **2016**, *102*, 211–220. [CrossRef]

43. Narumiya, M.; Nakada, N.; Yamashita, N.; Tanaka, H. Phase distribution and removal of pharmaceuticals and personal care products during anaerobic sludge digestion. *J. Hazard. Mater.* **2013**, *260*, 305–312. [CrossRef] [PubMed]

44. Samaras, V.G.; Stasinakis, A.S.; Mamais, D.; Thomaidis, N.S.; Lekkas, T.D. Fate of selected pharmaceuticals and synthetic endocrine disrupting compounds during wastewater treatment and sludge anaerobic digestion. *J. Hazard. Mater.* **2013**, *244*, 259–267. [CrossRef] [PubMed]

45. Barber, W.P.F. Thermal hydrolysis for sewage treatment: A critical review. *Water Res.* **2016**, *104*, 53–71. [CrossRef] [PubMed]

46. Tyagi, V.K.; Lo, S.-L. Microwave irradiation: A sustainable way for sludge treatment and resource recovery. *Renew. Sustain. Energy Rev.* **2013**, *18*, 288–305. [CrossRef]

47. Cirja, M.; Ivashechkin, P.; Schäffer, A.; Corvini, P.F.X. Factors affecting the removal of organic micropollutants from wastewater in conventional treatment plants (CTP) and membrane bioreactors (MBR). *Rev. Environ. Sci. Biol.* **2008**, *7*, 61–78. [CrossRef]

48. Guerra, P.; Kleywegt, S.; Payne, M.; Svoboda, M.L.; Lee, H.B.; Reiner, E.; Kolic, T.; Metcalfe, C.; Smyth, S.A. Occurrence and Fate of Trace Contaminants during Aerobic and Anaerobic Sludge Digestion and Dewatering. *J. Environ. Qual.* **2015**, *44*, 1193–1200. [CrossRef]

49. Trinh, T.; van den Akker, B.; Coleman, H.M.; Stuetz, R.M.; Drewes, J.E.; Le-Clech, P.; Khan, S.J. Seasonal variations in fate and removal of trace organic chemical contaminants while operating a full-scale membrane bioreactor. *Sci. Total Environ.* **2016**, *550*, 176–183. [CrossRef]

50. Hydromantis Inc. *Emerging Substances of Concern in Biosolids: Concentrations and Effects of Treatment Processes*; Final Report—Field Sampling Program, CCME Project #447-2009; Canadian Council of Ministers of the Environment: Winnipeg, MB, Canada, 2010.

51. Lee, H.B.; Peart, T.E. Organic contaminants in Canadian municipal sewage sludge. Part I. Toxic or endocrine-disrupting phenolic compounds. *Water Qual Res J Can* **2002**, *37*, 681–696. [CrossRef]

52. Nüchter, M.; Ondruschka, B.; Bonrath, W.; Gum, A. Microwave assisted synthesis—A critical technology overview. *Green Chem.* **2004**, *6*, 128–141. [CrossRef]

53. Ross, J.J.; Zitomer, D.H.; Miller, T.R.; Weirich, C.A.; McNamara, P.J. Emerging investigators series: Pyrolysis removes common microconstituents triclocarban, triclosan, and nonylphenol from biosolids. *Environ. Sci. Water Res. Technol.* **2016**, *2*, 282–289. [CrossRef]

54. Kor-Bicakci, G.; Ubay-Cokgor, E.; Eskicioglu, C. Effect of dewatered sludge microwave pretreatment temperature and duration on net energy generation and biosolids quality from anaerobic digestion. *Energy* **2019**, *168*, 782–795. [CrossRef]

55. Hosseini Koupaie, E.; Eskicioglu, C. Conventional heating vs. microwave sludge pretreatment comparison under identical heating/cooling profiles for thermophilic advanced anaerobic digestion. *Waste Manag.* **2016**, *53*, 182–195. [CrossRef] [PubMed]

56. Hamid, H.; Eskicioglu, C. Effect of microwave hydrolysis on transformation of steroidal hormones during anaerobic digestion of municipal sludge cake. *Water Res.* **2013**, *47*, 4966–4977. [CrossRef] [PubMed]

57. Guerra, P.; Kim, M.; Shah, A.; Alaee, M.; Smyth, S.A. Occurrence and fate of antibiotic, analgesic/anti-inflammatory, and antifungal compounds in five wastewater treatment processes. *Sci. Total Environ.* **2014**, *473*, 235–243. [CrossRef] [PubMed]

58. Chu, S.G.; Metcalfe, C.D. Simultaneous determination of triclocarban and triclosan in municipal biosolids by liquid chromatography tandem mass spectrometry. *J. Chromatogr. A* **2007**, *1164*, 212–218. [CrossRef] [PubMed]

59. Ahmad, M.; Abbott, T.; Eskicioglu, C. Effectiveness of single-stage and sequential sludge digestion on removal of recalcitrant pharmaceuticals and conventional pollutants. *Bioresour. Technol. Rep.* **2019**, *8*, 100326. [CrossRef]

60. Hosseini Koupaie, E.; Johnson, T.; Eskicioglu, C. Comparison of Different Electricity-Based Thermal Pretreatment Methods for Enhanced Bioenergy Production from Municipal Sludge. *Molecules* **2018**, *23*, 2006. [CrossRef]

61. *Pharmaceuticals and Personal Care Products in Water, Soil, Sediment, and Biosolids by HPLC/MS/MS, EPA-821-R-08-002*; U.S. Environmental Protection Agency (USEPA): Washington, DC, USA, 2007.

62. Kor-Bicakci, G. Effect of Microwave Pretreatment on Fate of Antimicrobials and Conventional Pollutants during Anaerobic Sludge Digestion and Biosolids Quality for Land Application. Ph.D. Thesis, Istanbul Technical University, Graduate School of Science Engineering and Technology, Istanbul, Turkey, December 2018.

63. Kor-Bicakci, G.; Abbott, T.; Ubay-Cokgor, E.; Eskicioglu, C. Effect of microwave pretreatment on removal of triclosan during anaerobic digestion of wastewater treatment sludge. In *Proceedings of the Water Environment Federation*; WEFTEC: Alexandria, VA, USA, 2017.

64. *Standard Methods*, 20th ed.; Standard Methods for the Examination of Water and Wastewater; American Public Health Association/American Water Works Association/Water Environment Federation: Washington, DC, USA, 2005.

65. van Huyssteen, J.J. Gas chromatographic separation of anaerobic digester gases using porous polymers. *Water Res.* **1967**, *1*, 237–242. [CrossRef]

66. Ackman, R.G. Porous polymer bead packings and formic acid vapor in the GLC of volatile free fatty acids. *J. Chromatogr. Sci.* **1972**, *10*, 560–565. [CrossRef]

Sample Availability: Samples of the compounds are not available from the authors.

 *molecules*

Article

Biogas Production from Physicochemically Pretreated Grass Lawn Waste: Comparison of Different Process Schemes

Georgia Antonopoulou [1,*], Dimitrios Vayenas [1,2] and Gerasimos Lyberatos [1,3]

1   Institute of Chemical Engineering Sciences, Stadiou, Platani, GR 26504 Patras, Greece; dvagenas@chemeng.upatras.gr (D.V.); lyberatos@chemeng.ntua.gr (G.L.)
2   Department of Chemical Engineering, University of Patras, GR 26500 Patras, Greece
3   School of Chemical Engineering, National Technical University of Athens, GR 15780 Athens, Greece
*   Correspondence: geogant@chemeng.upatras.gr; Tel.: +30-26-1096-5318

Received: 8 December 2019; Accepted: 9 January 2020; Published: 11 January 2020

Abstract: Various pretreatment methods, such as thermal, alkaline and acid, were applied on grass lawn (GL) waste and the effect of each pretreatment method on the Biochemical Methane Potential was evaluated for two options, namely using the whole slurry resulting from pretreatment or the separate solid and liquid fractions obtained. In addition, the effect of each pretreatment on carbohydrate solubilization and lignocellulossic content fractionation (to cellulose, hemicellulose, lignin) was also evaluated. The experimental results showed that the methane yield was enhanced with alkaline pretreatment and, the higher the NaOH concentration (20 g/100 gTotal Solids (TS)), the higher was the methane yield observed (427.07 L CH_4/kg Volatile Solids (VS), which was almost 25.7% higher than the BMP of the untreated GL). Comparing the BMP obtained under the two options, i.e., that of the whole pretreatment slurry with the sum of the BMPs of both fractions, it was found that direct anaerobic digestion without separation of the pretreated biomass was favored, in almost all cases. A preliminary energy balance and economic assessment indicated that the process could be sustainable, leading to a positive net heat energy only when using a more concentrated pretreated slurry (i.e., 20% organic loading), or when applying NaOH pretreatment at a lower chemical loading.

Keywords: grass lawn waste; anaerobic digestion; biochemical methane potential; pretreatment; whole slurry; separated fractions; alkali; acid; energy balance; economical assessment

---

1. Introduction

Anaerobic digestion (AD) is a mature and well-established technology worldwide for producing bioenergy in the form of methane. In order to improve its efficiency and reduce the process cost, many efforts have been made on finding alternative feedstocks, based on their availability and renewability [1]. Grass lawn (GL) waste coming from gardening or cuttings of sports fields, is nowadays an abundant carbon source, accounting for a significant fraction of organic municipal solid waste (biowaste) [2,3] and is generated even on a daily basis throughout the developed countries. Currently, this waste stream is usually burned, discarded or disposed in landfills [4], depending on the solid waste management strategy, which is followed by a specific country, causing environmental problems.

However, this type of waste is a valuable source of bioenergy, due to its high organic- and specifically carbohydrate-content [5], which remains unexploitable. Thus the possibility of using GL waste as alternative biomass feedstock for AD is quite appealing. Despite its potential use, the complex lignocellulosic structure of GL waste, limits the accessibility of microorganisms and enzymes, restricting thus its digestion. For this reason, an appropriate pretreatment method should be applied in order to remove the structural and compositional barriers and to improve the AD yields [1]. For GL waste, methods including chemical (through alkali or acid addition), physical (e.g., ultrasound, microwave, ionizing radiation), biological (e.g., enzyme and bacteria) or combined processes have been proposed so far, as pretreatment to enhance bioconversions, towards mainly biohydrogen production [5–7]. Up to now, limited studies have reported methods for enhancing the AD of GL waste. Tsapekos et al. [8] applied different mechanical pretreatment methods on ensiled meadow grass, or meadow grass without ensiling [9] to investigate their effect on biomass biodegradability and biochemical methane potential (BMP). Khor et al. [10] investigated the possibility of combining extrusion and $Ca(OH)_2$ pretreatment to improve storability, availability for biodegradation after storage and BMP of ensiled grass, while Yu et al. [11] investigated the effects of different pretreatments, including ozone, soaking aqueous ammonia (SAA), combined ozone and SAA and size reduction to enhance volatile fatty acid (VFA) and bio-methane production when GL was used as substrate. Finally, Antonopoulou et al. [12] applied SAA to enhance biodegradability and BMP of grass lawn, under different organic loadings.

The aim of the present study was to compare the effects of several pretreatment schemes on the structural and compositional characteristics of GL waste, as well as on the BMP under different process schemes. Specifically, acid and alkali pretreatment methods of GL waste were carried out and compared for the first time. In particular, three different inorganic acids (H_2SO_4, H_3PO_4 and HCl) and an alkali (NaOH) were tested at three different concentrations (2–20 g/100 gTS) and the impact of each pretreatment was assessed, in a comparative way, through techniques such as Scanning Electron Microscopy (SEM) and IR spectroscopy. Compositional analysis after all pretreatment methods, was used to assess the effect of different pretreatments on the lignocellulosic fractionation of GL waste.

An additional issue is whether it is preferable use the whole slurry resulting from pretreatment for the production of methane or whether it is worth to separate the solid and liquid fractions obtained from the pretreatment and use them separately for methane production. Use of the whole slurry instead of the separate fractions has the advantage of utilizing all the sugars of the pretreated slurry. In addition it leads to reduced process costs, since the step of separation and that of detoxification (to remove the inhibitors from the hydrolysate) are not needed [13].

Based on the individual pretreatment and BMP experiments, an integrated process for AD of GL waste is proposed, contributing to the reduction of a significant fraction of biowastes, while simultaneously producing energy in the form of methane.

2. Results and Discussion

2.1. Chemical Composition and Structure of GL before and after Pretreatment

The composition of GL waste used in this study was: total solids (TS) (%) = 92.2 ± 0.1, volatile solids (VS) (g/100 gTS) = 83.4 ± 0.1, cellulose (g/100 gTS) = 20.4 ± 0.1, hemicellulose (g/100 gTS) = 24.0 ± 2.0, lignin (g/100 gTS) = 12.3. ± 1.2, extractives (g/100 gTS) = 25.6 ± 3.1 and proteins (g/100 gTS) = 10.5 ± 0.5. Compared to the compositional analysis reported by other studies, the holocellulose content seems to be similar, while the lignin content is lower than that reported (20.39%. reported by Yu et al. [11]).

Figure 1a,b summarise the effect of different pretreatments (Figure 1a the acidic and Figure 1b the alkali ones) on the fractionation of biomass in terms of lignin, cellulose and hemicellulose. These values are expressed per kg of initial TS, taking into account the solid material recovery of biomass, which is presented in the right axis of both figures. It should be mentioned that during all types of pretreatment, the material recovery associated with the mass loss during pretreatment was less than 100% and the loss of biomass increased with pretreatment severity, leading to a higher material solubilisation. For example, the percentage material recovery (100 − loss of biomass (%)) after treatment with 2 g H_2SO_4/100 gTS was 72.4% and after treatment with 20 g H_2SO_4/100 gTS, it decreased to 49.9%. A material recovery ranging 64.1 and 94.5%, depending on the pretreatment conditions applied, was obtained by Antonopoulou et al. [14] who tested different acid and alkali pretreatment methods on sunflower straw biomass.

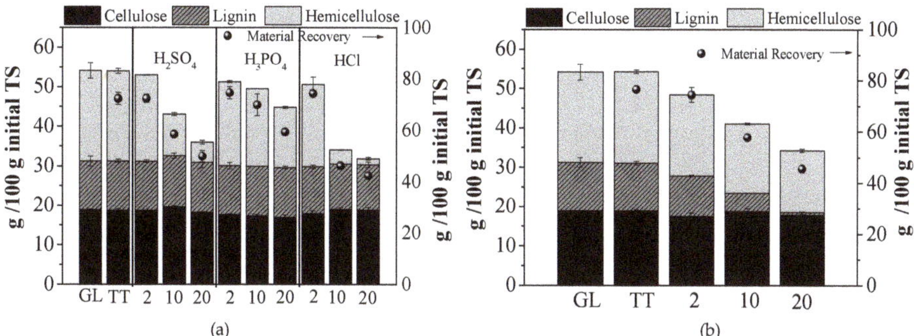

Figure 1. The effect of pretreatment on solid material recovery (right anex), as well as on the fractionation of grass lawn (GL) waste, in terms of cellulose, hemicellulose and lignin, during (**a**) thermal treatment (TT) at 120 °C for 1 h and acid (H_2SO_4, H_3PO_4, HCl) pretreatment, at the concentrations of 2, 10 and 20 g/100 gTS; (**b**) TT at 80 °C for 24 h and alkali (NaOH) pretreatment at the concentrations of 2, 10 and 20 g/100 gTS, respectively.

Acid pretreatment resulted in reduction of the hemicellulose fraction (due to its solubilization), and the removal of hemicellulose increased with the acid concentration (Figure 1a). Thus, pretreatment with 10 or 20 g HCl/100 gTS caused a reduction of hemicellulose by 84.37 and 93.4%, respectively, while when using H_2SO_4 or H_3PO_4 at the higher concentration of 20 g/100 gTS, the hemicellulose fraction was reduced by 77.9 and 33.8%, respectively. The acid pretreatment was not effective in removing cellulose and lignin, as confirmed also by other studies [1,13] reporting that under acidic conditions the main reaction that occurs is the hydrolysis of hemicellulose, especially xylan, while lignin is hardly solubilized, but is disrupted to a high degree, increasing cellulose susceptibility to enzymes.

As it can be seen from Figure 1b, alkali pretreatment was more effective in lignin breakdown, causing depolymerization and cleavage of lignin-carbohydrate linkages. The higher the NaOH concentration used, the higher was the lignin degradation observed. Specifically, when 2, 10 and 20 g NaOH/100 gTS were applied, a lignin removal of 16.7, 61.7 and 94.5%, respectively, was observed, indicating the effectiveness of the method for lignin decomposition. Under the same conditions, the hemicellulose removal efficiency was 10.3, 23.5 and 31.8% for 2, 10 and 20 g NaOH/100 gTS, respectively, while cellulose was not influenced at all. The fact that alkali pretreatment methods (such as soaking in NaOH or NH_3 based aquatic solutions), have been shown to be efficient in lignin removal, while the preservation of mainly cellulose has also been confirmed by other studies [1,12,15,16].

A t-Test of the lignocellulosic fractionation of GL waste, before and after H_2SO_4 pretreatment, showed that the average lignin and cellulose contained in GL were not affected significantly. For treatment with H_2SO_4 at all concentrations, the hemicellulose fraction before pretreatment was significantly higher than the respective fractions after pretreatment ($p = 0.0008$, $p = 0.0003$ and $p = 0.003 < 0.05$ for 2, 10 and 20 g/100 gTS, respectively). The same trend was observed for treatment with HCl and H_3PO_4. Regarding alkaline pretreatment, NaOH at all concentrations led to similar results, where statistical difference was found between hemicellulose and lignin, before and after pretreatment.

In Figure 2, representative Attenuated Total Reflection (ATR) spectra of raw, acid (H_2SO_4, H_3PO_4, HCl) at the higer concentration of 20 g/100 gTS and thermally pretreated, at 120 °C, GL waste, (Figure 2a), as well as the respective of thermally (80 °C) and alkaline pretreated (NaOH, 20 g/100 gTS (Figure 2b)) GL waste, are presented in a spectral range of 600 to 1800 cm^{-1}, in order to verify the chemical changes of the lignocellulosic material before and after pretreatment. The pretreated samples exhibited increased intensities in the regions of 1000–1200 cm^{-1} and 1500–1700 cm^{-1}.

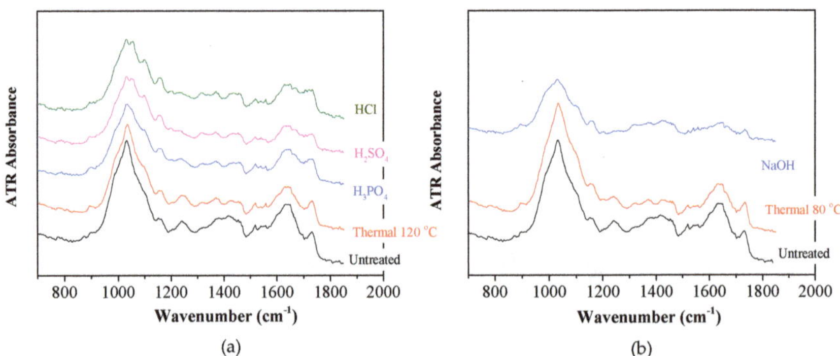

Figure 2. Attenuated Total Reflection (ATR) spectra of raw and (**a**) thermal treatment (TT) at 120 °C for 1 h and acid (H$_2$SO$_4$, H$_3$PO$_4$, HCl) pretreatment, at the concentration of 20 g/100 gTS; (**b**) TT at 80 °C for 24 h and alkali (NaOH) pretreatment at the concentration of 20 g/100 gTS, respectively.

The most important absorption bands commonly found in lignocellulosic biomass samples are observed at 894, 1043, 1242–1256, 1518, 1640 and 1730 cm^{-1} [17] and are associated with the three major lignocellulosic components. Fundamentally, cellulose is formed by glycosidic linkages and hydroxyl groups with a small amount of carboxyl, while hemicellulose and lignin are predominated by ether bond, with hemicellulose also characterized by a significant amount of carboxyl groups [18]. As shown in Figure 2a,b, the spectra of untreated and thermally treated GL waste are similar, indicating that thermal treatment without chemical agent addition did not significantly influence the lignocellulosic fraction, which was also confirmed by Figure 1a,b. The band at 894 cm^{-1} corresponding to C-H deformation/C-O-C stretching at β-1,4 glycosidic linkages, due to the amorphous part of cellulose, is intense in chemically pretreated samples, indicating the decrease in crystalline to amorphous fraction of cellulose, due to the different pretreatment methods. The signal of C-O, C-C and C-OH stretching vibrations at 1043 cm^{-1}, related to cellulose, hemicellulose and lignin [19] in the spectra of the chemically pretreated samples, corresponded to different peaks, compared to the thermally treated or raw GL waste, respectively. Moreover, the C-O vibrations of G rings of lignin at 1242–1256 cm^{-1}, the aromatic skeletal vibration of C=C bond of lignin at 1518 cm^{-1} [20] and the C=O stretching vibration in carbonyl of lignin [17] were represented by lower or no peaks, in the spectrum of alkaline treated sample, compared to the untreated one, as shown in Figure 2b. This could be attributed to the high reduction of lignin (94.5%) which took place due to alkaline pretreatment, as also confirmed by the characterization of the lignocellulosics (Figure 1b). Finally, the signal of ester bond due to C=O stretching in unconjugated ketone, carbonyl and ester groups related to xylan [21] is less intense in the acid treated samples, due to the high solubilization of hemicellulose, which took place under these pretreatment conditions.

In Figure 3, representative SEM images of: (a) raw, (b) alkali (c) acid with H$_2$SO$_4$ and (d) acid with HCl at 20 g/100 gTS, are presented. It is obvious that there is a different morphology after different pretreatment methods, compared to the raw sample. Specifically, treatment with 20 g H$_2$SO$_4$/100 gTS led to a different surface with disrupted parts, also containing pinholes and gaps as well as parts with a smoother outer layer. The smoother surface is evident in Figure 3b,d too, where it is obvious that HCl and NaOH pretreatment methods led to a different surface structure compared to the untreated GL waste. Similar images were also obtained from Yang and Wang [5] when using HCl (1% *w/w*) for 30 min at 100 °C as a pretreatment method of grass to enhance fermentative hydrogen production. Antonopoulou et al. [12] observed also a smoother surface of GL when applying SAA for 3 days at 22 °C, as a pretreatment method to enhance the BMP of this substrate. Also, Kang et al. [16] observed a deconstructed and more accessible surface after pretreatment of *Pennisetum hybrid*, a perennial grass with 2% NaOH at 35 °C for 24 h in order to enhance its anaerobic digestibility.

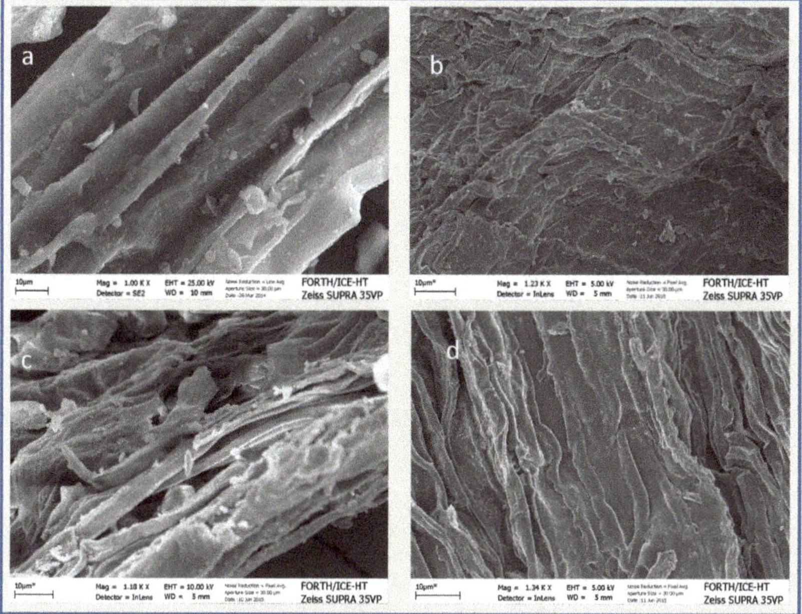

Figure 3. SEM images of raw (**a**), alkali (NaOH) (**b**) acid (H$_2$SO$_4$) (**c**) and acid (HCl) (**d**) pretreatment, at the concentration of 20 g/100 gTS, respectively.

2.2. BMP of GL Waste before and after Pretreatment

2.2.1. BMP of Untreated GL

The calculated methane production, after subtraction of the methane produced from blank experiments was 62.44 ± 0.36 mL, corresponding to a BMP of 260.65 ± 0.04 L CH$_4$/kg GL waste or 282.60 ± 0.04 L CH$_4$/kg TS or 339.86 ± 0.05 L CH$_4$/kg VS. The BMP of the GL waste used in the present study was comparable with that of ensiled meadow grass (372 ± 52 L CH$_4$/kg) [8] and GL (402.5 L CH$_4$/kg) [11], but higher than that of giant reed [22] or common reed [23] (188 L CH$_4$/kg). In any case, the BMP of the untreated GL waste is higher than that of other lignocellulosic feedstocks and this could be attributed to the low lignin content, since the latter is negatively correlated to the BMP [24]. Apart from the low lignin content, GL waste has high holocellulose (sum of cellulose and hemicellulose) and high extractives, content, which are possible sources for high methane productions.

2.2.2. BMP of the Whole Pretreatment Slurry

BMP experiments were conducted for the untreated GL waste, the whole pretreatment slurry (mixture of liquid and solids obtained after all pretreatment methods) as well as the separate fractions i.e., at the liquid (hydrolysate) and the solid fractions obtained after acid and alkali pretreatment methods used. In Figure 4, the effect of all pretreatment methods on the BMP of the whole slurry is presented, expressed as mL methane per g of initial VS. It is obvious that the kinetics of the process were not enhanced by the pretreatment, since more than 80–90% of the total biomethane of all experiments (even of the control- with untreated GL waste) was produced within about 13 days. It is also obvious that all alkali pretreatment methods affected positively the methane yield and the higher the NaOH concentration, the higher was the methane yield. Thus, treatment with 2, 10 and 20 g NaOH/100 gTS led to 389.0 ± 7.0, 397.7± 12.2 and 414.8 ± 26.5 L CH$_4$/kg VS, respectively, corresponded to a 14.74, 17.01 and 22.05% BMP increase, compared to the BMP of untreated GL waste. The increase of BMP can be attributed to the lignin reduction which occurred under alkaline pretreatment (Figure 1b) and was more intense at the higher NaOH concentration. During alkaline pretreatment, a saponification and cleavage of lignin- carbohydrate linkages has been reported to occur [25] rendering the structure smoother and increasing thus the biodegradability during the AD.

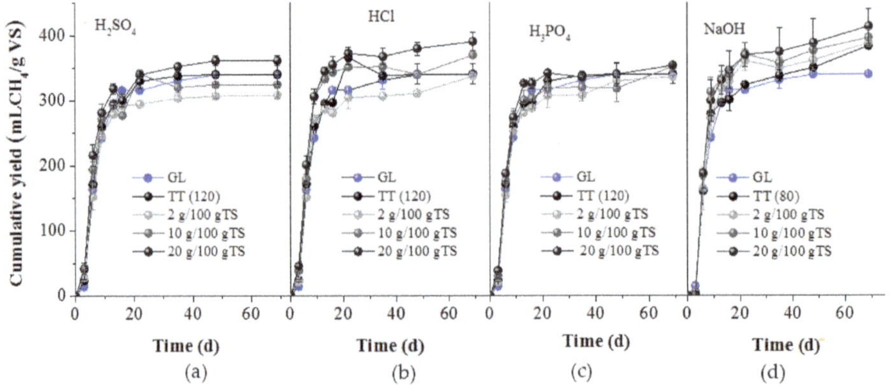

Figure 4. Cumulative methane yield of untreated (GL), thermally treated at 120 °C for 1h (TT (120)) or 80 °C for 24 h (TT (80)) and acid, with H_2SO_4 (**a**), HCl (**b**), H_3PO_4 (**c**) and alkali with NaOH (**d**), at the concentrations of 2, 10 and 20 g/100 gTS, respectively.

In the literature, application of an alkali (through the use of NaOH, $Ca(OH)_2$ or ammonia solution) for enhancing AD and/or the BMP of a lignocellulosic substrate, is commonly reported [1,26]. For instance, Jiang et al. [22] observed an improved enzymatic digestibility and biogas production from giant reed, after its treatment with NaOH and reusing the pretreatment leachate. Yu et al. [11] enhanced the specific methane yield of GL from 402.5 mL CH_4/gVS (untreated) to 481 mL CH_4/gVS, after treatment with SAA, while Kang et al. [16] improved the methane yield of *Pennisetum Hybrid* by 21%, after treatment with 2% NaOH (35 °C, 24 h).

Acid pretreatment enhanced slightly the BMP, i.e., a 4% increase was achieved through the addition of 20 g H_3PO_4/100 gTS, 6.9% due to 20 g H_2SO_4/100 gTS and 15.8% due to 20 g HCl/100 gTS. Recent studies report that acid pretreatments are appropriate for fermentative processes, such as biohydrogen or bioethanol production, due to the solubilization of hemicellulose and not for AD, since they have no effect on lignin [4,14].

2.2.3. BMP of the Solid and Liquid Fractions Obtained after Pretreatment

In Figure 5, the effect of pretreatment on the BMP of the solid fractions obtained after pretreatment is presented, expressed as mL methane per g vs. of the pretreated biomass. It is obvious that for all acid pretreatment methods, the lower chemical concentration of 2 g/100 gTS led to higher methane yields, which could be attributed to the higher holocellulose content in the solid fraction, available for biodegradation from the anaerobic sludge. Since treatment with 10 or 20 g acids/100 gTS led to high hemicellulose removal efficiency, which degraded towards xylose or arabinose and released in the hydrolysate, the biodegradable solid fraction was reduced.

NaOH pretreatment, on the other hand, enhanced significantly the BMP and the higher the NaOH concentration, the higher was the methane yield produced (345.1 ± 7.6, 365.9 ± 9.9 and 424.8 ± 1.5 L CH_4/kg $VS_{pretreated}$, for 2,10, and 20 g NaOH/100 g TS, respectively). This could be justified by the higher lignin removal that occurred under more severe conditions (higher NaOH concentration). Moreover, holocellulose was not affected significantly by alkali pretreatment and based on Figure 3b it can be concluded that only the surface structure was changed. This fact implies that an increase of the cellulose surface area, which was available for enzymatic attack in the subsequent digestion step, occurred, which is also consistent with other studies [12,13]. Apart from the methane yield,

NaOH pretreatment accelerated the kinetics of the process, since more than half of the ultimate methane yield was produced within the first 3 days for all NaOH concentrations.

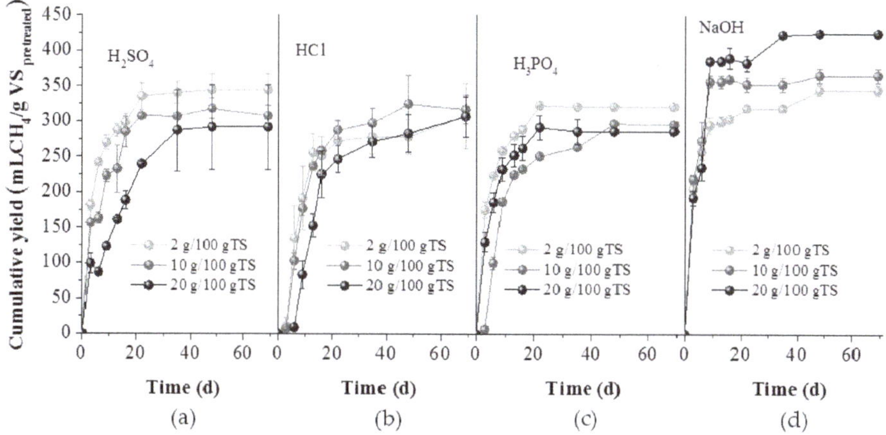

Figure 5. Cumulative methane yield of the solid fraction obtained after pretreatment of GL waste with H_2SO_4 (**a**), HCl (**b**), H_3PO_4 (**c**) and NaOH (**d**), at the concentrations of 2, 10 and 20 g/100 gTS, respectively.

In Figure 6, the BMP of the hydrolysates, expressed as mL methane per mL of hydrolysate, is presented. Contrary to the results from the experiments with the solid fractions, for NaOH and H_2SO_4, the higher the chemical concentration of the pretreatment agent, the higher were the methane yields obtained. Pretreatment with 10 or 20 g HCl/100 gTS and 20 gNaOH/100 gTS led to 9.73 ± 0.03, 8.09 ± 0.47 and 9.40 ± 0.97mL/mL$_{hydrolysate}$, respectively. These high values of the BMP could be attributed to the high organic and sugars content of the hydrolysates, which were released during hemicellulose solubilization, as also confirmed by the values of Table 1, in which the main sugar monomers (glucose, xylose and arabinose) as well as the concentration of soluble sugars, are presented.

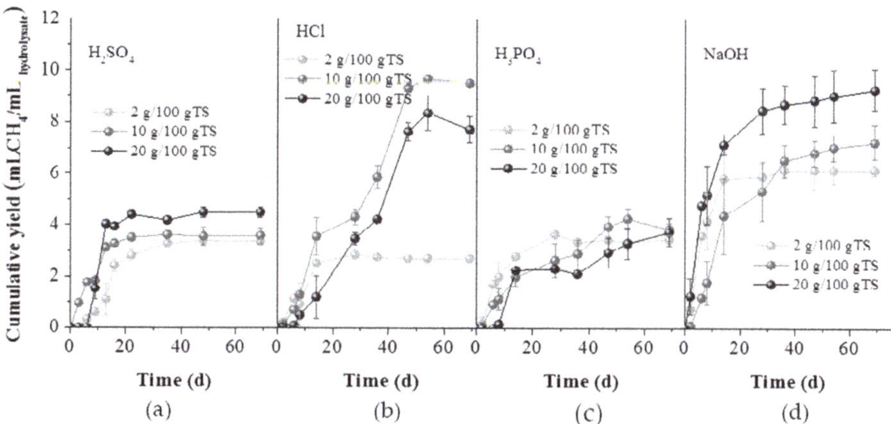

Figure 6. Cumulative methane yield of the hydrolysates obtained from GL waste pretreated with H_2SO_4 (**a**), HCl (**b**), H_3PO_4 (**c**) and NaOH (**d**), at the concentrations of 2, 10 and 20 g/100 gTS, respectively.

Table 1. Concentration of glucose, xylose, arabinose and sugars accompanied by their standard deviations, contained in the liquid fractions obtained after different pretreatment methods.

Pretreatment	Glucose (g/100 gTS)	Xylose (g/100 gTS)	Arabinose (g/100 gTS)	Sugars (g/100 gTS)
Thermal (120 °C)	1.43 ± 0.08	3.98 ± 0.10	-	6.36 ± 0.10
Thermal (80 °C)	1.12 ± 0.04	2.96 ± 0.12	-	5.21 ± 0.18
H_2SO_4,2 g/100 gTS	1.12 ± 0.01	3.29 ± 0.08	0.17 ± 0.01	6.71 ± 0.34
H_2SO_4, 10 g/100 gTS	2.59 ± 0.15	5.02 ±0.25	3.71 ±0.35	14.77. ± 0.11
H_2SO_4, 20 g/100 gTS	3.11 ± 0.03	10.42 ± 2.18	3.56 ± 0.91	15.12 ± 0.14
H_3PO_4, 2 g/100 gTS	1.82 ± 0.04	2.89 ± 0.17	0.52 ± 0.09	6.24 ± 0.03
H_3PO_4, 10 g/100 gTS	1.63 ± 0.03	3.07 ± 0.05	1.74 ± 0.06	7.69 ± 0.74
H_3PO_4, 20 g/100 gTS	1.73 ± 0.01	3.37 ± 0.02	2.66 ± 0.06	11.17 ± 1.78
HCl, 2 g/100 gTS	1.34 ± 0.01	3.16 ± 0.08	1.70 ± 0.10	8.08 ± 0.12
HCl, 10 g/100 gTS	2.33 ± 0.15	12.64 ± 0.25	4.05 ± 0.07	18.21± 0.69
HCl, 2 20/100 gTS	3.53 ± 0.03	13.28 ± 0.45	4.78 ± 0.23	19.03 ± 0.58
NaOH, 2 g/100 gTS	1.38 ± 0.08	2.12 ± 0.10	n.d.	5.34 ± 0.78
NaOH, 10 g/100 gTS	1.28 ± 0.08	4.52± 0.06	0.46 ± 0.05	6.80 ± 0.20
NaOH, 2 20/100 gTS	1.75 ± 0.05	5.87 ± 0.03	1.99 ± 0.01	8.58 ± 0.30

2.2.4. Comparison of the Methane Yields Obtained from Different Processes

The selection of an overall process scheme is crucial for the process economics. In Table 2, the methane yields of all fractions (whole biomass, solid and liquid fractions) obtained using all pretreatment methods, are presented. For comparison, the yields of the separate fractions have also been expressed in terms of mL CH_4/g $VS_{initial}$, taking into account the solid material recovery (loss of weight) due to pretreatment. Thus, for the solid fraction, the methane yield was calculated as:

$$CH_4 \text{ yield}\left(\frac{L}{kg\ VS_{initial}}\right) = CH_4 \text{ yield}\left(\frac{L}{kg\ VS_{pretreated}}\right) \times \text{Material Recovery}\left(\frac{kgVS_{pretreated}}{kgVS_{initial}}\right) \quad (1)$$

while for calculating the methane yield of the hydrolysates in terms of mL CH_4/g $VS_{initial}$, the fact that 100 mL of water were mixed with 5 g TS $_{initial}$ was taken into account, assuming also that no liquid was lost during separation.

Table 2. BMP of the whole pretreatment slurry and of the separated fractions obtained after all pretreatment methods.

	BMP (L/kg vs. Initial)			
Pretreatment	Whole Biomass	Solid Fraction	Liquid Fraction	Sum
Untreated GL	339.86 ± 1.75		-	
Thermal (120 °C)	340.72 ± 15.83	177.09± 1.84	86.51 ± 2.16	263.6
Thermal (80 °C)	383.70 ± 0.50	211.36± 1.28	105.49± 2.40	316.85
H_2SO_4,2 g/100 gTS	307.82 ± 1.62	203.93 ± 3.51	80.51 ± 1.29	284.44
H_2SO_4, 10 g/100 gTS	324.25 ± 20.32	174.74 ± 25.56	86.07 ± 7.21	260.81
H_2SO_4, 20 g/100 gTS	361.70 ± 7.04	141.18 ± 9.41	107.77 ± 4.93	248.95
H_3PO_4, 2 g/100 gTS	336.19 ± 3.24	211.39 ± 39.5	83.05 ± 4.28	294.44
H_3PO_4, 10 g/100 gTS	352.44 ± 4.07	177.62 ± 1.03	92.31 ± 5.50	269.93
H_3PO_4, 20 g/100 gTS	354.09 ± 1.062	146.99 ± 0.74	128.33 ± 13.56	275.32
HCl, 2 g/100 gTS	337.50 ± 2.68	222.37 ± 24.2	65.09 ± 1.24	287.46
HCl, 10 g/100 gTS	369.98 ± 6.81	164.72 ± 11.86	233.38 ± 0.83	398.1
HCl, 2 20/100 gTS	390.77 ± 13.92	134.23 ± 6.85	193.99 ± 11.30	328.22
NaOH, 2 g/100 gTS	388.13 ± 5.82	255.68 ± 3.34	147.42 ± 4.80	403.1
NaOH, 10 g/100 gTS	396.31 ± 11.68	214.70± 4.46	173.60± 16.15	388.3
NaOH, 2 20/100 gTS	413.50 ± 26.08	202.06 ± 4.70	225.01 ± 23.20	427.07

Comparing the BMP of the sum of both fractions, expressed in L/kg $VS_{initial}$ with the respective of the whole slurry at each pretreatment method, it is obvious that direct AD without separation of the pretreated biomass was favored in almost all cases. Only treatment with 20 gNaOH/100 gTS led to

427.07 L CH_4/kg VS, after separation, which corresponded to 25.7% enhancement of the BMP of the untreated GL waste. This value was only 3.3% higher than the BMP of the whole pretreated slurry (413.5 L CH_4/kg VS), under the same pretreatment conditions. However, the approach of using the whole slurry has the advantage of reduced process costs, since the step of separation is not needed [27]. Thus, taking into account these aspects, treatment with 20 g NaOH/100 gTS and direct AD of the whole slurry, seems to be the most promising scheme.

2.3. Energy and Cost Analysis

The experimental results showed that pretreatment with 20 g NaOH/100 gTS for 1d at 80 °C and direct AD of the whole slurry (NaOH-CH_4), led to higher methane yield, of 413.5 L/kg vs. or 346 L CH_4/kg TS. In the present study, a preliminary energy balance and economic assessment of the application of alkaline pretreatment was carried out, by comparing the extra cost (i.e., heating and chemical reagent) required for the pretreatment, with the extra energy in the form of methane due to pretreatment [28]. Thus, apart from the scenario in which alkali pretreatment was employed, the scenario of direct AD of GL waste, without pretreatment (CH_4 of GL) was also analyzed, for comparison reasons.

For realizing the process in full scale, a shredder for milling the grass to the proper size, a tank where the alkaline pretreatment will be carried out, an anaerobic digester for producing biogas from the whole slurry and a combined heat and power (CHP) unit, for biogas exploitation, should be involved.

The energy produced, as estimated by the BMP was found to be 3460 kWh/t TS for the alkali pretreated biomass and 2826 kWh/t TS, for the AD of untreated GL waste (taking into account the energy from methane as 10 kWh/m^3). From the energy yield, the surplus energy (thermal and electrical) produced by the CHP unit, was estimated and presented in Table 3, and compared with the energy demands of the whole process, such as heat energy requirement (HER), electrical energy for mixing the pretreatment tank and the cost of NaOH. Regarding the CHP generator, it was assumed that a typical unit produced 35% electricity and 50% heat (thermal efficiency).

Table 3. Energy analysis for CH_4 production from the alkaline pretreated GL (NaOH-CH_4) and for untreated GL waste (CH_4 of GL).

	NaOH-CH_4	CH_4 of GL
Energy from CH_4 (kWh/t TS)	3460	2826
Thermal energy produced (kWh/t TS)	1730	1413
Electrical energy produced (kWh/t TS)	1211	989.1
Energy produced (heat and electrical from CHP (kWh/t TS)	2941	2402.1

Regarding thermal energy gain, the surplus heat was calculated as the difference between the heat produced by the NaOH-CH_4 process and the process without alkaline pretreatment (CH_4 of GL) (317 kWh/t TS).

In the present study, a solids loading of 5% *w/v*, or 50 g TS/L was assumed. However, several researchers assessed the possibility of applying pretreatment methods at higher solid loading (>15% solids, *w/w*) [29,30]. Especially, in full scale AD the use of more concentrated slurries (e.g., 20% *w/v*) could be feasible. Thus, in the present analysis two different scenario were evaluated: the first scenario in which 50 g TS/L were used (as indicated from the experiments) and the second one, in which a more concentrated whole slurry of 200 g/L (20% *w/v*) was used (20g TS per 100 mL of water or aquatic solution of NaOH).

Regarding the pretreatment tank, addition of water or aquatic solution of NaOH should be carried, so as to reach the solids loadings (50 or 200 g TS/L) and a chemical loading of 20 g NaOH/100 g TS. The HER in kWh/t TS, for thermo-alkaline pretreatment (80 °C) of 1 ton TS of GL waste was estimated according to Equation (2) [28]:

$$HER = \frac{m \times Cp \times \left(T_{final} - T_{initial}\right)}{3600} \tag{2}$$

where m is the mass of water and substrate in kg; Cp the water specific heat (4.18 kJ/kg °C); $T_{initial}$ and T_{final} in °C is the initial and final temperature of the substrate suspension, assumed as 25 °C; and 80 °C, respectively.

The surplus heat was then compared to the thermal energy requirement for alkaline pretreatment (HER). At a solid loading of 50 gTS/L the surplus heat (317 kWh/t TS) was not sufficient to cover the heat requirement for alkaline pretreatment (1343 kWh/t TS). For the solid loading of 200 gTS/L, the net heat energy (NHE) of alkaline pretreatment was slightly negative (−68 kWh/t TS). Thus, assuming a heat energy recovery from the pretreatment step by almost 80% [31], a positive NHE was achieved for both solid loadings (48.4 and 240 kWh/t TS for solids loadings of 50 and 200 g TS/L, respectively) (Table 4).

Table 4. Energy analysis for NaOH-CH$_4$. The solids loadings of 50 and 200 g/L were considered.

Solid Loadings (gTS/L)	50	200
Thermal energy gain (kWh/t TS)[a]	317	317
Heat energy (HE) requirement (kWh/t TS)	1343	385
HE requirement with 80% of heat recovery (kWh/t TS)	268.6	77
Net heat energy (NHE) (kWh/t TS)[b]	−1026	−68
NHE with 80% of heat recovery (kWh/t TS)	48.4	240

[a] Thermal energy gain corresponds to the difference of heat energies produced by NaOH-CH$_4$ minus CH$_4$ of GL; [b] NHE is the difference between the thermal energy increase and the heat energy requirement for the alkaline pretreatment.

Regarding the requirements of electrical energy, only the mixing demands in the pretreatment tank were considered (10.5 kWh/t TS [32]). It should be emphasized that the electricity demands for GL waste grinding and milling were not considered, since these machines were also necessary for the AD of untreated GL. The net electrical energy required for alkaline pretreatment of GL waste at both solids loadings is presented in Table 5.

Table 5. Energy analysis and economical assessment for NaOH-CH$_4$.

	NaOH-CH$_4$
Electrical energy	
Electrical energy increase (kWh/t TS)[a]	221.9
Mixing pretreatment tank (kWh/t TS)	10.5
Net electrical energy (kWh/t TS)	211.4
Economic assessment	
NaOH cost (€/t TS)	82.4
Extra net gain (€/t TS),	52.85

[a] Electrical energy gain corresponds to the difference of electricity energies produced by NaOH-CH$_4$ process minus the CH$_4$ of GL.

For the economic assessment, the cost of chemicals from one side and the incomes from the sale to the public grid of electricity surplus (211.4kWh/t TS), from the other, should be compared (Table 5). As assumption an average price for biogas energy in European countries was considered (0.25 €/kWh). From the values of Table 5, it is obvious that the process should be sustainable either by using a lower alkali loading, or by selling with higher price in the public grid.

3. Materials and Methods

3.1. Biomass Used

GL was collected in the region of Attica, Greece, during gardening. It was initially air dried, then grinded with a house blender (Izzy X3, E560T3, Titanium, Crete, Greece) and milled with a lab grinder (A11 basic, IKA, Staufen, Germany) to powder, passing through a sieve with a pore size of 0.7 mm. Finally, it was air-dried at ambient temperature before being used for the experiments.

3.2. Pretreatment Methods Tested

For all pretreatment methods tested, the solids load was 5% *w/v*. Acid pretreatment was conducted at 121 °C for 1 h, by the use of three different inorganic acids (H_2SO_4, H_3PO_4 and HCl) at concentrations of 2, 10 and 20 g/100 g TS, respectively. Alkaline pretreatment was conducted at 80 °C for 24 h, by the use of NaOH at the same concentrations. For comparison, blank experiments, in which only thermal treatment (121 °C for 1 h or 80 °C for 24 h) without any chemical addition, were also carried out. After pretreatment, either the whole pretreatment slurry (liquid and solid fractions obtained after pretreatment) or the two fractions obtained after separation through filtering with 0.7 μm, were used for BMP, in batch reactors. A detailed physicochemical characterisation was also performed in the solid and liquid fractions, as described below.

3.3. BMP Experiments

BMP experiments were carried out in duplicate at 35 °C in serum bottles of 160 mL, using a working volume of 100 mL. The experiments were performed according to the modified protocol of Owen and Chynoweth [33] using as inoculum sludge from the anaerobic digester of the Patras wastewater treatment plant, treating municipal sewage sludge and operating at steady state at an hydraulic retention time (HRT) of 15 d. The main characteristics of the sludge were: pH: 8.13, total chemical oxygen demand (T.COD): 25.4 g/L, dissolved COD (d.COD): 0.81 g/L, total suspended solids (TSS): 23.03 g/L and volatile suspended solids (VSS): 14.13 g/L.

BMP tests were performed either at the whole slurry or at the separated fractions, obtained after pretreatment. For the experiments with the whole slurry, 20 mL mixed anaerobic culture, 76 mL water and 4 mL of the whole slurry at a solid loading of 5% *w/v*, were used. For the experiments with the solids obtained after pretreatment, 20 mL mixed anaerobic culture, 80 mL water and appropriate amounts of samples were added, in order to acquire the desirable TS content of 2g TS/L. For the experiments with the hydolysates, 20 mL mixed anaerobic culture were seeded with water and appropriate volumes of hydrolysates, so as their final COD concentration, being 2 g/L. For all experiments, the microbial culture was supplemented with 10 mL/L of a $(NH_4)_2HPO_4$ (7.21 g/L) solution, 10 mL/L of a $FeSO_4 \cdot 7H_2O$ (0.7g/L) solution and 10 mL/L of a trace metals solution [34]. Control experiments for checking the methanogenic biomass activity using glucose and cellulose, as well as blank experiments in order to determine the background gas productivity of the inoculum, were also carried out. The content of the vials was gassed with a mixture of N_2/CO_2 (80/20) in order to secure anaerobic conditions. The vials were sealed with butyl rubber stoppers and aluminum crimps and methane production was monitored as a function of time according to Owen and Chynoweth [33].

3.4. Analytical Methods

The analytical procedure for samples characterization in terms of their lignocellulosic content is presented in Antonopoulou et al. [14]. Briefly, raw samples were air-dried and then used for ethanol extraction (exhausted extraction for 24 h) [35] prior to the compositional analysis, which was performed according to the National Renewable Energy Laboratory (NREL)'s standard laboratory analytical procedure (LAP) [36]. Detection and quantification of sugar monomers (glucose, xylose and arabinose) were performed with HPLC-RI with an Aminex HPX-87H column (BioRad, Marnes-la-Coquette, France) at 60 °C and a Cation H micro-guard cartridge (RioRad) using H_2SO_4 0.006 N as an eluent at a flow rate of 0.7 mL/min. For the characterization of the pretreated samples, a separation of liquid and solid fractions was made, through filtering with 0.7 µm filters. The solid fractions were washed with water, air-dried and characterized as described above for the raw samples, but without performing an extraction process prior to the characterization.

The liquid fractions were used for soluble charbohydrates' content determination, according to Joseffson [37] and for the identification of monomeric sugars (glucose, xylose, arabinoze), using the method described above. The measurements of TS, VS, TSS and VSS as well as of d.COD and T.COD were carried out according to Standard Methods [38]. Raw and extractive-free samples were also used to determine Total Kjeldahl Nitrogen (TKN) according to Standard Methods [38] where the crude protein content was estimated by multiplying TKN by a factor of 6.25 [26]. The methane content of the produced biogas was quantified as described in Alexandropoulou et al. [26] while SEM images and IR spectra were obtained as described in Antonopoulou et al. [14].

3.5. Statistical Analysis

A two-sample t-test with a threshold *p*-value of 0.05 was applied in order to analyze statistically the effect of pretreatment on the lignocellulosic content of GL waste.

4. Conclusions

The experimental results obtained showed that the treatment with acids led to higher hemicellulose solubilization, while lignin removal from the solid matrix was achieved, when grass lawn waste (GL) was treated with NaOH. Higher acids concentrations led to higher solubilization of hemicellulose. The BMP of GL was enhanced with alkaline pretreatment and the higher the NaOH concentration, the higher was the methane yield observed. Comparing the BMP under different process schemes (whole or separated fractions) the experiments indicated that the use of the whole slurry was beneficial for the process yields and economy.

Author Contributions: G.A. performed the experiments, the experimental plan, the analysis of the results and wrote the manuscript. D.V. and G.L. supervised the work. All authors have read and agreed to the published version of the manuscript.

Funding: "This research was funded by the Greek General Secretariat for Research and Technology under "Supporting Postdoctoral Researchers Projects"—Pretreatment of lignocellulosic wastes for 2nd generation biofuels (POSTDOC_PE8(1756)) (post-doc fellowship of G. Antonopoulou).

Acknowledgments: The authors gratefully acknowledge Amaia Soto Beobide for ATR spectra.

Conflicts of Interest: The authors declare no conflict of interest.

References

1. Carrere, H.; Antonopoulou, G.; Passos, F.; Affes, R.; Battimelli, A.; Lyberatos, G.; Ferrer, I. Review of pretreatment strategies for the most common anaerobic digestion feedstocks: From lab-scale research to full-scale application. *Bioresour. Technol.* **2016**, *199*, 386–397. [CrossRef]
2. Yu, H.W.; Samani, Z.; Hanson, A.; Smith, G. Energy recovery from grass using two phase anaerobic digestion. *Waste Manag.* **2002**, *22*, 1–5. [CrossRef]
3. Yang, G.; Wang, J. Kinetics and microbial community analysis for hydrogen production using raw grass inoculated with different pretreated mixed culture. *Bioresour. Technol.* **2018**, *247*, 954–962. [CrossRef]
4. Cui, M.; Shen, J. Effects of acid and alkaline pretreatments on the biohydrogen production from grass by anaerobic dark fermentation. *Int. J. Hydrog. Energy* **2012**, *37*, 1120–1124. [CrossRef]
5. Yang, G.; Wang, J. Ultrasound combined with dilute acid pretreatment of grass for improvement of fermentative hydrogen production. *Bioresour. Technol.* **2019**, *275*, 10–18. [CrossRef] [PubMed]
6. Yang, G.; Wang, J. Pretreatment of grass waste using combined ionizing radiation- acid treatment for enhancing fermentative hydrogen production. *Bioresour. Technol.* **2018**, *255*, 7–15. [CrossRef] [PubMed]

7.   Sivagurunathan, P.; Kumar, G.; Mudhoo, A.; Rene, E.R.; Saratale, G.D.; Kobayashi, T.; Xu, K.; Kim, S.H.; Kim, D.H. Fermentative hydrogen production using lignocellulose biomass: An overview of pre-treatment methods, inhibitor effects and detoxification experiences. *Renew. Sustain. Energy Rev.* **2017**, *77*, 28–42. [CrossRef]

8.   Tsapekos, P.; Kougias, P.G.; Angelidaki, I. Biogas production from ensiled meadow grass; effect of mechanical pretreatments and rapid determination of substrate biodegradability via physicochemical methods. *Bioresour. Technol.* **2015**, *182*, 329–335. [CrossRef] [PubMed]

9.   Tsapekos, P.; Kougias, P.G.; Egelund, H.; Larsen, U.; Pedersen, J.; Trenel, P.; Angelidaki, I. Mechanical pretreatment at harvesting increases the bioenergy output from marginal land grasses. *Renew. Energy* **2017**, *111*, 914–921. [CrossRef]

10. Khor, W.C.; Vervaeren, H.; Rabaey, K. Combined extrusion and alkali pretreatment improves grass storage towards fermentation and anaerobic digestion. *Biomass Bioenergy* **2018**, *119*, 121–127. [CrossRef]

11. Yu, L.; Bule, M.; Ma, J.; Zhao, Q.; Frear, C.; Chen, S. Enhancing volatile fatty acid (VFA) and bio-methane production from lawn grass with pretreatment. *Bioresour. Technol.* **2014**, *162*, 243–249. [CrossRef] [PubMed]

12. Antonopoulou, G.; Gavala, H.N.; Skiadas, I.V.; Lyberatos, G. The effect of aqueous ammonia soaking pretreatment on methane generation using different lignocellulosic biomasses. *Waste Biomass Valorization* **2015**, *6*, 281–291. [CrossRef]

13. Antonopoulou, G.; Vayenas, D.; Lyberatos, G. Ethanol and hydrogen production from sunflower straw: The effect of pretreatment on the whole slurry fermentation. *Biochem. Eng. J.* **2016**, *116*, 65–74. [CrossRef]

14. Antonopoulou, G.; Dimitrellos, G.; Beobide, A.S.; Vayenas, D.; Lyberatos, G. Chemical pretreatment of sunflower straw biomass: The effect on chemical composition and structural changes. *Waste Biomass Valorization* **2015**, *6*, 733–746. [CrossRef]

15. Kucharska, K.; Rybarczyk, P.; Hołowacz, I.; Łukajtis, R.; Glinka, M.; Kaminski, M. Pretreatment of lignocellulosic materials as substrates for fermentation processes. *Molecules* **2018**, *23*, 2937. [CrossRef]

16. Kang, X.; Sun, Y.; Li, L.; Kong, X.; Yuan, Z. Improving methane production from anaerobic digestion of *Pennisetum* Hybrid by alkaline pretreatment. *Bioresour. Technol.* **2018**, *255*, 205–212. [CrossRef]

17. Sim, S.F.; Mohamed, M.; Lu, N.A.L.M.I.; Sarman, N.S.P.; Samsudin, S.N.S. Computer-assisted analysis of fourier transform infrared (FTIR) spectra for characterization of various treated and untreated agriculture biomass. *BioResources* **2012**, *7*, 5367–5380. [CrossRef]

18. Taherzadeh, M.J.; Karimi, K. Pretreatment of lignocellulosic wastes to improve ethanol and biogas production: A Review. *Int. J. Mol. Sci.* **2007**, *9*, 1621–1651. [CrossRef]

19. Yoo, C.G.; Kim, H.; Lu, F.; Azarpira, A.; Pan, X.; Oh, K.K.; Kim, J.S.; Ralph, J.; Kim, T.H. Understanding the physicochemical characteristics and the improved enzymatic saccharification of corn stover pretreated with aqueous and gaseous ammonia. *BioEnergy Res.* **2016**, *9*, 67–76. [CrossRef]

20. Bock, P.; Gierlinger, N. Infrared and Raman spectra of lignin substructures: Coniferyl alcohol, abietin, and coniferyl aldehyde. *J. Raman Spectrosc.* **2019**, *50*, 778–792. [CrossRef]

21. Shi, J.; Li, J. Metabolites and chemical group changes in the wood-forming tissue of *Pinus Koraiensis* under inclined conditions. *BioResources* **2012**, *7*, 3463–3475.

22. Jiang, D.; Ge, X.; Zhang, Q.; Zhou, X.; Chena, Z.; Keener, H.; Li, Y. Comparison of sodium hydroxide and calcium hydroxide pretreatments of giant reed for enhanced enzymatic digestibility and methane production. *Bioresour. Technol.* **2017**, *244*, 1150–1157. [CrossRef] [PubMed]

23. Lizasoain, J.; Rincon, M.; Theuretzbacher, F.; Enguídanos, R.; Nielsen, P.J.; Potthast, A.; Zweckmair, T.; Gronauer, A.; Bauer, A. Biogas production from reed biomass: Effect of pretreatment using different steam explosion conditions. *Biomass Bioenergy* **2016**, *95*, 84–91. [CrossRef]

24. Monlau, F.; Sambusiti, C.; Barakat, A.; Guo, X.M.; Latrille, E.; Trably, E.; Steyer, J.P.; Carrère, H. Predictive models of biohydrogen and biomethane production based on the compositional and structural features of lignocellulosic materials. *Environ. Sci. Technol.* **2012**, *46*, 12217–12225. [CrossRef]

25. Tarkow, H.; Feist, W.C. A mechanism for improving the digestibility of lignocellulosic materials with dilute alkali and liquid ammonia. In *Cellulases and Their Applications*; Hajny, G.J., Reese, E.T., Eds.; ACS Publications: Washington, DC, USA, 1969; Volume 95, pp. 197–218. [CrossRef]

26. Alexandropoulou, M.; Antonopoulou, G.; Ntaikou, I.; Fragkou, E.; Lyberatos, G. Fungal pretreatment of willow sawdust and its combination with alkaline treatment for enhancing biogas production. *J. Environ. Manag.* **2017**, *203*, 704–713. [CrossRef]
27. Jung, Y.H.; Kim, I.J.; Kim, H.K.; Kim, K.H. Dilute acid pretreatment of lignocellulose for whole slurry ethanol fermentation. *Bioresour. Technol.* **2013**, *132*, 109–114.
28. Monlau, F.; Kaparaju, P.; Trably, E.; Steyer, J.-P.; Carrère, H. Alkaline pretreatment to enhance one-stage CH_4 and two-stage H_2/CH_4 production from sunflower stalks: Mass, energy and economical balances. *Chem. Eng. J.* **2015**, *260*, 377–385. [CrossRef]
29. Modenbach, A.A.; Nokes, S.E. The use of high-solids loadings in biomass pretreatment: A review. *Biotechnol. Bioeng.* **2012**, *109*, 430–442. [CrossRef]
30. Schell, D.J.; Farmer, J.; Newman, M.; McMillan, J.D. Dilute-sulfuric acid pretreatment of corn stover in pilot-scale reactor: Investigation of yields, kinetics, and enzymatic digestibilities of solids. *Appl. Biochem. Biotechnol.* **2003**, *105–108*, 69–85. [CrossRef]
31. Dhar, B.R.; Nakhla, G.; Ray, M.B. Techno-economic evaluation of ultrasound and thermal pretreatments for enhanced anaerobic digestion of municipal waste activated sludge. *Waste Manag.* **2012**, *32*, 542–549. [CrossRef]
32. Pavlostathis, S.G.; Gosset, G.M. Alkaline pretreatment of wheat straw for increasing anaerobic biodegradability. *Biotechnol. Bioeng.* **1985**, *27*, 334–344. [CrossRef] [PubMed]
33. Owens, J.M.; Chynoweth, D.P. Biochemical methane potential of municipal solid waste (MSW) components. *Wat. Sci. Technol.* **1993**, *27*, 1–14. [CrossRef]
34. Skiadas, I.V.; Lyberatos, G. The periodic anaerobic baffled reactor. *Water Sci. Technol.* **1998**, *38*, 401–408. [CrossRef]
35. Sluiter, A.; Ruiz, R.; Scarlata, C.; Sluiter, J.; Templeton, D. *Determination of Extractives in Biomass*; Laboratory Analytical Procedure, National Renewable Energy Laboratory: Golden, CO, USA, 2008.
36. Sluiter, A.; Hames, B.; Ruiz, R.; Scarlata, C.; Sluiter, J.; Templeton, D.; Crocker, D. *Determination of Structural Carbohydrates and Lignin in Biomass*; Laboratory Analytical Procedure, National Renewable Energy Laboratory: Golden, CO, USA, 2008.
37. Joseffson, B. Rapid spectrophotometric determination of total carbohydrates. In *Methods of Seawater Analysis*; Grasshoff, K., Ehrhardt, M., Kremling, K., Eds.; Verlag Chemie GmbH: Weinheim, Germany, 1983; pp. 340–342.
38. APHA; AWWA; WPCF. *Standard Methods for the Examination of Water and Wastewater*; Franson, M.A., Ed.; American Public Health Association: Washington, DC, USA, 1995.

Article

Biogas Production from Sunflower Head and Stalk Residues: Effect of Alkaline Pretreatment

Marinela Zhurka [†], Apostolos Spyridonidis, Ioanna A. Vasiliadou and Katerina Stamatelatou *

Department of Environmental Engineering, Democritus University of Thrace, 67100 Xanthi, Greece; maria123zhurka@hotmail.com (M.Z.); aspyri@env.duth.gr (A.S.); ioavasil@env.duth.gr (I.A.V.)
* Correspondence: astamat@env.duth.gr; Tel.: +30-25410-79315
† Current affiliation: Chemical and Materials Engineering Group, School of Engineering, University of Aberdeen, Aberdeen AB24 3FX, UK.

Academic Editors: Ivet Ferrer, Cigdem Eskicioglu, Georgia Antonopoulou and Audrey Battimelli
Received: 5 December 2019; Accepted: 25 December 2019; Published: 31 December 2019

Abstract: Sunflower residues are considered a prominent renewable source for biogas production during anaerobic digestion (AD). However; the recalcitrant structure of this lignocellulosic substrate requires a pretreatment step for efficient biomass transformation and increased bioenergy output. The aim of the present study was to assess the effect of alkaline pretreatment of various parts of the sunflower residues (e.g., heads and stalks) on their methane yield. Experimental data showed that pretreatment at mild conditions (55 °C; 24 h; 4 g NaOH 100 g^{-1} total solids) caused an increase in the biochemical methane potential (BMP) of both heads and stalks of the sunflower residues as determined in batch tests. The highest methane production (268.35 ± 0.11 mL CH_4 g^{-1} volatile solids) was achieved from the pretreated sunflower head residues. Thereafter; the effect of alkaline pretreatment of sunflower head residues was assessed in continuous mode; using continuous stirred-tank reactors (CSTRs) under two operational phases. During the first phase; the CSTRs were fed with the liquid fraction produced from the pretreatment of sunflower heads. During the second phase; the CSTRs were fed with the whole slurry resulting from the pretreatment of sunflower heads (i.e., both liquid and solid fractions). In both operating phases; it was observed that the alkaline pretreatment of the sunflower head residues had a negligible (phase I) or even a negative effect on biogas production; which was contradictory to the results of the BMP tests. It seems that; during alkaline pretreatment; this part of the sunflower residues (heads) may release inhibitory compounds; which induce a negative effect on biogas production in the long term (e.g., during continuously run digesters such as CSTR) but not in the short-term (e.g., batch tests) where the effect of the inoculum may not permit the inhibition to be established.

Keywords: anaerobic digestion; lignocellulosic biomass; NaOH pretreatment; bioreactor experiments; biochemical methane potential; inhibition

---

1. Introduction

Lignocellulosic substrates are considered important feedstocks for the production of second-generation biofuels (e.g., H_2, CH_4). Agricultural residues comprise a renewable resource which does not compete with plant cultivation for food, since they consist of the nonedible parts of the plants (leaves, stalks), which are usually burned on the fields creating environmental issues related with biomass incineration in the open air [1]. This is also the case with sunflower residues (stalks, leaves, heads) which are produced in large quantities and left in the fields after seed harvesting. In the absence of any alternative reuse, sunflower residues may be a promising renewable resource for bioenergy production via anaerobic digestion [2]. However, the presence of lignin is apparently the most important factor affecting the biodegradability of lignocellulosic materials. Therefore, the recalcitrant structure of lignocellulosic substrates consisting of holocelluloses (cellulose, hemicelluloses) that are embedded in the lignin network, requires the application of pretreatment, to increase the accessibility of holocelluloses to bacteria during anaerobic digestion [3].

Pretreatment methods, such as mechanical, thermo-chemical, chemical, and biological, have been applied to improve biogas production from anaerobic digestion of lignocellulosic residues [4,5]. Among them, dilute-acid and alkaline chemical pretreatments are the most widely used for sunflower residues, being low-cost and effective processes [5–7]. Dilute-acid pretreatment, hydrolyzes hemicelluloses into monomeric sugars, improving cellulose conversion and increasing enzyme accessibility [8]. By alkaline pretreatment, the cleavage of ester bonds mainly occurs in lignin/phenolics–carbohydrate complexes, resulting in partial lignin removal [3]. On the other hand, it should be emphasized, that applying pretreatment on plant biomass, lignin derived compounds are released which may be inhibitory to anaerobic microorganisms [6]. Acid pretreatment causes the hydrolyzation of the hemicellulosic fraction into furaldehydes [5-hydroxymethylfurfural (5-HMF) and furfural] and aliphatic acids (i.e., formic and acetic acid) [9]. Moreover, Antonopoulou et al. [7] reported that high NaOH concentration during pretreatment of sunflower straw biomass, resulted in a high concentration of phenolics. These compounds might have an inhibitory or even toxic effect on anaerobic microorganisms and therefore, should be always taken into consideration. Monlau et al. [6,10] applied various chemical pretreatments on sunflower straw studying their influence on lignin removal and evaluating the biochemical methane potential (BMP). They reported that alkaline pretreatment at 55 °C, for 24 h with 4 g NaOH 100 g^{-1} total solids, was the most suitable for enhancing the anaerobic digestion process and methane potential. Monlau et al. [2], applied this mild NaOH pretreatment in continuous experiments and noticed an increase of 26% in methane yield. However, it is not well understood how the different parts of the residues (i.e., stalks and heads) can respond to the pretreatment scheme at different modes of the anaerobic digestion process (batch and continuous).

The case study considered in this work was of the sunflower residues, abundant in Greece and especially in the area of East Macedonia and Thrace. FAOSTAT recorded an area of 90,600 hectares of sunflower seeds cultivated in Greece, with an estimated yield of 2433.8 kg/ha in 2017 [11]. These agricultural areas produce large quantities of sunflower residues; Searle and Malins [12] estimated that the ratio of residue to harvested grain is 1.77 for the case of sunflower crops. Therefore, they could be used as feedstock in biogas plants which have been developed in Greece over the recent decade. The biochemical methane potential of an organic material is usually evaluated through batch tests. On the other hand, experimental results coming from continuously run digesters are scarce but more trustworthy since they simulate the conditions prevailing in real biogas plants (which operate on a continuous mode) and reveal the long-term response of the anaerobic biomass to a particular feedstock. To this end, the effect of alkaline pretreatment on the biogas production from sunflower residues was studied in two different types of digesters (batch and continuous), focusing mostly on the heads. Consequently, the results of the presented study can be considered to be complementary to those reported in Monlau et al. [2], which is the only reference (to our knowledge) focusing on the digestion of sunflower stalks in continuous mode at mesophilic conditions, after alkaline pretreatment under the same conditions. These results show how an estimation based on the typical biomethane potential tests can be proven wrong by experiments run in continuous mode.

2. Results and Discussion

2.1. Sunflower Heads and Stalks Characteristics

Heads and stalks, prior to any analysis, were dried in an oven resulting in a low humidity level of ca. 10% for both residues. The chemical oxygen demand (COD) concentration was higher for stalks by 20.8% in comparison to heads, which is in agreement with the higher volatile solid (VS) content of stalks (87.7% of total solids; TS for heads and 79.9% of TS for stalks). On the contrary, total Kjeldahl nitrogen (TKN) and lipids of heads were more than two times higher compared to stalks (Table 1).

Table 1. Composition of the sunflower heads and stalks (raw) as well as their solid and liquid fractions (SF and LF respectively) after alkaline pretreatment.

	Raw	SF Pretreated	LF Pretreated	Raw	SF Pretreated	LF Pretreated
Humidity (%)	9.46 ± 0.4	1.3 ± 1.8	-	8.1 ± 0.4	2.5 ± 1.0	-
Total solids; TS (% WM)	90.5 ± 0.4	98.7 ± 17.6	-	91.9 ± 0.3	97.5 ± 1.0	-
Volatile solids; vs. (% WM)	72.4 ± 0.8	87.8 ± 10.4	-	80.6 ± 0.2	83.3 ± 11.5	-
VS (% TS)	79.9 ± 0.5	88.9 ± 0.5	-	87.7 ± 0.1	85.5 ± 0.3	-
Chemical oxygen demand; COD (SF g kg^{-1}TS or LF g L^{-1})	890 ± 26	967 ± 28	8.3 ± 0.3	1075 ± 34	1067 ± 56	6.2 ± 0.1
TKN (g kg^{-1}TS)	13.3 ± 0.3	13.8 ± 0.3	-	6.3 ± 0.0	5.6 ± 0.2	-
Lipids (g kg^{-1}TS)	2.5 ± 0.1	-	-	0.99 ± 0.06	-	-
Phenols (mg Gallic Acid L^{-1})	-	-	214 ± 4	-	-	112 ± 2

WM: wet matter, DM: dry matter.

The latter could be attributed to some seed residues in the heads, which are rich in lipids and proteins [13]. Klason lignin content of raw stalks reported here (20.36% VS, Table 2) was lower compared to the results of Monlau et al. (29.7% VS) [6]. Klason lignin concentration of stalks (20.36% VS) was much higher in comparison to heads (10.48% VS, Table 2), providing a more rigid structure for stalks and reducing their biodegradability. Elemental analysis showed that both, heads and stalks, contained a significant amount of Ca, Mg, K, Si, and Fe (Table 3). Nevertheless, trace elements such as Co, Mo, Ni, and W which are considered to be essential for the process of anaerobic digestion [14], were either detected in small concentrations or were absent.

Table 2. Lignin content of raw and pretreated sunflower residues.

	Klason Lignin
	(% Raw VS)
Heads raw matter	10.48 ± 0.28
Heads pretreated	7.19 ± 0.19
Stalks raw matter	20.36 ± 0.66
Stalks pretreated	18.74 ± 0.24

2.2. Pretreatment Effect on the Biochemical Methane Potential (BMP) of Sunflower Heads and Stalks—Batch Experiments

The BMPs of all fractions were expressed as mL CH$_4$ per g of vs. prior to pretreatment (called as "raw") for both pretreated and untreated sunflower heads and stalks, so that comparisons were possible. Control tests were conducted with the untreated sunflower head or stalk (raw). In all cases, the methane volume produced from the blank tests was subtracted to exclude the contribution of the inoculum in the BMP. The alkaline pretreated sunflower head or stalk were separated into their respective liquid and solid fractions (LF and SF respectively). Each fraction was tested separately, but since the BMP of each fraction was expressed per vs. of the raw material it came from, the BMPs of both LF and SF could be summed up and, therefore, express the total BPM of the pretreated head or stalk. The results are shown in Figure 1. In all cases, the LF yielded the lowest methane but at the highest rate. This was attributed to the less, but more readily available organic matter present in the LF as compared to the SF.

Table 3. Elemental analysis of raw stalks and heads.

Component	Concentration (ppm)		Component	Concentration (ppm)	
	Stalks	**Heads**		**Stalks**	**Heads**
Al	1570	600	Ga	1	0
Ca	53,770	27,790	La	10	1
Fe	990	490	Mo	0	0
Mg	14,170	10,540	Nd	1	1
P	2210	2080	Ni	12	4
K	30,760	98,780	Rb	35	185
Si	19,120	2870	Sc	0	0
Na	2770	130	Sr	119	95
S	6670	5780	Th	0	0
Ti	100	50	U	0	0
As	0	0	Sb	0	0
Cd	0	0	Hg	0	0
Cr	18	0	Br	165	253
Cu	31	29	Cs	3	4
Pb	109	162	Bi	0	0
Mn	234	22	Sm	10	11
V	0	0	W	0	1
Zn	111	21	Zr	0	0
Ba	128	58	Cl	18,944	21,637
Ce	5	3	Y	0	0
Co	5	5	Nb	0	0

Focusing on the untreated residues (raw heads and stalks), it can be seen that the BMP of raw stalk was 65% lower ($p = 1.4 \times 10^{-5}$, <0.05) compared to the raw head residues, i.e., 127.98 ± 5.19 and 210.56 ± 1.97 mL CH$_4$ g^{-1} raw VS, respectively (Table 4). This was not expected since the heads contained less VS, resulting in less COD (Table 1). However, their different lipid content (2.5 g kg^{-1} TS heads vs. 0.99 g kg^{-1} TS stalks) and protein content (13.3 g TKN kg^{-1} TS heads vs. 6.3 g TKN kg^{-1} TS stalks) could explain the higher BMP of the heads (lipids and proteins yield higher methane than other organic compounds such as carbohydrates). Similarly, after alkaline pretreatment, the methane yield of sunflower stalks was estimated to be 168.17 ± 6.87 mL CH$_4$ g^{-1} raw VS, which was significantly lower ($p = 2.2 \times 10^{-5}$, <0.05) than the BMP of pretreated sunflower heads (268.47 ± 3.38 mL CH$_4$ g^{-1} raw VS) (Table 4). In fact, it was 37% lower. The methane yield achieved from the pretreated heads was 60% higher than the yield from the pretreated stalks (Table 4).

In general, alkaline pretreatment as resulted from the BMP tests enhanced the methane yield by 31% and 28% for the stalk and head residues, respectively. This can be correlated with the lignin removal (Table 2) observed after alkaline pretreatment of head and stalk residues (31% and 8%, respectively). Similar results were obtained by Monlau et al. [15] who applied alkaline pretreatment under the same conditions and observed a high lignin removal of 22%, after pretreatment at 55 °C. Moreover, the lignin removal achieved in this study for sunflower heads is comparable with the removal achieved under higher temperature conditions in other studies. Antonopoulou et al. [7] reported 20% and 36% of lignin removal using 2 and 20 g NaOH 100 g^{-1} TS, respectively, at 80 °C.

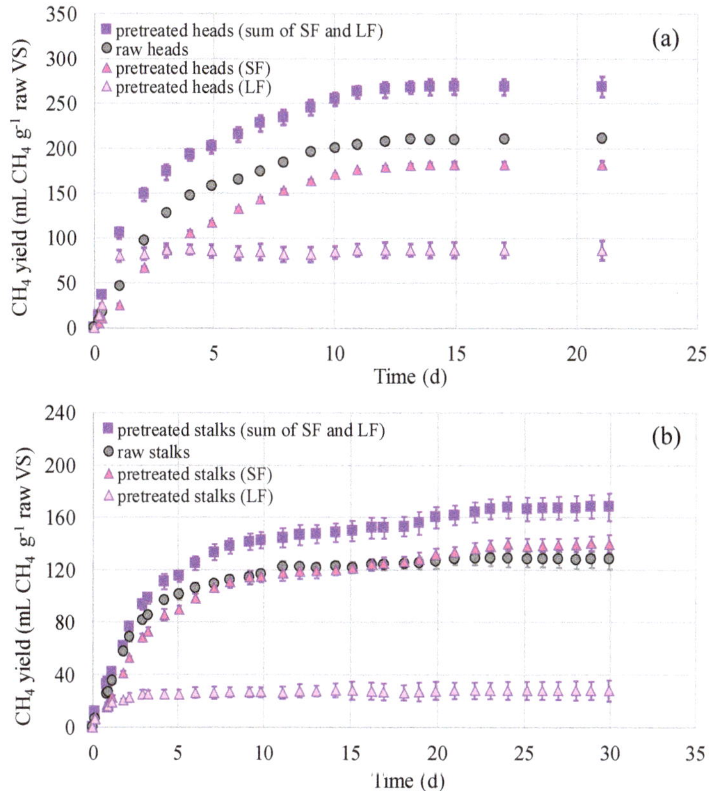

Figure 1. Biochemical methane potential (BMP) profiles of pretreated and untreated sunflower (**a**) heads and (**b**) stalks. Values correspond to means of triplicates of independent values ± standard deviations (error bars). SF: solid fraction, LF: liquid fraction.

Table 4. Biochemical methane potential (BMP) of raw and pretreated sunflower residues. SF: solid fraction, LF: liquid fraction.

Sunflower Residues	BMP (mL CH4 g^{-1} Raw VS) Mean ± SD (± CI)			
	Raw	Pretreated Residues (Sum of SF and LF)	Pretreated Residues (SF)	Pretreated Residues (LF)
Heads	210.56 ± 1.97 (±2.23)	268.47 ± 3.38 (±3.83)	182.01 ± 3.17 (±3.59)	86.46 ± 6.44 (±7.29)
Stalks	127.98 ± 5.19 (±5.88)	168.17 ± 6.87 (±7.77)	140.11 ± 4.62 (±5.23)	28.06 ± 5.47 (±6.19)

Finally, it is noteworthy that Monlau et al. [6] achieved higher methane potential from sunflower stalks (259 ± 6 mL CH$_4$ g^{-1} raw VS) as compared to this study (Table 4), after pretreatment at 55 °C with 4% NaOH for 24 h. In another study, Monlau et al. [10] found a methane potential of 262 mL CH$_4$ g^{-1} raw vs. from sunflower stalks using identical alkaline pretreatment. The BMP of the untreated sunflower stalks was also higher than this study (193 mL CH$_4$ g^{-1} raw vs. > 128 mL CH$_4$ g^{-1} raw VS). Furthermore, Hesami et al. [16] reported much higher values, compared to this study, BMPs from sunflower stalks after hydrothermal (180 °C, 60 min) and isopropanol-based organosolv pretreatment (160 °C, 30 min, 1% H$_2$SO$_4$), of 234 and 278 mL CH$_4$ g^{-1} VS, respectively. However, the methane production yield obtained from the digestion of untreated stalks was lower (124 mL CH_4 g^{-1} VS) than this study. The differences in BMPs as reported in various research studies should be expected, due to the differences in the sunflower variety, the geographical area and the practices used to grow the crop [17]. This is in agreement with Monlau et al. [10], who reported that the alkaline pretreatment (55 °C, 24 h, 4 g NaOH 100 g^{-1} TS) of stalks of different sunflower varieties resulted in different lignin reduction, varying from 23.3% to 36.3% VS. Moreover, the time the residues were left in the field after harvesting and prior to collection is also a determining factor. In the case of the present study, the residues were collected approximately 2–3 weeks after harvesting.

2.3. Continuous Stirred Tank Reactors (CSTRs)–Continuous Experiments

BMP tests indicated that pretreated head residues were the most promising substrate for methane production. Therefore, continuous experiments were performed to evaluate the long-term effect of alkaline pretreatment on biogas production from sunflower head residues. During the first operational phase, all three CSTRs were fed on the liquid fraction coming from the pretreated sunflower head (CSTR1 and CSTR2) or the untreated sunflower head under the conditions summarized in Table 5.

Table 5. Operating conditions of Continuous Stirred Tank Reactors (CSTRs) for both phases (values are given as the average ± standard deviations; numbers in parenthesis express the 95% confidence intervals on the mean values).

	CSTR1	CSTR2	CSTR3
	Operating Phase I: Liquid Fraction Only		
NaOH (g 100 g^{-1} TS)	4	4	0
Hydraulic Retention Time; HRT (d)	21 ± 7	15 ± 4	16 ± 6
Organic Loading Rate; OLR (mg L^{-1} d^{-1})	657 ± 243 (± 82)	791 ± 224 (± 75)	624 ± 316 (± 108)
	Operating Phase II: Whole Slurry		
NaOH (g 100 g^{-1} TS)	4	8	0
HRT (d)	25 ± 4	25 ± 1	25 ± 1
OLR (mg L^{-1} d^{-1})	2079 ± 159 (± 47)	1900 ± 81 (± 25)	1970 ± 0 (± 0)

The organic loading rates (OLRs) and biogas production rates (BPRs) versus time are shown in Figure 2a–c. The experimental data showed a significantly higher biogas production rate during the operation of CSTR2 (205 ± 23 mL L^{-1} d^{-1}) than CSTR1 (161 ± 26 mL L^{-1} d^{-1}) ($p = 0.005$, <0.05) (Table 6), due to the higher OLR in CSTR2. However, the methane yield achieved by CSTR2 (0.193 ± 0.035 L CH_4 g^{-1} VS) was comparable with the yield achieved by CSTR1 (0.187 ± 0.025 L CH_4 g^{-1} VS) ($p = 0.71$, >0.05). The concentration of phenolic compounds determined in both CSTRs during operation was below 250 mg L^{-1} (data not shown).

On the other hand, despite the alkaline pretreatment of the feeding substrate in CSTR2, the biogas production rate of CSTR2 (205 ± 23 mL L^{-1} d^{-1}) was not significantly higher than the control CSTR3 (179 ± 12 mL L^{-1} d^{-1}) (Table 6) that operated at the same HRT (16 ± 6 d) ($p = 0.67$, >0.05). In order to elucidate if any inhibition occurred in the bioreactors, 1 g of acetic acid per L of reactor was added in all three CSTRs, while the feeding with the liquid fraction of the sunflower heads continued. This disturbance occurred on the 37th day when the biogas had been stabilized in all three CSTRs. Acetic acid remained constant with some temporal increases in the CSTR1 and CSTR2 (Figure 2d,e). This indicated that no more acetic acid than that produced from the digestion of the liquid fraction could be degraded, which revealed that these reactors were kinetically limited. On the other hand, the acetic acid was degraded in the CSTR3 within 5.5 days (Figure 2f). To alleviate the inhibitory effect that averted acetic acid degradation, the feeding of CSTR1 and CSTR2 was stopped and these reactors operated at batch mode until acetic acid concentration was minimized. Even at batch mode, the degradation of acetic acid was much slower in CSTR1 (84 mg L^{-1} d^{-1}) and CSTR2 (50 mg L^{-1} d^{-1}) than in CSTR3 (350 mg L^{-1} d^{-1}).

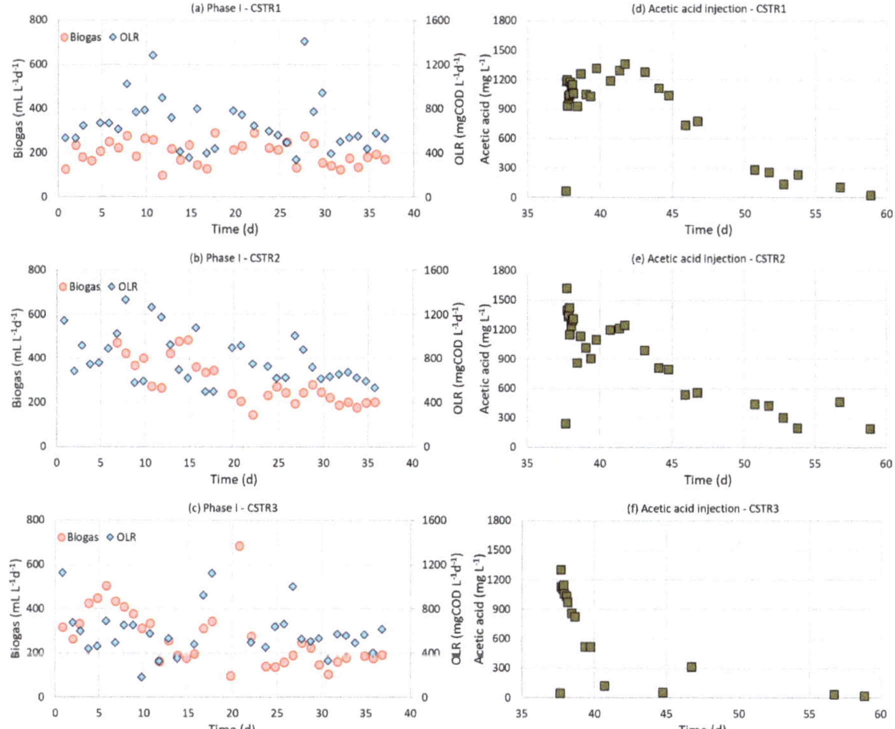

Figure 2. Biogas production rate compared to organic loading rate in three CSTRs fed on sunflower heads during operational phase I (**a**) CSTR1 (NaOH pretreated, HRT = 21 ± 7 d), (**b**) CSTR2 (NaOH pretreated, HRT = 15 ± 4 d), (**c**) CSTR3 (untreated, HRT = 16 ± 6 d) and acetic acid degradation profiles after adding acetate on the 37th day in (**d**) CSTR1, (**e**) CSTR2, (**f**) CSTR3.

Table 6. Biogas production rate (BPR) and operational characteristics during the continuous mode of the three CSTRs (Mean ± SD (± CI)).

Phase I	VSS (g L^{-1})	pH	BPR (mL L^{-1} d^{-1})
CSTR 1	3.2 ± 0.4 (± 0.5)	7.38 ± 0.04	161 ± 26 (± 19)
CSTR 2	3.7 ± 0.4 (± 0.4)	7.42 ± 0.06	205 ± 23 (± 17)
CSTR 3	2.8 ± 0.3 (± 0.3)	7.30 ± 0.07	179 ± 12 (± 11)
Phase II	VSS (g L^{-1})	pH	BPR (mL L^{-1} d^{-1})
CSTR 1	15.1 ± 1.3 (± 1.8)	8.49 ± 0.18	unstable
CSTR 2	16.1 ± 1.6 (± 1.8)	7.22 ± 0.16	unstable
CSTR 3	16.0 ± 2.0 (± 2.8)	7.21 ± 0.05	505± 52 (± 42)

The CSTRs' operation was continued with feeding of the pretreated sunflower head as a whole (without separating it into liquid and solid fractions). This enabled a direct comparison with the results of Monlau et al. [2], who also fed the CSTRs with the pretreated sunflower stalks as a whole. Moreover, it could be possible that feeding with a medium rich in solids would allow the slow release of organic matter due to hydrolysis, which may be beneficial. The solid matrix of the feeding material could also provide a more suitable microenvironment for the microorganism to cope with adverse conditions. Blika et al. [18] reported that the presence of solids in olive-mill wastewater was a crucial factor for the stability of the AD process, that might have resulted in adsorption of toxic hydrophobic compounds onto the solids, such as long-chain fatty acids. Therefore, during phase II, the three CSTRs were fed with an increased OLR of ac. 2000 mg L^{-1} d^{-1} (Table 5). All CSTRs were operated under an HRT of 25 days, while CSTR2 was fed on sunflower heads pretreated at a higher dose of NaOH (8 g 100 g^{-1} TS compared to 4 g 100 g^{-1} TS). Both CSTR 1 and CSTR2 were unstable, while CSTR3 exhibited a stable response in terms of biogas production rate (Figure 3). The biogas and methane yield were 0.26 ± 0.03 L g^{-1} vs. and 0.15 ± 0.015 L CH_4 g^{-1} vs. in CSTR3, respectively. Monlau et al. [2] found that the methane yield of sunflower stalks without alkaline preferment was 0.152 L g^{-1} vs. (which agrees with the methane yield of the non-treated heads obtained in this study), while after pretreatment it was increased by 25.6% to 191 mL g^{-1} vs. (which does not agree with the results of the present study, since alkaline pretreatment caused instability in the digesters). The inhibitory conditions prevailed in the CSTR1 and the CSTR2 are also proven by the increasing trend of acetic acid profiles in these digesters which is more intense in the case of CSTR2 (Figure 3d,e). This indicates that inhibition is related with the NaOH pretreatment which was harsher in the feeding of CSTR2 (a double dose of NaOH, i.e., 8 g 100 g^{-1} TS, was applied). Eventually, CSTR3, which was fed with raw sunflower head residues, achieved the highest biogas production rate (505 ± 52 mL L^{-1} d^{-1}) as compared to CSTR1 and CSTR2.

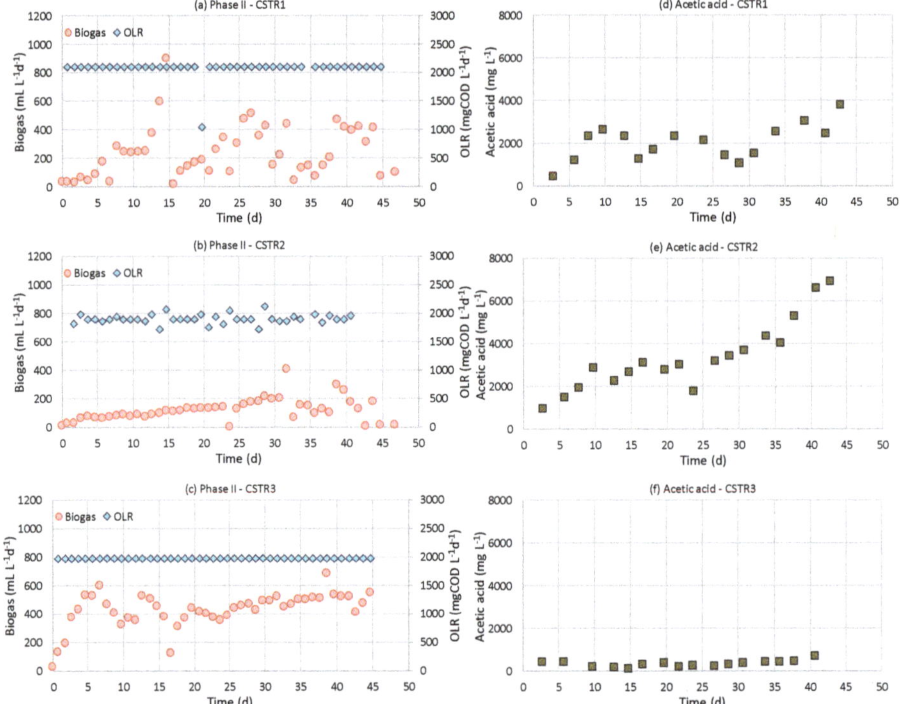

Figure 3. Biogas production rate compared to organic loading rate in three CSTRs fed on sunflower heads during operational phase II (**a**) CSTR1 (NaOH pretreated, 4% NaOH), (**b**) CSTR2 (NaOH pretreated, 8% NaOH), (**c**) CSTR3 (untreated) and acetic acid degradation profiles in (**d**) CSTR1, (**e**) CSTR2, (**f**) CSTR3.

In another work, Polat et al. [19] studied the methane production of a lab-scale fermenters fed with a slurry (2% TS *w/v*) of alkali-pretreated sunflower heads (2 and 5 g NaOH 100 g^{-1} TS, 24 h, ambient temperature) at various HRTs (8–15 d) at thermophilic conditions. They found a positive effect of the alkaline pretreatment. The biogas and methane yields of sunflower heads without alkaline pretreatment were 0.199 L g^{-1} vs. and 0.125 L CH$_4$ g^{-1} VS, respectively, for an HRT of 15 days. These values are lower than the yields found in this study for the control reactor (0.26 ± 0.03 L biogas g^{-1} vs. and 0.15 ± 0.015 L CH$_4$ g^{-1} vs. in CSTR3), but it should be taken into account that the HRT was longer i.e., 25 d). However, when compared to the yields they obtained at a dose of 5 NaOH 100 g^{-1} TS, the yields were 20% higher (0.249 L biogas g^{-1} vs. and 0.154 L CH$_4$ g^{-1} VS). It should be noted, that the solid concentration of the feeding slurry in Polat et al. [19] was lower (2%) than in this work (4.6%). Although the NaOH dose was a little higher (5 NaOH 100 g^{-1} TS) than the dose used in CSTR2 (4 NaOH 100 g^{-1} TS), the feeding in Polat et al. [19] contained less TS which may have induced a lower concentration of inhibitors. Moreover, the pretreatment in Polat et al. [19] was conducted at even milder conditions with respect to temperature. This might be an additional reason for less inhibitors' production during alkaline pretreatment.

Regarding the causes of process low efficiency or instability while the digesters were fed on alkaline pretreated sunflower heads, it is usually considered that the solubilization or degradation of lignin results in secondary by-products such as phenolic or other compounds which may cause inhibition (i.e., vanillin, syringaldehyde) [2]. There are contradictory reports in the literature about their inhibitory level. For example, furfural and 5-hydroxymethylfurfural (HMF) which originate from the dehydration of pentoses and hexoses, as well as syringaldehyde and vanillin resulting from the lignin polymers' degradation, were not found inhibitory (even at high concentrations, 1 g L^{-1}), and, moreover, were converted to biogas during BMP tests [20]. On the other hand, Kayembe et al. [21] evaluated the inhibitory effect of phenolic monomers on methane production by acetoclastic methanogens (archaea) in batch assays, correlating the inhibition with the hydroxyl groups' number of their aromatic structures. They suggested that monomers as phenol and resorcinol, exhibit inhibition of ca. 1.2 and 1.7 g L^{-1} IC50, respectively, corresponding to 50% inhibition of methanogenic activity. However, the phenolics measured in the liquid fractions of this study (Table 1) were low enough compared to the above IC50 values. This indicates that phenolics as measured through the Folin–Ciocalteu method are not the cause of inhibition, but this does not exclude the presence of other potential inhibitors resulting as derivatives from the sunflower heads after alkaline pretreatment.

Apart from the inhibitory compounds probably released during pretreatment, the higher sodium ion concentration, might also be a cause for methanogenic bacteria inhibition. Antonopoulou et al. [22] observed a threshold of NaOH (1% *w/v* NaOH, 0.5% TS *w/v*) concentration above which, inhibition or toxicity of methanogens might occur. This corresponds to a sodium cation concentration of 5750 mg L^{-1}. Moreover, Polat et al. [19] reported an unstable behavior of continuously run digesters fed on sunflower heads when the solid concentration of the feed was 5%. They correlated this instability with the excessive alkali dosage (NaHCO$_3$) required to adjust the very low pH of the feeding medium. A concentration range of sodium ion between 3500 and 5500 mg Na$^+$ L^{-1} has been considered to cause moderate inhibition [23]. In the presented study, the sodium ion concentration entering the digesters was estimated to be 1058 and 2116 mg Na$^+$ L^{-1} in the case of pretreatment with 4 g and 8 g NaOH per 100 g TS, respectively (4.6% TS was the influent concentration). Therefore, alkali pretreatment, at these NaOH doses applied in this work, does not seem to have caused inhibition through the sodium anion.

Finally, the contradictory results for the methane yields derived from batch and continuous experiments in the present study reveal that batch tests may not give representative results since they are affected by the activity of the inoculum. A highly active inoculum may achieve yield high ratios of methane produced per mass of substrate. On the contrary, under continuous operation, the inoculum is constantly exposed to the feeding conditions which increases the concentration of inhibitory compounds inside the reactor if these are recalcitrant and are not consumed. Therefore, alkaline pretreatment seems to have a positive effect due to organic matter solubilization and destruction of the lignocellulosic matrix, but also a negative effect due to inhibitory compounds probably released as lignin degrades [2]. The positive effect was prominent in the short term during the BMP tests, while the negative effect prevailed in the long term during continuous operation of the digesters.

3. Materials and Methods

3.1. Sunflower Residues Collection and Pretreatment

Sunflower residues were collected, from a farm located in northern Greece (Xanthi city) after seed harvest. The residues were separated into stalks and heads after seed removal. The heads and stalks were dried at 40 °C in an oven (MMM-Venticell) and then milled separately to obtain a particle size of ca. 0.15–1.4 mm using a cutting mill (Retsch SM100). The particle size distribution is shown in Figure 4. The alkaline pretreatment conditions (55 °C, 24 h, 4 g NaOH 100 g^{-1} TS) were selected after Monlau et al. [6]. A loading of heads or stalks equal to 4.6 g dry matter was added into glass bottles (120 mL) with a working volume of 100 mL. The bottles were continuously stirred in a thermostatically controlled incubator at 55 °C. After pretreatment, the pH value was adjusted to 7 with 6N HCl and the whole slurry was separated into liquid and solid fractions by filtration through a 2.5 mm mesh filter. The solid fractions were dried at 40 °C for 24 h before analysis. The main characteristics of the raw and pretreated fractions of the sunflower residues are summarized in Table 1 as average values from triplicate analysis. Table 2 shows the lignin content (Klason lignin) of raw and pretreated heads and stalks. Klason lignin or acid-insoluble lignin is the insoluble residue portion obtained after removing the ash by concentrated acid hydrolysis of the plant tissues [24].

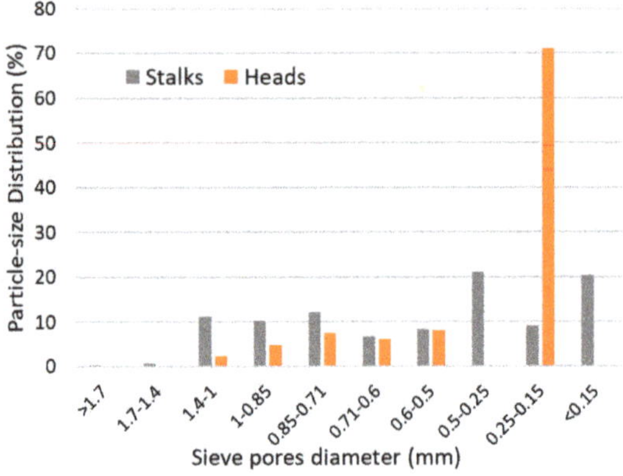

Figure 4. Graphical presentation of particle-size distribution of milled stalks and heads after sieving.

3.2. Biochemical Methane Potential

In order to evaluate the biochemical methane potential (BMP) of the pretreated residues, the solid and liquid fractions of the heads and the stalks were separately digested in batch anaerobic serum bottles (110 mL, working volume 97.5 mL). All BMP tests were conducted in triplicate at 37 °C. Each bottle was inoculated with 85 mL of anaerobic sludge (24.3 g vs. L^{-1}) as inoculum, taken from a full-scale anaerobic digester fed on cow manure and a mixture of various wastes. The inoculum was kept for about 5 days before the batch digestion test at 37 °C for degassing. Then 12.5 mL of the liquid fraction or 1.07 g of the solid fraction were added into serum bottles to achieve an organic loading of 0.04 or 0.55 g COD g^{-1} vs. inoculum, respectively. Since the liquid fractions contained readily biodegradable COD, the COD loading was lower in these tests. On the other hand, the solid fractions contained organic matter, a part of which was biodegradable, and therefore, the COD added was higher. Control tests were run with raw (untreated) sunflower heads and stalks (organic loading of 0.55 g COD g^{-1} vs. inoculum). Blank tests containing only the inoculum and water to reach the same final volume were also set up. No addition of macro elements or trace elements took place. After the preparation of serum bottles, degasification with N_2/CO_2 mixture (80/20) was carried out for 1 min to obtain anaerobic conditions and the bottles were sealed with rubber stoppers. The headspace of each bottle was connected with a NaOH (6N) displacement apparatus to trap CO_2. Methane production was monitored via the volume displacement method and was expressed at standard conditions of temperature and pressure (STP).

3.3. Continuously Stirred Tank Reactors: Set-up and Operation

Experiments were conducted in three identical continuously stirred tank reactors (CSTRs) (Figure 5). Each CSTR had 3.2 L and 2.5 L total and working volume, respectively. All reactors were constructed from two Plexiglas cylinders concentrically configured to allow for the recirculation of hot water within the in-between void space, thus maintaining the digester's temperature at 35 ± 1 °C through a controller. The CSTRs were mixed at ca. 85 rpm by a mechanical stirrer attached onto the top of the reactors. The biogas and the effluent were removed from the same tube (which determined the level of the mixed liquor in the CSTR) into a flask and they were separated. This flask allowed the solids to settle down while the supernatant overflowed through a port placed in the middle. The solids were removed daily from the bottom of the flask. Moreover, the biogas was released from the upper part of the flask and was led to a volumetric measurement device consisting of a U-tube filled with oil and a 3-way valve solenoid. The valve was activated by a level sensor which allowed for the displacement of oil by the biogas entering the U-tube and the release of biogas after the oil reached a certain level (Figure 5). The biogas production rate (BPR) is reported at STP [25].

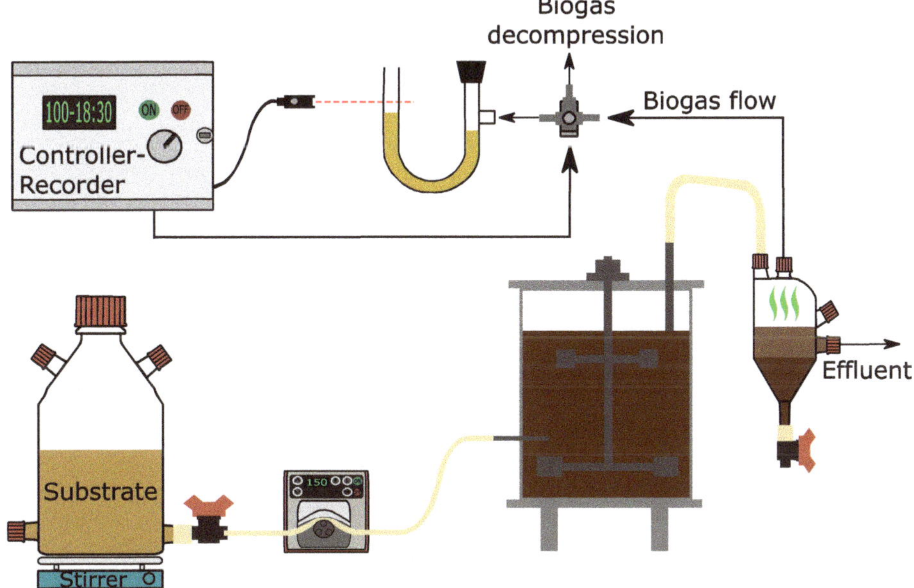

Figure 5. Experimental configuration used for the continuous tests.

The three CSTRs were inoculated with biomass taken from a full-scale anaerobic digester treating various animal wastes. The operation can be distinguished in two main phases. As shown in Table 5, during the first operating phase, all three CSTRs were fed on the liquid fraction obtained from the sunflower residues' heads as described in Section 2.1 in the presence of NaOH (mild alkaline pretreatment; CSTR 1 and CSTR 2) or in the absence of NaOH (control experiment; CSTR3). For the hydraulic retention time (HRT) it was attempted to maintain equality for CSTR2 and CSTR3, while the

HRT was kept higher in the CSTR1 to study the effect of the feeding rate. Since feeding was conducted three times a day via peristaltic pumps set by timers, there was an inevitable variation in the feeding rate which was recorded daily. As a result of the different HRT and the liquid fraction's characteristics (in the presence or absence of NaOH), the organic loading rate (OLR) was higher in the CSTR 2 (lower HRT) and lower in the CSTR3 (the absence of NaOH resulted in lower extraction of the organic matter in the liquid fraction) The liquid fraction was supplemented with 10 ml L^{-1} from nutrient and trace elements solutions consisting of 8.2 g L^{-1} K_2HPO_4 (solution A), 40.9 g L^{-1} NH_4Cl (solution B), 33 g L^{-1} $FeCl_3.4H_2O$ (solution C), and other trace elements (solution D; [25]) which were stored in a different bottle from Fe salt to avoid insoluble salt formation.

After the three CSTRs reached a quasi-steady biogas production rate under the operating conditions of phase I, an impulse disturbance of acetate was imposed, while the CSTRs continued to be fed with the sunflower substrate. Specifically, 1 g L^{-1} acetate was spiked simultaneously into all three CSTRs on the 37th day, increasing instantly the concentration of acetate inside the reactors. The aim of this disturbance was to study the response of the acetoclastic methanogens prevailing under the imposed operating conditions.

After depletion of acetate, the second operating phase started to introduce the whole slurry biomass (without separating it into liquid and solid fractions) after alkaline pretreatment in two different NaOH doses (CSTR1 and CSTR2) or without alkaline pretreatment (CSTR3; control experiment). Similarly, to the first operating phase, nutrient and trace element solutions were added. The feeding was conducted manually to avoid clogging problems due to the solids. The HRT resulting was 25 d for all three CSTRs. The organic loading rate (OLR) resulted as shown in Table 5. All feeding mixtures introduced into the CSTRs were adjusted to pH 7 with 6 N HCl.

3.4. Statistics

All data obtained were statistically analyzed to calculate average values (Equation (1)) and standard deviations (Equation (2)).

$$\overline{Y} = \frac{\sum_{i=1}^{n} Y_i}{n} \tag{1}$$

$$SD = \sqrt{\frac{\sum_{i=1}^{n}\left(Y_i - \overline{Y}\right)^2}{n-1}} \tag{2}$$

where $\overline{Y}$ is the average of the measured values, SD is the standard deviation, Y_i are the experimental values, and n is the number of data.

The 95% Confidence Interval (CI) on the mean was computed as follows (Equation (3)):

$$CI = \overline{Y} \pm Z_{95} \cdot \frac{SD}{\sqrt{n}} \tag{3}$$

where Z_{95} is the upper $(1 - 0.95)/2$ critical value for the standard normal distribution.

An F-test was performed to determine the homogeneity of variances, followed by analysis of variance method (two-sample *t*-test, *p* 0.05), to assess the effect of pretreatment of sunflower residues on anaerobic digestion performance using Microsoft Excel software.

3.5. Analytical Methods

The characterization of sunflower heads and stalks as well as the samples from all bioreactors was performed in the laboratory of Wastewater Management and Treatment Technologies of the Department of Environmental Engineering at Democritus University of Thrace (DUTH, Greece). Chemical oxygen demand (COD), total Kjeldahl nitrogen (TKN), total solids (TS), volatile solids (VS) and lipids were determined according to Standard Methods [26]. The pH was measured with a pH meter (HANNA, HI 83141, HANNA INSTRUMENTS HELLAS, Athens, Greece). Phenols' concentration was measured according to Cindric et al. [27]. Klason lignin content was determined according to the Laboratory Analytical Procedure (LAP) [28]. The concentrations of volatile fatty acids (VFAs) were measured in a gas chromatograph (Perkin Elmer, Waltham, MA, US) equipped with a capillary free fatty acid phase (FFAP) column, a flame ionization detector, and helium as carrier gas, while the oven temperature program was 50 °C to 200 °C with 10 °C min^{-1} rate. A gas analyzer equipped with CO_2 and CH_4 infrared sensors (GasCard NG model of the Edinburgh Sensors) (Edinburgh Instruments Ltd, Livingston, UK) was used for determining biogas composition.

For the elemental analysis of stalks and heads, the samples were pelletized in the Atlas 25 ton Manual Hydraulic Press (Specac, Kent, UK) with Atlas™ 40 mm Evacuable Pellet Die (Specac, Kent, UK) by using a 2 g subsample with boric acid as substrate (H_3BO_3 99.9%, Socachim, Belguim). Analysis was conducted using a Wavelength dispersive X-ray fluorescence spectrometer (ZSX Primus II, Rigaku Corporation, Tokyo, Japan), which was calibrated using the NIST standards: 1646a, 1648a, 2584, and 2710a.

4. Conclusions

Results presented herein showed that the sunflower head and stalk residues after alkaline pretreatment had a positive effect on the biogas production as evaluated during BMP tests. On the other hand, CSTR operation with pretreated sunflower heads, indicated the opposite. Experimental observations revealed that alkaline pretreatment obviously causes inhibitory conditions, which could not be detected in the batch tests where, mostly, the short-term response of the microorganisms is shown. This indicates, that BMP tests cannot always reveal the real effect of treatment conditions, since the microbial biomass is not continuously exposed to the positive or adverse effects of these conditions. Since sunflower heads, after alkaline pretreatment, cause instability to continuously operated anaerobic digesters, they should be excluded from lignocellulosic mixtures which are subjected to alkaline pretreatment, even under mild conditions. On the contrary, if not pretreated, they seem to support stability in the digesters (with a methane yield of 0.15 ± 0.015 L CH_4 g^{-1} VS) and could be used as feedstock.

Author Contributions: This work was carried out in collaboration between all authors. Authors M.Z. and A.S. performed the experimental investigation. Author I.A.V.; performed the data curation and the statistical analysis and wrote the first draft of the manuscript. Corresponding Author K.S.; designed the study, performed the supervision, the writing—review and the editing. All authors have read and agreed to the published version of the manuscript.

Funding: This research was funded by the project "INVALOR" (MIS 50002495) which is implemented under the Action "Reinforcement of the Research and Innovation Infrastructure", funded by the Operational Programme "Competitiveness, Entrepreneurship and Innovation" (NSRF 2014–2020) and co-financed by Greece and the European Union (European Regional Development Fund).

Conflicts of Interest: The authors declare no conflict of interest.

References

1. Ren, N.; Wang, A.; Cao, G.; Xu, J.; Gao, L. Bioconversion of lignocellulosic biomass to hydrogen: Potential and challenges. *Biotechnol. Adv.* **2009**, *27*, 1051–1060. [CrossRef] [PubMed]
2. Monlau, F.; Kaparaju, P.; Trably, E.; Steyer, J.P.; Carrere, H. Alkaline pretreatment to enhance one-stage CH4and two-stage H_2/CH_4 production from sunflower stalks: Mass, energy and economical balances. *Chem. Eng. J.* **2015**, *260*, 377–385. [CrossRef]
3. Monlau, F.; Barakat, A.; Trably, E.; Dumas, C.; Steyer, J.P.; Carrère, H. Lignocellulosic materials into biohydrogen and biomethane: Impact of structural features and pretreatment. *Crit. Rev. Environ. Sci. Technol.* **2013**, *43*, 260–322. [CrossRef]
4. Taherzadeh, M.J.; Karimi, K. Pretreatment of lignocellulosic wastes to improve ethanol and biogas production: A review. *Int. J. Mol. Sci.* **2008**, *9*, 1621–1651. [CrossRef] [PubMed]
5. Kainthola, J.; Kalamdhad, A.S.; Goud, V.V. A review on enhanced biogas production from anaerobic digestion of lignocellulosic biomass by different enhancement techniques. *Process Biochem.* **2019**, *84*, 81–90. [CrossRef]

6. Monlau, F.; Barakat, A.; Steyer, J.P.; Carrere, H. Comparison of seven types of thermo-chemical pretreatments on the structural features and anaerobic digestion of sunflower stalks. *Bioresour. Technol.* **2012**, *120*, 241–247. [CrossRef]

7. Antonopoulou, G.; Dimitrellos, G.; Beobide, A.S.; Vayenas, D.; Lyberatos, G. Chemical Pretreatment of Sunflower Straw Biomass: The Effect on Chemical Composition and Structural Changes. *Waste Biomass Valorization* **2015**, *6*, 733–746. [CrossRef]

8. Cao, S.; Pu, Y.; Studer, M.; Wyman, C.; Ragauskas, A.J. Chemical transformations of Populus trichocarpa during dilute acid pretreatment. *RSC Adv.* **2012**, *2*, 10925–10936. [CrossRef]

9. Monlau, F.; Sambusiti, C.; Barakat, A.; Quéméneur, M.; Trably, E.; Steyer, J.P.; Carrère, H. Do furanic and phenolic compounds of lignocellulosic and algae biomass hydrolyzate inhibit anaerobic mixed cultures? A comprehensive review. *Biotechnol. Adv.* **2014**, *32*, 934–951. [CrossRef]

10. Monlau, F.; Aemig, Q.; Barakat, A.; Steyer, J.P.; Carrère, H. Application of optimized alkaline pretreatment for enhancing the anaerobic digestion of different sunflower stalks varieties. *Environ. Technol. (United Kingdom)* **2013**, *34*, 2155–2162. [CrossRef]

11. Faostat Food and Agriculture Organization of the United Nations Crop Statistics. Available online: http://www.fao.org/faostat/en/#data/QC (accessed on 1 December 2019).

12. Searle, S.; Malins, C. *Availability of Cellulosic Residues and Wastes in the EU*; International Council on Clean Transportation: Washington, DC, USA, 2013.

13. Gonzalez-Perez, S.; Vereijken, J.M. Review Sunflower proteins: Overview of their physicochemical, structural and functional properties. *J. Sci. Food Agric.* **2007**, *87*, 2173–2191. [CrossRef]

14. Molaey, R.; Bayrakdar, A.; Sürmeli, R.Ö.; Çalli, B. Anaerobic digestion of chicken manure: Mitigating process inhibition at high ammonia concentrations by selenium supplementation. *Biomass Bioenergy* **2018**, *108*, 439–446. [CrossRef]

15. Monlau, F.; Trably, E.; Barakat, A.; Hamelin, J.; Steyer, J.P.; Carrere, H. Two-stage alkaline-enzymatic pretreatments to enhance biohydrogen production from sunflower stalks. *Environ. Sci. Technol.* **2013**, *47*, 12591–12599. [CrossRef] [PubMed]

16. Hesami, S.M.; Zilouei, H.; Karimi, K.; Asadinezhad, A. Enhanced biogas production from sunflower stalks using hydrothermal and organosolv pretreatment. *Ind. Crops Prod.* **2015**, *76*, 449–455. [CrossRef]

17. Amon, T.; Amon, B.; Kryvoruchko, V.; Machmüller, A.; Hopfner-Sixt, K.; Bodiroza, V.; Hrbek, R.; Friedel, J.; Pötsch, E.; Wagentristl, H.; et al. Methane production through anaerobic digestion of various energy crops grown in sustainable crop rotations. *Bioresour. Technol.* **2007**, *98*, 3204–3212. [CrossRef]

18. Blika, P.S.; Stamatelatou, K.; Kornaros, M.; Lyberatos, G. Anaerobic digestion of olive mill wastewater. *Global Nest J.* **2009**, *11*, 364–372.

19. Polat, H.; Selçuk, N.; Soyupak, S. Effect of pretreatments on the semicontinuous anaerobic digestion of sunflower heads. *Energy Sources* **1992**, *14*, 391–403. [CrossRef]

20. Barakat, A.; Monlau, F.; Steyer, J.P.; Carrere, H. Effect of lignin-derived and furan compounds found in lignocellulosic hydrolysates on biomethane production. *Bioresour. Technol.* **2012**, *104*, 90–99. [CrossRef]

21. Kayembe, K. Inhibitory Effects of Phenolic Monomers on Methanogenesis in Anaerobic Digestion. *Br. Microbiol. Res. J.* **2013**, *3*, 32–41. [CrossRef]

22. Antonopoulou, G.; Lyberatos, G. Effect of pretreatment of sweet sorghum biomass on methane generation. *Waste Biomass Valoriza.* **2013**, *4*, 583–591. [CrossRef]

23. McCarty, P.L. Anaerobic Waste Treatment Fundamentals. *Public Works* **1964**, *95*, 91–94.

24. Chen, H.; Chen, H. Lignocellulose biorefinery feedstock engineering. In *Lignocellulose Biorefinery Engineering*, 1st ed.; Woodhead Publishing: Cambridge, UK, 2015; pp. 37–86.

25. Spyridonidis, A.; Skamagkis, T.; Lambropoulos, L.; Stamatelatou, K. Modeling of anaerobic digestion of slaughterhouse wastes after thermal treatment using ADM1. *J. Environ. Manag.* **2018**, *224*, 49–57. [CrossRef] [PubMed]

26. APHA; AWWA; WEF. *Standard Methods for the Examination of Water and Wastewater*, 20th ed.; American Public Health Association/American Water Works Association/Water Environment Federation: Washington, DC, USA, 1999.

27. Cindrić, I.J.; Kunštić, M.; Zeiner, M.; Stingeder, G.; Rusak, G. Sample preparation methods for the determination of the antioxidative capacity of apple juices. *Croatica Chem. Acta* **2011**, *84*, 435–438. [CrossRef]

28. Sluiter, A.; Hames, B.; Ruiz, R.O.; Scarlata, C.; Sluiter, J.; Templeton, D.; Energy, D. *Determination of Structural Carbohydrates and Lignin in Biomass*; Biomass Analysis Technology Team Laboratory Analytical Procedure: Golden, CO, USA, 2004; pp. 1–14.

Sample Availability: Samples of the compounds not available from the authors.

Article

Soft Microwave Pretreatment to Extract *P*-Hydroxycinnamic Acids from Grass Stalks

Aurélie Bichot [1], Mickaël Lerosty [1], Laureline Geirnaert [1], Valérie Méchin [2], Hélène Carrère [1], Nicolas Bernet [1], Jean-Philippe Delgenès [1] and Diana García-Bernet [1,*]

[1] Univ Montpellier, INRA, 102 Avenue des Etangs, CEDEX, 11100 Narbonne, France; aurelie.bichot@supagro.fr (A.B.); mickael.lerosty@inra.fr (M.L.); laureline.geirnaert@utc.fr (L.G.); helene.carrere@inra.fr (H.C.); nicolas.bernet@inra.fr (N.B.); jean-philippe.delgenes@inra.fr (J.-P.D.)

[2] INRA Institut Jean-Pierre Bourgin, CEDEX, 78026 Versailles, France; valerie.mechin@inra.fr

* Correspondence: diana.garcia-bernet@inra.fr

Academic Editor: Ivet Ferrer
Received: 30 September 2019; Accepted: 25 October 2019; Published: 28 October 2019

Abstract: The aim of this article is to provide an analysis of microwave effects on ferulic and coumaric acids (FA and CA, respectively) extraction from grass biomass (corn stalks and miscanthus). Microwave pretreatment using various solvents was first compared to conventional heating on corn stalks. Then, microwave operational conditions were extended in terms of incident power and treatment duration. Optimal conditions were chosen to increase p-hydroxycinnamic acids release. Finally, these optimal conditions determined on corn stalks were tested on miscanthus stalks to underlie the substrate incidence on p-hydroxycinnamic acids release yields. The optimal conditions—a treatment duration of 405 s under 1000 W—allowed extracting 1.38% FA and 1.97% CA in corn stalks and 0.58% FA and 3.89% CA in miscanthus stalks. The different bioaccessibility of these two molecules can explain the higher or lower yields between corn and miscanthus stalks.

Keywords: microwave pretreatment; grass biomass; p-hydroxycinnamic acids extraction

---

1. Introduction

Lignocellulosic biomass is a key resource for the sustainable development of the bioeconomy, and its effective implementation would allow a decisive matter and energy co-valorization. First, it is available on all continents in different forms and species. Second, the volumes of available resources are very large, represent an interesting carbon source, and come from numerous sustainable sources avoiding land-use conflicts: agricultural and agri-food wastes, wastes from land-use planning, forest woody residues, perennial crops on polluted soils (miscanthus), algae, or municipal solid wastes [1]. Nevertheless, the use of lignocellulosic biomass is curtailed because of high processing costs and difficulties related to obtaining high extraction yields. Nowadays, many applications (food and chemical industry, materials) are being explored from agricultural residues and miscanthus, as they represent an alternative source of many high-value molecules (sugars, ethanol, acetic acid, butyric acid, xylitol, propionic acid, oil, etc.) and fibers [2].

Among them, p-hydroxycinnamic acids are increasingly attracting the attention of scientists and industry because of their interesting features. Indeed, due to their anti-oxidant properties, p-hydroxycinnamic compounds can be used as food preservatives and have a preventive role in some types of cancer [3]. Ferulic, coumaric, syringic, and caffeic acids are the most commonly found acids in cereals, whether in grains or stems [4]. They are covalently bounded to cell walls and difficult to access. Indeed, p-coumaric acid is principally esterified on lignin S units, and ferulic acid is both etherified to lignin and esterified to hemicelluloses [5–7]. Figure 1 graphically represents the different chemical bonds that form the complex three-dimensional biomass network.

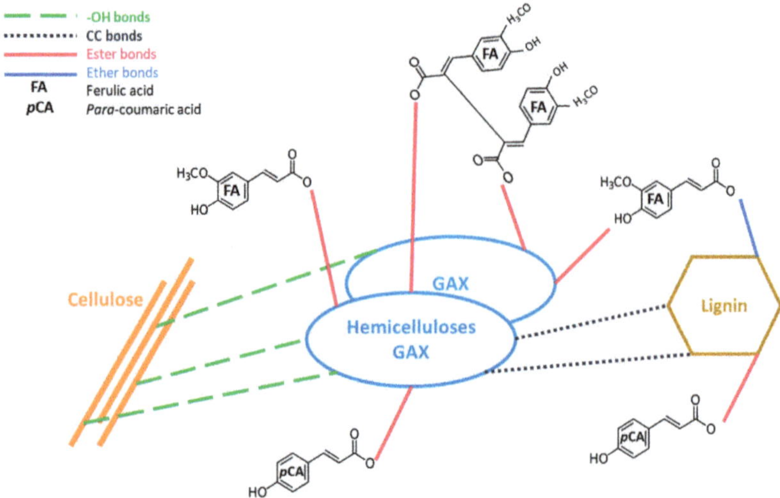

Figure 1. Global organization of grass cell wall.

Many factors can explain *p*-hydroxycinnamic acids' accessibility and lignocellulosic biomass resistance to degradation: polymers composition and/or organization, protein and acetyl group abundance, specific surface area, or a combination of all these factors [8]. Three main polymers—cellulose, hemicelluloses, and lignin—compose lignocellulosic biomass [9]. While lignin is believed to play an important role in cellulose protection from hydrolysis to sugar monomers by forming a physical barrier and adsorbing enzymes [10], other key factors such as hemicelluloses rate, cellulose crystallinity, acetylation, or porosity are supposed to hinder cellulose hydrolysis by forming a complex and solid lignocellulose network [11]. Thus, applying an efficient pretreatment to break down lignocellulosic biomass in order to favor bioconversions and facilitate access to molecules of interest is mandatory to obtain an economically sustainable process.

Many pretreatments have been tested in order to reduce lignocellulose recalcitrance to *p*-hydroxycinnamic acids release, including: biological, chemical, or physical treatment [12,13]. The present article is focusing on microwave pretreatment, which is a thermal treatment that has gained interest over the past 10 years due to its numerous advantages. Indeed, microwave treatment heats material directly without any contact, shortens reaction time by reaching high temperatures faster than conventional heating systems, increases yields and purity by reducing the formation of side products and inhibitors, diffuses homogenous microwave irradiation inside the cavity, thus allowing high replication, consumes less solvent, and allows a quick control of the operating parameters [14,15]. Due to these potential benefits, many recent studies are devoted to the application of this technology to improve *p*-hydroxycinnamic acids and sugars release from lignocellulosic biomass and/or waste activated sludge biodegradability [16]. A systemic approach is proposed: after testing large range of operating parameters (solvent, duration, power, power density) on a specific biomass (corn stalks), the conditions that lead to the highest *p*-hydroxycinnamic acids yields were extended on another lignocellulosic biomass.

2. Results and Discussion

2.1. Raw Matter Composition

Before performing any treatment, dry matter content (DM) was estimated according to the NREL (National Renewable Energy Laboratory from US Department of Energy) protocol. Table 1 presents parietal composition of raw materials expressed in % DM. Parietal composition corresponds to all the elements that are part of the biomass cell wall: hemicellulose, cellulose, lignin, and ash. Van Soest protocol was used to determine hemicelluloses, cellulose, acid detergent lignin (ADL), and ash in a gravimetrical way, and NREL protocol was used to determine Klason lignin. A clear difference can be seen in the composition of the two studied biomass samples, particularly in terms of wall proportion. Soluble content in miscanthus stalks is very low, and 95% DM of the mass consists of the cell wall, of which cellulose and lignin represent a significant part. Miscanthus is lignified to a greater extent than corn stalks (twice as much), and its cellulose content is two times higher. It appears that the compositions of the two biomasses differ significantly, which could explain the differences in behavior following microwave treatment.

Table 1. Raw matter characteristics: parietal composition analyzed by Van Soest method and Klason lignin (mean of triplicate and standard error) and alkali extraction for ferulic acid (FA) and coumaric acids (CA) (mean of duplicate and standard error). ADL: acid detergent lignin, DM: dry matter.

		Corn Stalks F98902	Miscanthus Stalks GIB Genotype
Soluble Content (%)		35.67 ± 1	4.81 ± 1
Parietal Composition using Van Soest Method (%DM)	Cell wall	64.33 ± 1	95.19 ± 1
	Cellulose	28.87 ± 0.5	55.75 ± 1.6
	Hemicellulose	26.06 ± 1	21.75 ± 1.1
	ADL	8.18 ± 1.7	16.86 ± 1.6
	Ash	1.22 ± 0.3	0.83 ± 0.5
Klason Lignin (% DM)		15.00 ± 0.5	24.34 ±0.2
p- **Hydroxycinnamic Acids (mg/g DM)**	Ferulic acid	4.2 ± 0.9	2.1 ± 0.5
	Coumaric acid	13.1 ± 2.6	6.5 ± 2.3

Biomass composition (*p*-hydroxycinnamic acids, parietal polymers) was in agreement with values found in the literature. Provan et al. [17] worked on phenolic compounds in maize straw (2 g CA/100 g DM, 0.6 g FA/100 g DM) or wheat straw (0.5 g CA/100 g DM, 0.01 g FA/100 g DM). In terms of parietal polymers, the composition can vary significantly among species, but generally admitted values vary between 27%–40% DM cellulose, 25%–34% DM hemicellulose, and 9%–15% DM lignin for corn stalks, and 28%–49% DM cellulose, 24%–32% DM hemicellulose, and 15%–28% DM lignin for miscanthus stalks [11].

2.2. Effect of Solvent and Treatment on p-Hydroxycinnamic Acids Release

Preliminary experiments were performed on corn stalks sample F98902 using one microwave condition (500 W for 270 s) and different solvents, in order to select the best one for the pretreatment of biomass with the microwave pilot. Indeed, solvent is one of the most important parameters in microwave pretreatments, as it affects the solubility of the target components, and it interacts differently with microwaves depending on its polarity. Polar solvents absorb microwaves, allowing reaction media to reach high temperatures quickly, and provoking cell disruption, thus facilitating the release of molecules of interest; while non-polar solvents are transparent to microwaves [15,18,19]. In the literature, many solvents have been tested, including non-polar solvents that do not absorb waves and allow waves to be concentrated on material. To our knowledge, no scientists have worked with non-polar solvents in the microwave pretreatment of biomass studies, but hexane is commonly used mixed with a polar solvent, such as water, for the treatment of polluted soils [20]. Concerning polar solvents, water and ethanol mixtures are the most commonly used. With ethanol 50% (*w:w*) and 30 W, Carniel et al. [18] extracted 3.74 mg gallic acid equivalent/g *Physalis angulate* in less than a minute. Oufnac et al. [21] extracted with methanol up to 467 µg catechin equivalent/g wheat bran, which is twice as high as the extraction yield obtained using conventional heating.

In this study, only polar solvents were tested, including water, alkaline water (NaOH), acidic water (H_2SO_4), and water/ethanol mixture. The operating conditions were the following (Table 2): 270 s of microwave treatment under 500 W (0.179 Wh/g), 360 s of conventional heating, or control without any heating. pH was measured after one hour of soaking, just before treatment. No significant change in pH was observed after the treatment (Table 2). The antioxidant activities were not studied in this study and considered equivalent between samples, meaning that the biological activity was assessed as not having a significant impact on the biomass during the treatments. The maximum temperature was 99 °C for microwave pretreatments in water, 80 °C for microwave pretreatment in ethanol, and between 66 °C and 75 °C for conventional pretreatments, depending on solvents. Therefore, conventional processing involved a much longer heating time, which is consistent with the bibliographic references: Del Rio et al. [22] realized a biomass thermal treatment for 20 min to reach 120 °C.

2.2.1. Composition Analysis of Pretreated Biomass

Biomass composition was characterized using Van Soest and Klason methods (Figure 2). According to Figure 2A, parietal polymers contents are not statistically different between raw matter and pretreated biomass: standard deviations ranging from 0.1% to 2.3% intersect and do not lead to significant differences in parietal composition due to treatments (ANOVA results not shown). For all treatments tested, ADL represents 7.6% DM, cellulose represents 29% DM, hemicellulose represents 26% DM, and total soluble content (from treatment and from NDF Van Soest first step) represents 36.7% DM. Corn stalks' parietal composition is not significantly modified by the treatments tested.

The lignin content estimated by two different analytical methods is presented in Figure 2B. From statistical analysis, the treatment or the solvent do not statistically affect lignins (ADL and Klason). Lignin is a very resistant polymer to be impacted by the tested soft microwaves conditions. By looking in greater detail, Klason lignin content is almost always twice as high as ADL content [23], and the differences between the two calculation methods vary between 5.4% and 9.4% (excepted for ethanol microwave treatment, for which the difference is much smaller with 1.7%). As explained earlier, this difference could be interpreted as the part of the lignin linked by β-O-4 bonds and solubilized in 72% H_2SO_4. As this difference does not vary with the type of treatment employed, it means that β-O-4 bonds are not affected by the treatments, and it can be assumed that the lignin internal structure remains the same.

The results differ from those obtained by Choudhary et al. [24]. When sorghum was pretreated with alkaline water up to 0.2 g NaOH/g DM in 20 mL water for 4 min and 1000 W, 58.9% lignin was removed, and 54.9% of hemicellulose was removed, but the cellulose content never varied. In our study, the minor changes in biomass structure can be explained by the moderate temperature reached, which never exceeded 100 °C, by the low incident power during the tests (500 W). Moreover, in Choudhary et al. [24], the NaOH concentration was high compared to our study, and explains lignin degradation and hemicellulose solubilization.

The total mass recovered after treatments of 10 g was around 8 g (Figure 2A). Acid, alkaline, or ethanol did not modify the solubility of components: in all cases tested, 2 g of raw matter (out of 10 g used) were solubilized during the treatment.

Table 2. Impact of solvents and treatments on FA and CA release in liquid phase after treatments (mg/g DM) with 200/10 (*w/w*) as liquid/solid ratio. Each treatment was performed in duplicate, CA and FA measurements were duplicate for each test, and the standard deviation was expressed as a percentage.

Solvent	Reactant Concentration (mg/g liquid)	Reactant Mass (mg/g DM)	Initial pH	Final pH	Microwave Heating (mg/g DM)		Final Microwave Temperature (°C)	Conventional Heating (mg/g DM)		Final Conventional Temperature (°C)	Control (mg/g DM)	
					CA	FA		CA	FA		CA	FA
Water	/	/	5.3	5.1	0.201 ± 0.4%	0.047 ± 0.1%	99	0.125 ± 0.1%	0.051 ± 0.2%	69	0.085 ± 0.3%	0.057 ± 0.2%
Acid	0.34	7.3	3.95	3.7	0.113 ± 1.2%	0.030 ± 0.3%	99	0.071 ± 0.6%	0.027 ± 0.3%	66	0.060 ± 0.2%	0.025 ± 0.4%
Base	0.12	27	8.5	8.3	0.247 ± 0.1%	0.056 ± 0.1%	99	0.230 ± 0.5%	0.081 ± 0.3%	71	0.173 ± 1.4%	0.093 ± 1.6%
Ethanol	500	/	5.5	5.5	0.298 ± 0.6%	0.037 ± 0.4%	81	0.174 ± 0.1%	0.035 ± 0.3%	75	0.159 ± 0.5%	0.038 ± 0.5%

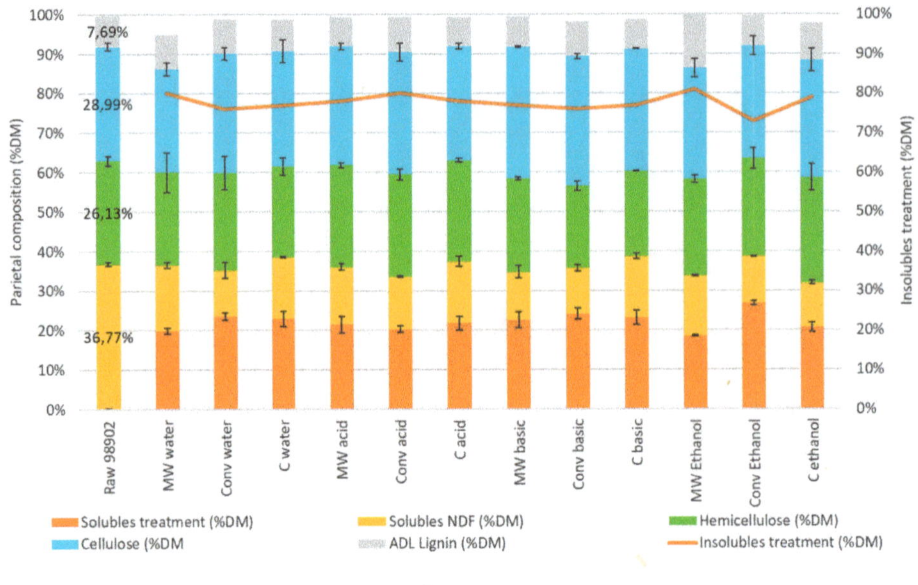

A

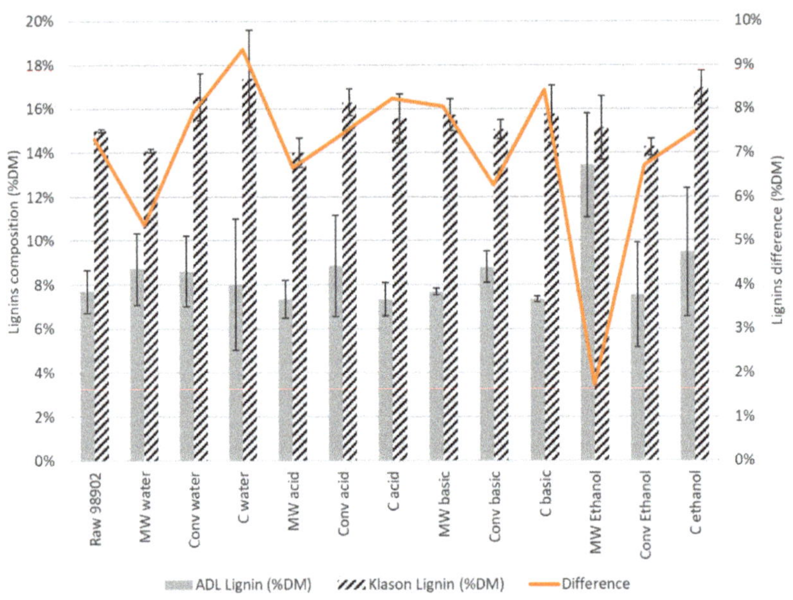

B

Figure 2. Biomass parietal composition depending on treatment (%DM) with Van Soest method (**A**). The figures in bold (**A**) are the percentages of insoluble content following the treatments (the mass of solid recovered after drying at 40 °C); the ash content is not represented. Comparison between Klason lignin and ADL (**B**). The figures in bold (**B**) are the differences between the two lignins measurements. Means of triplicates.

2.2.2. *p*-Hydroxycinnamic Acids Release

According to the literature, an increased extraction of *p*-hydroxycinnamic acids in the presence of NaOH was expected due to the ester linkages first between ferulic acid and hemicelluloses and second between *p*-coumaric acid and S lignin units, which correspond to the cleavage of the ester bonds and α-aryl ether bonds rupture linking *p*-hydroxycinnamic acids to the cell wall [25]. In contrast, acid treatment should break the glycosidic bonds and favor sugar solubilization while leaving the ester bonds intact [26]. An analysis of the *p*-hydroxycinnamic acids released in the liquid phase during the different pretreatments performed is summarized in Table 2.

From Table 2, it appears that the FA and CA recovered in the liquid fraction after treatments varied slightly with the experimental conditions tested in this study. The highest CA extraction values were observed with ethanol and alkaline water as solvent and microwave heating as treatment (0.298 mg CA/g DM and 0.247 mg CA/g DM, respectively). In case of FA, alkaline water appeared to be the most effective solvent without the microwave heating further increasing efficiency.

In the chosen conditions, alkaline treatment ensured the highest yields compared to the other solvents tested. Moreover, the NaOH concentration was similar to the one employed by Mussatto et al. [27]. Indeed, Mussatto et al. [27] carried out a conventional treatment of acid pretreated brewer's spent grain. They extracted up to 4.27 mg FA/g DM and 4.08 mg CA/g DM after 90 s at 120 °C using 2% NaOH (*w/v*). In that case, no microwave treatment was investigated, but the temperature and the previous acid pretreatment were the main parameters affecting FA and CA release.

The total initial amounts of FA and CA in raw corn stalks were 4 mg/g DM and 13 mg/g DM, respectively. Therefore, *p*-hydroxycinnamic acids release yields did not exceed 2.3% with all the tested solvents in the studied conditions (500 W, 270 sec). Moreira et al. [28] found approximately the same extraction yields from brewer's spent grain: one gram treated with 20 mL of NaOH 0.75% for 15 min and 100 °C in a microwave oven permitted approaching a yield of 1.31% of FA. These relatively low yields are compared to the high ones obtained with 2 mL of NaOH 2N protocol on 20 mg of dry matter for a night: in this latter case, the percentage of NaOH is much higher and reaches 88% g/g DM. Even if no heating is applied during the 24 h of reaction, the reagent concentration is so high that the ester bonds are largely degraded. In the present study, such high NaOH concentrations are not desired in order to develop an environmentally friendly process.

Moreover, in our study, according to ANOVA (Table 3), the solvent had more impact on FA release than the treatment itself (p = 1.13×10^{-7} and p = 0.0704 for solvent and treatment effects respectively). On the contrary, the release of CA was significantly dependent on solvent (p = 5.07×10^{-7}) and on treatment (p = 2.7×10^{-7}). Finally, the interaction between the solvent and treatment did not have a significant effect on *p*-hydroxycinnamic acids release (p > 0.05%).

Table 3. ANOVA parameters to the significant effects on FA and CA extraction depending on treatment and solvent (CA and FA measurements realized in duplicate for each test).

Factor	CA					FA				
	d.f.	SS	MS	F	p	d.f.	SS	MS	F	p
Treatment	2	0.0380	0.0190	39.23	2.7×10^{-7} *	2	0.0005	0.0002	3.086	0.0704
Solvent	3	0.0747	0.0249	51.31	5.07×10^{-9} *	3	0.0082	0.0027	34.506	1.13×10^{-7} *
Residuals	18	0.0087	0.0004			18	0.0014	0.0001		

d.f. = degrees of freedom, SS = sum of squares, MS = mean of squares, * significant at 95% confidence level.

Another ANOVA was performed to determine the effect of each type of treatment and solvents: conventional heating did not significantly favored FA extraction (p = 0.286) compared to the control. Microwave treatment had a significant negative effect on FA release with t value = −2.479. This could be explained by a negative effect of waves on biomass structure and *p*-hydroxycinnamic acids organization. Acid solvent as well was significantly disadvantageous for the extraction of FA (t value = −4.394) compared to water alone. When water was used as solvent during microwave treatment, Diaz et al. [29]

demonstrated that neither lignin nor cellulose were modified, and that only a small amount of hemicellulose content was reduce; this may explain the negative effects on *p*-hydroxycinnamic acid release in our case with water. A similar or lower yield is naturally expected with acid as solvent.

From Table 2, for both FA and CA, the alkaline solvent allows an increase in acids release compared to other solvents for the same treatment, because a break-up of ester bonds linking *p*-hydroxycinnamic acids to the cell wall is favored with alkali [26].

Concerning *p*-hydroxycinnamic acids release, microwave pretreatment efficiency is variable according to the literature: recovered molecules yields vary a lot depending on the solvent and raw matter. In this way, Carniel et al. [18] demonstrated that to release a maximum amount of total phenolic compounds (TPC), the most important parameter is the solvent and using ethanol 50% in volume allowed to obtain 3.74 mg TPC/g DM after extraction from *Physalis angulate* seeds. Concerning maize, this substrate is often cited in the literature, but usually the microwave is used to degrade the cell wall and gain access to polysaccharides, and only the evolution of parietal polymers composition is analyzed. Wheat bran was treated by microwave (500 W, 20 min 100 °C) with methanol by Oufnac et al. [21], but even if the incident power and final temperature were the same as in this study, the results are difficult to compare: treatment duration was more than four times longer, and efficiency is measured by catechin equivalent release (467 µg catechin equivalent/g wheat bran). Catechin is part of the flavonoids molecules family and is responsible for the color of the plant. Since this molecule is not linked to cell walls and is hydrolysable, it reacts to microwaves differently than *p*-hydroxycinnamic acids that are covalently linked to cell walls.

In order to better understand microwave effects on *p*-hydroxycinnamic acids release, it was decided to focus on water, which is considered as a suitable extraction solvent: Moreira et al. [28] found it to be the second best solvent to extract polyphenols behind NaOH at 150 °C for 25 min. Moodley and Kana [30] demonstrated that NaOH concentration has more impact than microwave duration on sugars release. Thus, by working with water, NaOH effects are constrained, which allows better understanding the thermal and non-thermal microwave effects. Finally, using water as solvent is environmentally friendly and appropriate from an industrial application point of view: water is low cost compared to other solvents, and wastewater management is much easier without chemical reagent addition.

Therefore, the optimization of microwave treatment on corn stalks with water as solvent was performed using a response surface methodology (RSM). The best conditions were then applied to miscanthus.

2.3. Response Surface Methodology (RSM) Analysis to Determine Better Microwave Parameters

RSM analysis was used to determine the optimal operating conditions in terms of FA and CA release from corn stalks. Moreover, the lignocellulosic biomass structure was also analyzed through the Van Soest and Klason lignin methods.

2.3.1. Experimental Design

Using the conditions described in 3.9, values for the center points were highly repeatable with a standard deviation of 0.3% and 0.2% for CA and FA, respectively. This provided an idea of the experimental error over the entire design of the experiments. Each *p*-hydroxycinnamic acid analysis was performed in duplicate, and the results are shown in Figure 3 and Table 4. The polynomial equations for FA (Equation (1)) and CA (Equation (2)) are the following:

$$Y_{FA} = 9,26 - 1.17X_1 - 0.36X_2 + 1.17X_1X_2 + 0.03X_1^2 + 1.89X_2^2 \quad R^2 = 0.749; \ F-value = 58.521 \quad (1)$$

$$Y_{CA} = 70.50 + 12.37X_1 + 11.52X_2 - 2.20X_1X_2 - 5.81X_1^2 + 16.09X_2^2 \quad R^2 = 0.995; \ F-value = 0.003 \quad (2)$$

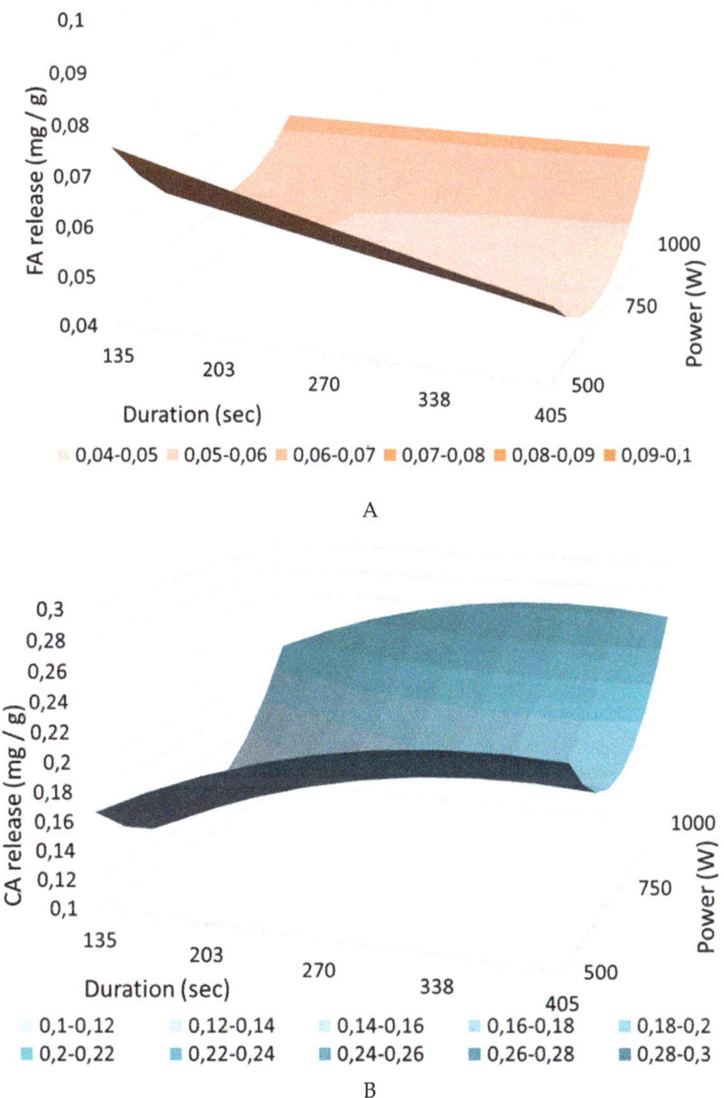

Figure 3. Response surface plotplan for FA (**A**) and CA (**B**) release.

As indicated in Joglekar and May [31], if the R^2 of a model is superior to 0.8, the model matches the experimental data. The R^2 obtained in the case of CA is high, which corresponds to an excellent regression: the model explains 99.5% of the experimental data. Moreover, the F-value calculated for CA (0.003) is very much lower than the 95% confidence level Fischer parameter (18.51), thus meaning that the model is relevant in explaining CA release results. On the contrary, the small R^2 (0.749), and the high F-value (58.521) obtained in case of FA indicated that the model does not fit the data, and results cannot be easily exploited using the model.

Using Pareto charts, the effects of each individual term of the polynomial equations are presented on Figure 4 for FA release (Figure 4A) and CA release (Figure 4B). A term is considered to have a significant impact on acid release if it exceeds the black vertical bar ($p < 0.05\%$) corresponding to 3.18 in the study conditions. Analyzing each term of the polynomial equation permits highlighting interactions between parameters. In the case of FA (Figure 4A), none of the parameters significantly impacted FA release, which can be due to the design itself not fitting the data. Maybe FA acid release is not impacted by power nor by treatment duration: temperature could be another parameter to test, especially under pressed conditions. In that case, care must be taken not to obtain significantly negative interactions between new parameters. Another hypothesis is that the studied domain is too restrictive to obtain a significant trend on FA release, but expanding the range may not be a good option from an industrial point of view, as increasing power or processing time consumes a lot of energy and will certainly not achieve high yield. Thus, no optimum is visible on the three-dimensional (3-D) representation (Figure 3A). On the contrary, for the CA Pareto chart (Figure 4B), all coefficients have a significant impact on CA release except concerning interaction between power and duration. Another method to find the optimum of the model would have been to perform an optimization from close to close by fixing a variable (for example, the duration) and varying the other (the incident power) variables, as they have no significant effect on each other. All the other variables have a significant positive or negative impact on CA release (t-value >3.18), which supports the choice to use this model with these variables and this domain. Incident power and duration both have a significant positive impact on CA release, and this impact is similar for both variables (both coefficients close to 0.02). Thus, by increasing the incident power and duration, it is expected that the release of CA increases, and especially with regard to the increase in incident power, as its quadratic interaction is high. These comments on CA release are also visible on the response surface 3-D plot (Figure 3B), where it is noteworthy that as the power or the duration increases, the value of released CA steps up. Finally, it would be wise to choose processing times among the high tested values and a sufficiently high power to obtain good yields while respecting a possible and viable economic process.

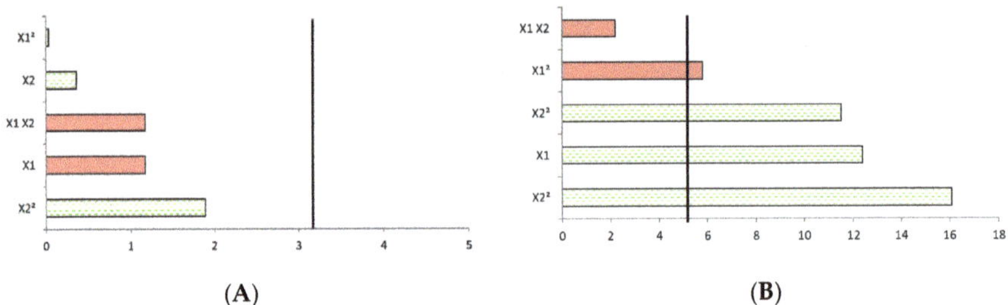

Figure 4. Pareto diagram for FA (**A**) and CA (**B**). X1 corresponds to duration, X2 corresponds to incident power, red surfaces indicate a negative impact, green surfaces indicate a positive impact, the vertical line corresponds to the value to be exceeded so that the response is significantly impacted by the parameter ($p < 0.05\%$, corresponding to 3.18 in the design developed)

It would be interesting to test stronger conditions, for example 1500 W and a duration of 810 s. However, these conditions were experimentally difficult to reach above a certain power and duration; the material boiled strongly in the reactor, and liquid escaped. For example, this boiling phenomenon was observed in the case of NaOH at too-high power levels or with water with a longer treatment duration. In addition, a longer reaction time would only be advantageous from an industrial point of view if it would greatly increase *p*-hydroxycinnamic acid yields. Working under pressure to reach higher temperatures is also a possible solution, but the destruction of products is an important trend to be controlled [15]. Finally, over time, some inhibitors may form, creating problems for the eventual rest of the process in the case of bioprocesses [32].

2.3.2. *p*-Hydroxycinnamic Acids Release Depending on Incident Power and Treatment Duration

According to Table 4, CA release was always higher than FA release: under the most severe conditions (1000 W and 405 s), 1.38% FA and 2.0% CA were released. This observation is valid for all operating conditions. The initial high presence of CA (13.1 mg CA/g and 4.2 mg CA/g) in raw corn stalks can explain the higher CA released yields. Moreover, FA release cannot be optimized by varying the power and duration, as these parameters have no impact according to the 3-D plots (Figure 3). This is consistent with Pinela's analysis [33]: On tomato fruits, the incident microwave power had no effect on *p*-hydroxycinnamic acids release, and a Box–Behnken design was carried out using four independent variables (duration, temperature, ethanol concentration, and liquid–solid ratio). With a treatment time of 3 min, a temperature of 140 °C, and no ethanol, 23 mg CA/g was removed. Tomatoes are naturally richer in *p*-hydroxycinnamic acids than corn stalks, and molecules are more accessible. Carniel et al. [18] reached the same conclusion using *Physalis angulate*: incident power and time had no effect on the total phenolic compound released, but the parameters range tested were very tight: between 10 and 30 W for 40 to 60 s. Moreover, the Folin–Ciocalteu colorimetric method was used to determine the total phenolic compounds, which is a different method from the one used in the present study, and the detected molecules are not the same. The Folin–Ciocalteu colorimetric method permits detecting all phenolic compounds contrary to HPLC, which detects selectively *p*-hydroxycinnamic acids.

Table 4. Ferulic acid (FA) and coumaric acid (CA) release yields using the experimental design. Yields are expressed in %: $mg_{phenolic\ acid\ release}/mg_{initial\ phenolic\ acid}$, mean of duplicate.

Assay	T Final (°C)	FA (mg/g)	FA Yield (%)	CA (mg/g)	CA Yield (%)
1	63	0.080 ± 2%	1.90	0.166 ± 1.1%	1.3
2	98	0.058 ± 0.1%	1.38	0.216 ± 0.1%	1.6
3	99	0.056 ± 0.8%	1.33	0.219 ± 1.1%	1.7
4	99	0.071 ± 0.1%	1.69	0.260 ± 0.1%	2.0
5	98	0.059 ± 0.9%	1.40	0.229 ± 2.8%	1.7
6	98	0.058 ± 0.7%	1.38	0.260 ± 1.9%	2.0
7/8/9	98	0.048 ± 0.2%	1.14	0.176 ± 0.3%	1.3

ANOVA was performed to determine the significant impact of time and/or power on FA and CA release (Table 5). According to Table 5, CA release was significantly impacted by both time and power: CA release could be optimized depending on these parameters. With 405 or 270 s treatment time and 1000 W, CA release significantly increased, and reached 0.260 mg/g. This result is higher than the one obtained with NaOH without microwave treatment (0.173 mg/g). Thus, CA release increased with microwave treatment, and the results obtained with Equation (2) were confirmed as CA release having increased with time exposure and microwave power. On the contrary, FA release was not significantly impacted by the different power tested nor by the treatment duration. This is consistent with the results obtained in the previous section. To conclude, it appeared that 1000 W and 405 s were good options for microwave treatment with feasible conditions.

Table 5. ANOVA parameters to the significant effects on FA and CA extraction depending on treatment duration (sec) and power (W). Phenolic acid analysis in duplicate for each test condition.

Factor	CA					FA				
	d.f.	SS	MS	F	p	d.f.	SS	MS	F	p
Duration (sec)	1	0.011	0.01	16.59	3.6×10^{-4} *	1	4.2×10^{-4}	0	2.65	0.115
Power (W)	1	0.009	0.009	14.15	8.3×10^{-4} *	1	4.6×10^{-5}	0	0.28	0.597
Residuals	27	0	6.9×10^{-4}			27	0	0		

2.3.3. *p*-Hydroxycinnamic Acids Release Depending on Power Density

An important point to elucidate was the effect of absorbed energy density (Wh/g) on *p*-hydroxycinnamic acids release: lignocellulosic biomass may react differently under various energy densities. The different energy densities tested are presented in Table 6, and a summary for each energy density is represented in Figure 5. The corresponding temperature rise graphs are shown in Figure 6. The maximum temperature of 100 °C is reached twice as fast at 1000 W as at 500 W. It is important to note that with a treatment time of 135 s and under 500 W, the reaction medium does not reach 100 °C but rather only 63 °C, which may explain the ineffectiveness of this treatment in terms of phenolic acid release. On the contrary, during 1000-W treatments, the temperature of 100 °C is reached earlier, and the material remains at 100 °C for a longer period compared to 500 W.

From Figure 5, concerning CA, the higher the energy density, the higher the *p*-hydroxycinnamic acid release ($p = 3.44 \times 10^{-5}$). Nevertheless, in the case of FA, it was not as simple, and no trend could be noticed ($p = 0.274$). These results are complicated to confirm, because to our knowledge, no other literature article has studied the energy density applied to the product. As a partial conclusion, to release more coumaric acid, it would be necessary to increase the energy density applied to the material either by increasing the incident power or by varying the amount of material introduced into the reactor. In all the cases tested, whatever the time, the power, or energy density, yields obtained were low (<2%). A different hypothesis could explain these low yields, especially in the case of FA.

Table 6. Energetic aspects for duration and incident power tested.

Total Mass (g)	Pi (W)	Pid (W/g)	Duration (s)	Eid (Wh/g)
210	500	2.38	135	0.089
210	1000	4.76	135	0.179
210	500	2.38	270	0.179
210	1000	4.76	270	0.357
210	500	2.38	405	0.268
210	1000	4.76	405	0.536
210	750	3.57	270	0.268

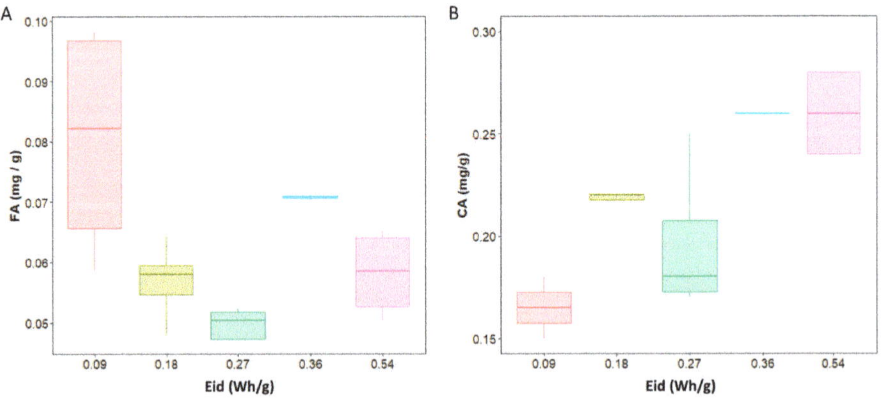

Figure 5. Ferulic (**A**) and coumaric (**B**) acids release depending on incident energy density.

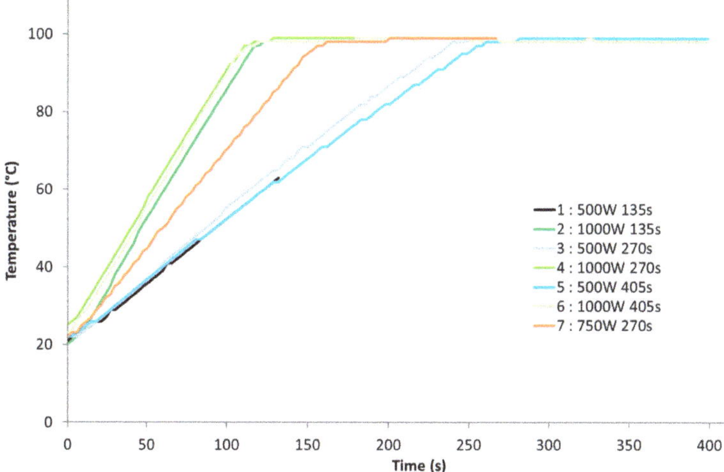

Figure 6. Temperature rise during response surface methodology (RSM) experiments.

A hypothesis is that the free acids liberated during the treatment were degraded between 80 °C and 100 °C. Tests with known ferulic and coumaric acids concentrations diluted in ethanol and submitted to the following operational conditions were performed: 500 W and 1000 W for 540 s (results not shown). A temperature of 80 °C was reached after 200 s of treatment. FA and CA concentrations detected after HPLC were similar to the ones injected: p-hydroxycinnamic acids were not destroyed or evaporated during treatment. This was consistent with the study of Liazid et al. [34] demonstrating that ferulic acid degradation did not occur at temperatures lower than 130 °C.

Another hypothesis takes into account the chemical bonds between acids and walls. FA can be present in a variety of forms in the cell [35]. FA may be present in free form or covalently linked to lignin by ether links and to hemicelluloses by ester links forming a bridge between those two parietal polymers. FA could also be only linked to hemicelluloses by ester links (Figure 1). We suppose that ether links are more difficult to access because they are "hidden" in lignin polymers. Moreover, ether links are also difficult to break, as they are not hydrolyzed after a night with NaOH 2N. The FA release measurement was carried out on the liquid after treatment without any additional alkaline treatment: the small amounts could represent only free FA or even simple esterified FA, because etherified FA was protected by parietal components [17]. In addition, the treatment conditions implemented in this study are very soft (short duration and moderate temperature) to allow these bonds to be broken: only a few minutes in water compared to a full night in sodium hydroxide. On the contrary, CA is only linked to parietal components (both hemicellulose and lignin) by esters links: these covalent links are more easily accessible, as they are not trapped in the complex structure of the wall, and they are easier to hydrolyze and release [35]. Finally, CA are more abundant in raw corn stalks than FA. Since CA are more extracted (in mg/g) than FA and the extracting yields are similar among FA and CA, an effect of abundance in the release of CA could be hypothesized. By analyzing the composition of the walls, a modification could be detected in comparison with the raw material that would explain these low yields.

2.3.4. Effect of Operational Conditions on the Pretreated Biomass Composition

The pretreated biomass was analyzed with the Van Soest method (Figure 7). Figure 7 represents the biomass composition after the RSM plan. It appears that biomass composition was not modified by the different operational conditions tested ($p > 5\%$) when compared to the composition of the raw material. Similar results in the Klason lignin method (results not shown) confirmed that the polymer composition was not modified during microwave processing. Despite the harsher microwave conditions compared to the previous part, the composition of the biomass is still not affected. This means that the same conclusions as previously can be assumed: the conditions tested are still not effective in breaking ester and ether links and releasing phenolic acids in the liquid phase. A more precise analysis of the links between lignin units by thioacydolysis would make it possible to highlight a change in the structural organization of the plant. It must be notified that there was always an average of two grams of solubilized material during processing, corresponding to a large part of the material solubilized during the first stage of the Van Soest method on raw corn stalks.

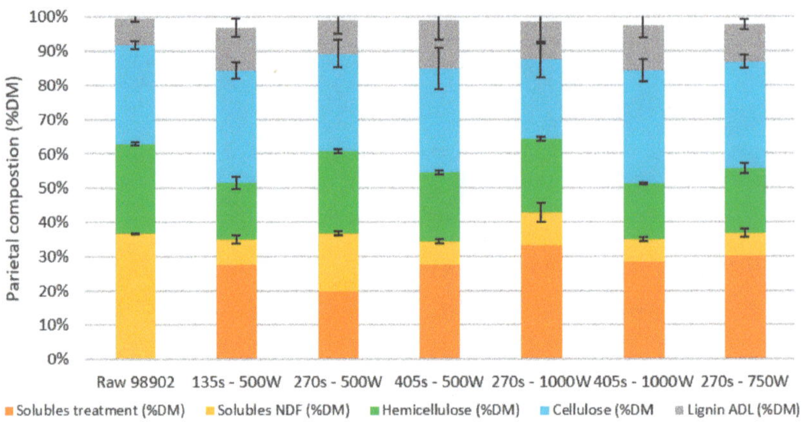

Figure 7. Pretreated biomass composition after RSM plan (mean of duplicate analysis).

2.4. Microwave Optimum Conditions Applied to Miscanthus

Based on the results from the previous paragraphs, the microwave conditions that released the most coumaric acid were chosen, as no optimization was performed for ferulic acid release. The best conditions (1000 W for 405 s) were applied to another biomass in order to find out if similar yields could be achieved (Table 7). Miscanthus was chosen because in France, it benefits the energy crop subvention in order to implement "land under industrial set-aside". Miscanthus, similar to corn stalk, is a poaceae biomass that is very different both in terms of cell wall content (%NDF) and wall composition. Miscanthus GIB was submitted to 1000 W for 405 s with the same operational conditions as those described in Section 2.3. Control and conventional heating were also performed in duplicate on miscanthus.

Table 7. CA and FA yields after optimum microwave conditions for corn and miscanthus stalks, and after control and conventional treatment for miscanthus stalks. FA and CA yields are expressed in %: $mg_{phenolic\ acid\ release}/mg_{initial\ phenolic\ acid}$.

		FA Recovery Yield (%)	CA Recovery Yield (%)
MW 1000 W–405 s	Corn stalks	1.38 ± 0.1	1.97 ± 0.0
	Miscanthus stalks	0.58 ± 0.2	3.89 ± 0.2
Conventional heating	Miscanthus stalks	0.53 ± 0.0	1.94 ± 0.4
Control (no heating)	Miscanthus stalks	0.47 ± 0.0	1.46 ± 0.0

From Table 7, for miscanthus stalks, 0.58% FA and 3.89% CA were recovered in comparison with corn stalks with 1.38% FA and 1.97% CA. The FA yield from miscanthus is much lower than the one from corn: more than 50%. Raw miscanthus is poorer in FA than raw corn (Table 1), which can explain the low amount of FA released during the treatment. However, we have previously demonstrated that microwave treatment, under the tested conditions, has only a limited impact on ferulic acid release, which limits the interpretation of this result.

On the contrary, CA yield in miscanthus is twice as high as in corn, in spite of the very low initial content of CA in raw matter (0.65 mg CA/g miscanthus). As explained earlier, these differences may be explained by the chemical bonds impacted, and miscanthus CA must certainly be present in free form in cells. Moreover, by comparing miscanthus microwave extraction yields with miscanthus conventional or control extraction yields, it appeared that the microwave treatment promoted the CA extraction, but the FA release remained unchanged, which was in agreement with the previous parts: CA being significantly released following the microwave treatment compared to FA.

In the tested conditions, only 0.8 g out of the 10 g of miscanthus were solubilized during the treatment, contrary to corn (2 g), which can be explained by the high initial NDF percentage in miscanthus stalks compared to NDF in corn stalks (95% DM and 65% DM, respectively). The final miscanthus composition was the following, according to the Van Soest method: 4.3% (±0.1%) soluble content, 18% (±3%) hemicellulose, 53% (±3%) cellulose, and 15.9% (±0.8%) ADL, which was not significantly different from the initial composition. These results differed from Boonmanumsin et al. [36], who experimented with microwaves combined with NH_4OH (1% *w/v*) on miscanthus for 15 min and 300 W. *p*-hydroxycinnamic acids were not analyzed; instead, they examined monomeric sugars. In optimum conditions, up to 25.6 g sugars/100g biomass were released (principally under xylose form). The extracting solvent, the final temperature, and the duration were the main differences between the two experiments. It is always difficult to compare results from different microwave studies as various parameters may differ at the same time (duration, solvent, temperature, pressure, ratio, substrates) and may affect microwave pretreatment efficiency. As a matter of fact, general statements cannot be made for all plant materials because of their diversity in structure and composition. Processing conditions must be defined and optimized taking into account the lignocellulosic matrix and the sought-after molecules or applications [33]. Strictly speaking about *p*-hydroxycinnamic acids release, most of the studies are focusing on plant stalks, but rhizome cut in small pieces has also been investigated in the case of Scirpus holoschoenus [37]. When 20 mL of acetone 56% was added to 1 g rhizome and put under 600 W for 70 sec, up to 30 mg of gallic acid equivalent/g were extracted. Similarly, Galan et al. [38] extracted 162 mg gallic acid equivalent/g from sea buckthorn leaves after 450 s of treatment. Finally, 62 µg FA/g was extracted from Soybeen seeds after 75-W treatment for 10 min. These results highlight the importance of the raw material used during the treatments and the variability of the results obtained. The results of microwave processing are dependent on the matrix.

3. Materials and Methods

3.1. Raw Biomass

Corn stalks named F 98902 were supplied by INRA IJPB (Versailles-Grignon unit, Versailles Cedex, 78026, France) and were harvested in September 2016.

Miscanthus stalks (*M. x giganteus Britannique*, noted GIB) were supplied by INRA AgroImpact (Estrées Mons esperimental unit, Péronne, 80203, France, 49_53 N, 3_00 E) [39] and were harvested in February 2017.

Air-dried samples were coarsely crushed (Viking crusher model GE 220, STIHL, Stuttgart, Germany) and sent to the laboratory in Narbonne. Then, samples were finely ground to 1 mm using a Fritsch Pulverisette 19 grinder and sieved to retain only particles between 200 and 1000 µm. Ground and sieved samples, called "raw substrates" in the following, were kept in closed boxes in ambient air before used.

3.2. Chemicals

All chemicals were purchased from Merck. High-purity water (Merck Millipore Quantum TEX) was used for all pretreatments and analyses.

3.3. Microwave Pretreatment (MW)

Microwave pretreatments were performed with a Minilabotron 2000 microwave pilot (SAIREM, FRANCE), operating at atmospheric pressure, 2.45 GHz with a maximum power of 2 kW. All pretreatments were carried out at constant incident power level in open vessel.

According to the preliminary results (not shown), a total mass of 210 g was determined to be optimal for waves absorption based on microwave pilot configuration. Ten grams of prepared raw material with known dry matter content were transferred to a 500-mL glass reactor. Then, 200 g of liquid was added. The tested liquids (solvents) were: water, acidic water (pH = 4 obtained by adding 0.75 mmol H_2SO_4 to 200 mL water), alkaline water (pH = 8.5 obtained by adding 6.7 mmol NaOH to 200 mL water), or an ethanol/water mixture (50% *w/w*). This solid:liquid ratio (1:21 *w:w*) allowed correct magnetic stirring during treatments. Indeed, adequate stirring is essential for microwave treatment [15]. Samples underwent 1 h of pre-soaking at ambient temperature before microwave treatment. pH was measured at the end of this hour of contact.

Then, the reactor was closed using a glass cover connected to a refrigerant limiting solvent evaporation. A fiber-optic temperature sensor was used to monitor the temperature of the reaction medium: the maximum temperature could not exceed 100 °C in the case of aqueous solvents and 80 °C in the case of ethanol treatment because of the boiling points of these solvents. All treatments were performed at constant power and at atmospheric pressure (open vessel), in order to avoid excess temperature increase and the thermal degradation of molecules (in particular *p*-hydroxycinnamic acids) [39].

Pretreatments were performed in duplicate (excepted for the center point in the RSM plan performed in triplicate). After treatment, the reactor was air-cooled to room temperature for 15 min before weighing and pH measurement. Then, the reaction mixture was filtered through a 200-μm sieve. The solid was washed with 250 mL of deionized water to remove chemicals and by-products inhibiting enzymatic hydrolysis. The solid fraction was placed at 40 °C for seven days to dry. Then, dry matter content was measured to determine the amount of solubilized matter during processing and the solid recovery yield (% of g pretreated biomass/g raw matter). The supernatant was filtered through a cellulose filter (2.7 μm) and stored at −20 °C until further analysis.

3.4. Energy Calculations

Incident power varied between 500 W, 750 W, and 1000 W for 135 s, 270 s, and 405 s (Table 6). Temperature, incident power, and reflected power were recorded every three seconds during the treatment. Absorbed power was obtained using Equation (3). The total absorbed power was the sum of all the absorbed power recorded every three seconds; see Equation (4). Absorbed energy was calculated from Equation (5) by multiplying the total Pa by 3, since it was only recorded every 3 s.

$$\text{Absorbed Power (Pa)} = \text{Incident Power (Pi) - Reflected Power (Pr) in W} \tag{3}$$

$$\text{Total Pa} = \sum \text{Pa in W} \tag{4}$$

$$\text{Absorbed Energy density (Ead)} = \left(\sum \text{Pa.3}\right)/(3600.\text{Total mass}) \text{ in Wh/g} \tag{5}$$

3.5. Conventional Heating Treatment (Conv)

A 500-mL Schott bottle containing the same reaction mixture as the microwave-treated sample was immersed in a heat-stabilized oil bath at 100 °C for 360 s, which is the reaction time necessary to reach a temperature close to 100 °C. As the bottle was closed, pressure could slightly increase during treatment without exceeding 1.3 bars. This pressure, close to atmospheric pressure, was considered to have no impact during conventional treatment. The separating protocol was the same as that described before. The conventional test allowed comparing conventional and microwave heating modes. The temperature and pH were recorded during processing.

3.6. Control treatment (C)

A control treatment (soaked biomass sample without any heating) was also carried out. Liquid and solid phases were separated after an hour of contact using the method described before. Pretreatment effects on biomass were evaluated by *p*-hydroxycinnamic acids release and biomass structure changes.

3.7. Ferulic and Coumaric Acids Analysis

Initial FA and CA amounts were determined by HPLC (method described below). A mild alkaline extraction for a night using 2 mL of NaOH 2N with 20 mg of raw matter permitted releasing esterified *p*-hydroxycinnamic acids [13,26,27]. These initial amounts (in mg/g) were used to calculate the release yields (Equation (6)).

Ferulic acid (FA) and p-coumaric acid (CA) released in the liquid phase after pretreatments were quantified in duplicate. *P*-hydroxycinnamic compounds were analyzed by HPLC using a HPLC-DAD Waters system: autosampler 717, multisolvent delivery system 600, Diode Array Detector 2996. Then, *p*-hydroxycinnamic acids were detected at 320 nm, and the peak areas were calculated by Empower3 software (Waters). The mobile phases consisted of ultrapure water/formic acid–95/5 (Solvent A) and acetonitrile/ultrapure water/formic acid–80/15/5 (Solvent B). The flow rate was 1 mL/min, and the injection volume was 10 µl. Separation was performed at 30 °C on a Waters Atlantis T3 Column, 100 Å, 5 µm, 4.6 mm × 250 mm (C18) equipped with a C18–4 × 3 mm Security Guard Cartridge (Phenomenex, France).

Then, the results—which were obtained in mg/l—were transformed to g/g DM using the collected liquid volume (l) after treatment (Equations (6) and (7)):

$$\text{Collected liquid} = \text{Initial V} - \text{evaporation V} - \text{swelling V in L} \tag{6}$$

$$\text{FA released} = \text{FA (mg/L)} \times \text{Collected liquid}/10\text{gDM in mg/g} \tag{7}$$

Swelling was previously measured (results not shown) and corresponded to 1 mL/g DM and 0.9 mL/g DM for corn stalks and miscanthus stalks, respectively. The evaporation volume was measured by taking the mass difference between the beginning and the end of the treatment. The evaporated volume did not depend on the biomass used, but instead only on the power and duration of the treatment, as well as on the solvent. On average, for water microwave treatments, the evaporated volume can be calculated with Equation (8) with $R^2 = 0.91$:

$$\text{f(duration} \times \text{Incident power)} = 7449.4 \times \text{evaporation V} + 83002 \tag{8}$$

For ethanol microwave treatments, the evaporation volume was twice that obtained in water under the same conditions. For control and conventional heating, evaporation was considered negligible, as it was less than 10 mL. In any case, these mass changes in the microwave reactor over time were not taken into account in the calculations of power or energy density.

FA extraction yield was calculated as the quotient of the liberated FA/CA mass released by the total FA/CA mass in the raw substrates (equation (9) for FA).

$$\text{FA yield} = \text{FA released (mg/g)}/\text{FA initial (mg/g) in \%} \tag{9}$$

3.8. Biomass Composition Analysis

The impact of the pretreatment on biomass structure was evaluated using the Van Soest method [40] and Klason lignin protocol [41].

The Van Soest protocol allowed evaluating the evolution in parietal polymers content after treatment. This method is based on the mass sequential partitioning of cell walls, from most extractible to less extractible, with successive extractions using different solvents (water, neutral detergent solution, acid detergent solution, and acid 72%). After the first Van Soest step, a part of neutral detergent fiber (considered as parietal residue) was used to measure the Klason lignin content using the NREL protocol. ADL (acid detergent lignin) lignin (%DM) is the lignin obtained with the Van Soest method, and corresponds to lignin insoluble in 72% H_2SO_4 after a first extraction step in an acid detergent solution (ADS). Klason lignin is obtained from parietal residue and corresponds to lignin insoluble in 72% and in 4% H_2SO_4; the result is then reported in dry matter percentage. As explained in Hatfield and Fukushima [42], ADS can solubilize cellulose and the most accessible part of lignin (about 50%), explaining why Klason lignin is much higher than ADL, reaching twice its value. Moreover, some proteins could condense and increase the Klason lignin mass, even if this explanation is improbable in the case of grass. Finally, it was determined on a maize cell wall that Klason lignin and ADL represent two different lignin measurements: therefore, it is essential to always compare values from the same analysis [43]. The lignin part solubilized in ADS fraction was found to be correlated with β-O-4 lignin bonds. Thus, according to Zhang et al. [43], comparing Klason lignin and ADL lignin values could reveal a variation in chemical bonds within lignin.

3.9. Statistical Analysis and Response Surface Methodology Method

All the statistical tests were performed using R software (version 3.4.0). Microwave effects on biomass composition and on *p*-hydroxycinnamic acids release were analyzed with ANOVA and considered significant when p-value < 0.05 with residuals distributed according to a normal law. Biomass composition with Van Soest or Klason lignin were repeated twice for each sample obtained from a pretreatment. In addition, *p*-hydroxycinnamic analysis was performed in duplicate from the liquid phase of each pretreatment.

RSM allows obtaining robust results with as few experiments as possible to understand the effects of each variable on the response and find optimal conditions [44]. The plan chosen was a full factorial plan, combining mathematical and statistical techniques to model a problem and optimize the response [45]. The experimental domain was determined based on the literature, the results from experimental assays, and taking into account the experimental limits of the Minilabotron microwave pilot. Since we chose to work in an open vessel and thus without pressure, temperature was not chosen as a parameter. The two chosen parameters ($z = 2$) were the reaction time (X1 in seconds) and the incident power (X2 in Watt). The number of experiments (N) was determined with Equation (10):

$$N = z^2 + z + C = 2^2 + 2 + 3 = 9 \tag{10}$$

where C represents the number of center points, which was repeated three times. The experimental plan is summarized in Table 8.

Table 8. Pretreatment conditions for the experimental design and results (mean of two measures from assays 1 to 6).

Assay	Parameter X_1: Duration (sec)		Parameter X_2: Power (W)	
	Code	Value	Code	Value
1	−1	135	−1	500
2	−1	135	1	1000
3	0	270	−1	500
4	0	270	1	1000
5	1	405	−1	500
6	1	405	1	1000
7/8/9	0	270	0	750

Experimental data can be adjusted to a second-order model, following Equation (11):

$$Yi = b_0 + b_1\,X_1 + b_2\,X_2 + b_{12}\,X_1\,X_2 + b_{11}\,X_1{}^2 + b_{22}\,X_2{}^2 \tag{11}$$

where Yi is the experimental response that corresponds to FA and CA release in the liquid phase (mg/g), Xi represents the studied parameters, b_0 is the average response, b_i represents the linear coefficients, b_{12} is the average effect of interaction factor, and b_{ij} represents the quadratic coefficients. The model was validated or not using a Fisher test for each studied response (FA and CA). For the calculation of the different coefficients, refer to Witek-Krowiak et al. [46]. The solution of the response equations was performed using a Microsoft Excel spreadsheet.

4. Conclusions

The present study was designed to determine the effects of microwave irradiation on grass stalks as a pretreatment for biomass deconstruction and *p*-hydroxycinnamic acids release. Different microwave experimental conditions were set up for various incident powers, durations, and solvents. Experiments have shown that microwave power and duration had a low impact on biomass parietal composition and on FA release. CA release yields remained low, but by increasing power and duration, they could progress according to RSM. It would be interesting to implement an experimental design to understand grass biomass behavior under microwaves and to find out their real effect (radiation vs heat).

Author Contributions: Conceptualization, J.-P.D., V.M. and D.G.-B.; methodology, A.B., M.L. and L.G.; formal analysis, A.B.; investigation, A.B., M.L. and L.G.; writing—original draft preparation, A.B.; writing—review and editing, J.-P.D., D.G.-B., V.M., N.B. and H.C.

Funding: This research received no external funding.

Acknowledgments: The authors would like to thank the NH Verre company (Nicolas Holfeltz) for its collaboration in designing the microwave reactor. The authors also thank Yannick Sire from INRA Pech Rouge for *p* hydroxycinnamic acids analysis and Jordan Siera for his invaluable help in the exploitation of the RSM results. The authors are gratefully acknowledged for the Ph.D. Grant allocated by the GAIA Ph.D. school to Aurélie BICHOT.

Conflicts of Interest: The authors declare no conflict of interest.

References

1. Fatma, S.; Hameed, A.; Noman, M.; Ahmed, T.; Shahid, M.; Tariq, M.; Sohail, I.; Tabassum, R. Lignocellulosic Biomass: A Sustainable Bioenergy Source for the Future. *Protein Pept. Lett.* **2018**, *25*, 148–163. [CrossRef]
2. Apprich, S.; Tirpanalan, Ö.; Hell, J.; Reisinger, M.; Böhmdorfer, S.; Siebenhandl-Ehn, S.; Novalin, S.; Kneifel, W. Wheat bran-based biorefinery 2: Valorization of products. *LWT* **2014**, *56*, 222–231. [CrossRef]
3. Naczk, M.; Shahidi, F. Extraction and analysis of phenolics in food. *J. Chromatogr. A* **2004**, *1054*, 95–111. [CrossRef]
4. Xing, Y.M.; White, P.J. Identification and function of antioxidants from oat greats and hulls. *J. Am. Oil Chem. Soc.* **1997**, *74*, 303–307. [CrossRef]
5. Grabber, J.H.; Quideau, S.; Ralph, J. *p*-coumaroylated syringyl units in maize lignin: Implications for beta-ether cleavage by thioacidolysis. *Phytochemistry* **1996**, *43*, 1189–1194. [CrossRef]
6. Kondo, T.; Mizuno, K.; Kato, T. Cell wall-bound para-coumaric and ferulic acids in italian ryegrass. *Can. J. Plant. Sci.* **1990**, *70*, 495–499. [CrossRef]
7. Helm, R.F.; Ralph, J. Lignin hydroxycinnamyl model compounds related to forage cell-wall structure. 2. Ester-linked structures. *J. Agric. Food Chem.* **1993**, *41*, 570–576. [CrossRef]
8. Bichot, A.; Delgenès, J.-P.; Méchin, V.; Carrère, H.; Bernet, N.; García-Bernet, D. Understanding biomass recalcitrance in grasses for their efficient utilization as biorefinery feedstock. *Rev. Environ. Sci.* **2018**, *17*, 707–748. [CrossRef]

9. Ragauskas, A.; Williams, C.K.; Davison, B.H.; Britovsek, G.; Cairney, J.; Eckert, C.A.; Frederick, W.J.; Hallett, J.; Leak, D.J.; Liotta, C.L.; et al. The Path Forward for Biofuels and Biomaterials. *Science* **2006**, *311*, 484–489. [CrossRef] [PubMed]

10. Chang, V.S.; Holtzapple, M.T. Fundamental Factors Affecting Biomass Enzymatic Reactivity. *Twenty-First Symposium on Biotechnology for Fuels and Chemicals* **2000**, *6*, 5–37.

11. Weijde, T.V.D.; Alvim Kamei, C.L.; Torres, A.F.; Vermerris, W.; Dolstra, O.; Visser, R.G.F.; Trindade, L.M. The potential of C4 grasses for cellulosic biofuel production. *Frontiers in Plant Science* **2013**, *4*, 107. [PubMed]

12. Dupoiron, S.; Lameloise, M.-L.; Bedu, M.; Lewandowski, R.; Fargues, C.; Allais, F.; Teixeira, A.R.S.; Rakotoarivonina, H.; Rémond, C. Recovering ferulic acid from wheat bran enzymatic hydrolysate by a novel and non-thermal process associating weak anion-exchange and electrodialysis. *Sep. Purif. Technol.* **2018**, *200*, 75–83. [CrossRef]

13. Buranov, A.U.; Mazza, G. Extraction and purification of ferulic acid from flax shives, wheat and corn bran by alkaline hydrolysis and pressurized solvents. *Food Chem.* **2009**, *115*, 1542–1548. [CrossRef]

14. Bundhoo, Z.M. Microwave-assisted conversion of biomass and waste materials to biofuels. *Renew. Sustain. Energy Rev.* **2018**, *82*, 1149–1177. [CrossRef]

15. Zhang, H.-F.; Yang, X.-H.; Wang, Y. Microwave assisted extraction of secondary metabolites from plants: Current status and future directions. *Trends Food Sci. Technol.* **2011**, *22*, 672–688. [CrossRef]

16. Alqaralleh, R.M.; Kennedy, K.; Delatolla, R. Microwave vs. alkaline-microwave pretreatment for enhancing Thickened Waste Activated Sludge and fat, oil, and grease solubilization, degradation and biogas production. *J. Environ. Manag.* **2019**, *233*, 378–392. [CrossRef] [PubMed]

17. Provan, G.J.; Scobbie, L.; Chesson, A. Determination of phenolic acids in plant cell walls by microwave digestion. *J. Sci. Food Agric.* **1994**, *64*, 63–65. [CrossRef]

18. Carniel, N.; Dallago, R.M.; Dariva, C.; Bender, J.P.; Nunes, A.L.; Zanella, O.; Bilibio, D.; Luiz Priamo, W. Microwave-Assisted Extraction of Phenolic Acids and Flavonoids from Physalis angulata. *J. Food Process Eng.* **2017**, *40*, e12433. [CrossRef]

19. Favretto, L. *Basic Guidelines for Microwave Organic Chemistry Applications*; Milestone: Bergamo, Italy, 2014.

20. Eskilsson, C.S.; Björklund, E. Analytical-scale microwave-assisted extraction. *J. Chromatogr. A* **2000**, *902*, 227–250. [CrossRef]

21. Oufnac, D.S.; Xu, Z.; Sun, T.; Sabliov, C.; Prinyawiwatkul, W.; Godber, J.S. Extraction of Antioxidants from Wheat Bran Using Conventional Solvent and Microwave-Assisted Methods. *Cereal Chem. J.* **2007**, *84*, 125–129. [CrossRef]

22. Del Rio, A.V.; Palmeiro-Sanchez, T.; Figueroa, M.; Mosquera-Corral, A.; Campos, J.L.; Mendez, R. Anaerobic digestion of aerobic granular biomass: effects of thermal pre-treatment and addition of primary sludge. *J. Chem. Technol. Biotechnol.* **2014**, *89*, 690–697. [CrossRef]

23. Jung, H.-J.G.; Varel, V.H.; Weimer, P.J.; Ralph, J. Accuracy of Klason Lignin and Acid Detergent Lignin Methods As Assessed by Bomb Calorimetry. *J. Agric. Food Chem.* **1999**, *47*, 2005–2008. [CrossRef] [PubMed]

24. Choudhary, R.; Umagiliyage, A.L.; Liang, Y.; Siddaramu, T.; Haddock, J.; Markevicius, G. Microwave pretreatment for enzymatic saccharification of sweet sorghum bagasse. *Biomass-Bioenergy* **2012**, *39*, 218–226. [CrossRef]

25. Lam, T.B.T.; Iiyama, K.; Stone, B.A. Distribution of free and combined phenolic acids in wheat internodes. *Phytochemistry* **1990**, *29*, 429–433.

26. Mathew, S.; Abraham, T.E. Ferulic Acid: An Antioxidant Found Naturally in Plant Cell Walls and Feruloyl Esterases Involved in its Release and Their Applications. *Crit. Rev. Biotechnol.* **2004**, *24*, 59–83. [CrossRef]

27. Mussatto, S.I.; Dragone, G.; Roberto, I.C. Ferulic and p-coumaric acids extraction by alkaline hydrolysis of brewer's spent grain. *Ind. Crop. Prod.* **2007**, *25*, 231–237. [CrossRef]

28. Moreira, M.M.; Morais, S.; Barros, A.A.; Delerue-Matos, C.; Guido, L.F. A novel application of microwave-assisted extraction of polyphenols from brewer's spent grain with HPLC-DAD-MS analysis. *Anal. Bioanal. Chem.* **2012**, *403*, 1019–1029. [CrossRef]

29. Díaz, A.B.; Moretti, M.M.D.S.; Bezerra-Bussoli, C.; Nunes, C.D.C.C.; Blandino, A.; Da Silva, R.; Gomes, E. Evaluation of microwave-assisted pretreatment of lignocellulosic biomass immersed in alkaline glycerol for fermentable sugars production. *Bioresour. Technol.* **2015**, *185*, 316–323. [CrossRef]

30. Moodley, P.; Kana, E.G. Development of a steam or microwave-assisted sequential salt-alkali pretreatment for lignocellulosic waste: Effect on delignification and enzymatic hydrolysis. *Energy Convers. Manag.* **2017**, *148*, 801–808. [CrossRef]

31. Joglekar, A.M.; May, A.T. Product excellence through design of experiments. *Cereal Foods World* **1987**, *32*, 857.

32. Carrión-Prieto, P.; Martín-Ramos, P.; Hernández-Navarro, S.; Sánchez-Sastre, L.F.; Marcos-Robles, J.L.; Martín-Gil, J. Furfural, 5-HMF, acid-soluble lignin and sugar contents in *C. ladanifer* and *E. arborea* lignocellulosic biomass hydrolysates obtained from microwave-assisted treatments in different solvents. *Biomass-Bioenergy* **2018**, *119*, 135–143. [CrossRef]

33. Pinela, J.; Prieto, M.A.; Carvalho, A.M.; Barreiro, M.F.; Oliveira, M.B.P.; Barros, L.; Ferreira, I.C.; Barreiro, F. Microwave-assisted extraction of phenolic acids and flavonoids and production of antioxidant ingredients from tomato: A nutraceutical-oriented optimization study. *Sep. Purif. Technol.* **2016**, *164*, 114–124. [CrossRef]

34. Liazid, A.; Palma, M.; Brigui, J.; Barroso, C.G.; Lovillo, M.P. Investigation on phenolic compounds stability during microwave-assisted extraction. *J. Chromatogr. A* **2007**, *1140*, 29–34. [CrossRef] [PubMed]

35. de Oliveira, D.M.; Finger-Teixeira, A.; Mota, T.R.; Salvador, V.H.; Moreira-Vilar, F.C.; Molinari, H.B.C.; Craig Mitchell, R.A.; Marchiosi, R.; Ferrarese-Filho, O.; Dantas dos Santos, W. Ferulic acid: A key component in grass lignocellulose recalcitrance to hydrolysis. *Plant Biotechnol. J.* **2015**, *13*, 1224–1232. [CrossRef] [PubMed]

36. Boonmanumsin, P.; Treeboobpha, S.; Jeamjumnunja, K.; Luengnaruemitchai, A.; Chaisuwan, T.; Wongkasemjit, S. Release of monomeric sugars from Miscanthus sinensis by microwave-assisted ammonia and phosphoric acid treatments. *Bioresour. Technol.* **2012**, *103*, 425–431. [CrossRef]

37. Oussaid, S.; Madani, K.; Houali, K.; Rendueles, M.; Díaz, M. Optimized microwave-assisted extraction of phenolic compounds from Scirpus holoschoenus and its antipseudomonal efficacy, alone or in combination with Thymus fontanesii essential oil and lactic acid. *Food Bioprod. Process.* **2018**, *110*, 85–95. [CrossRef]

38. Galan, A.-M.; Calinescu, I.; Trifan, A.; Winkworth-Smith, C.; Calvo-Carrascal, M.; Dodds, C.; Binner, E. New insights into the role of selective and volumetric heating during microwave extraction: Investigation of the extraction of polyphenolic compounds from sea buckthorn leaves using microwave-assisted extraction and conventional solvent extraction. *Chem. Eng. Process.* **2017**, *116*, 29–39. [CrossRef]

39. Thomas, H.L.; Pot, D.; Latrille, E.; Trouche, G.; Bonnal, L.; Bastianelli, D.; Carrère, H. Sorghum Biomethane Potential Varies with the Genotype and the Cultivation Site. *Waste Biomass Valorization* **2019**, *10*, 783–788. [CrossRef]

40. Goering, H.U.; Van Soest, P.J. Forage fiber analyses (apparatus, reagents, procedures, and some applications). In *No. 379. US Agricultural Research Service*; US government printing office: Washington, DC, USA, 1970.

41. Dence, C.W. The Determination of Lignin. In *Methods in Lignin Chemistry*; Springer: Berlin, Germany, 1992; pp. 33–61.

42. Hatfield, R.; Fukushima, R.S. Can lignin be accurately measured? *Crop Sci.* **2005**, *45*, 832–839. [CrossRef]

43. Zhang, Y.; Culhaoglu, T.; Pollet, B.; Melin, C.; Denoue, D.; Barrière, Y.; Baumberger, S.; Méchin, V. Impact of Lignin Structure and Cell Wall Reticulation on Maize Cell Wall Degradability. *J. Agric. Food Chem.* **2011**, *59*, 10129–10135. [CrossRef]

44. Oliveira, J.; Alves, M.; Costa, J.; Alves, M. Optimization of biogas production from Sargassum sp. using a design of experiments to assess the co-digestion with glycerol and waste frying oil. *Bioresour. Technol.* **2015**, *175*, 480–485. [CrossRef] [PubMed]

45. Thomas, H.L.; Seira, J.; Escudie, R.; Carrere, H. Lime Pretreatment of Miscanthus: Impact on BMP and Batch Dry Co-Digestion with Cattle Manure. *Molecules* **2018**, *23*, 13. [CrossRef] [PubMed]

46. Witek-Krowiak, A.; Chojnacka, K.; Podstawczyk, D.; Dawiec, A.; Pokomeda, K. Application of response surface methodology and artificial neural network methods in modelling and optimization of biosorption process. *Bioresour. Technol.* **2014**, *160*, 150–160. [CrossRef] [PubMed]

Sample Availability: Samples of the compounds are not available from the authors.

 *molecules*

Article

Matrix Discriminant Analysis Evidenced Surface-Lithium as an Important Factor to Increase the Hydrolytic Saccharification of Sugarcane Bagasse

Ana Sílvia de Almeida Scarcella [1], Alexandre Favarin Somera [2],
Christiane da Costa Carreira Nunes [3], Eleni Gomes [3], Ana Claudia Vici [2],
Marcos Silveira Buckeridge [4] and Maria de Lourdes Teixeira de Moraes Polizeli [2,*]

1   Departamento de Bioquímica e Imunologia, Faculdade de Medicina de Ribeirão Preto, Universidade de São Paulo. Bandeirantes Av., 3900, 14049-900 Ribeirão Preto, São Paulo, Brazil; asascarcella@yahoo.com.br
2   Departamento de Biologia, Faculdade de Filosofia, Ciências e Letras de Ribeirão Preto, Universidade de São Paulo. Bandeirantes Av., 3900, 14040-901 Ribeirão Preto, São Paulo, Brazil; afsomera@gmail.com (A.F.S.); acvici@gmail.com (A.C.V.)
3   Departamento de Biologia, Instituto de Biociências Letras e Ciências Exatas, Universidade Estadual Paulista Júlio de Mesquita Filho. Cristovão Colombo Street, 2265, 15054000 São José do Rio Preto, São Paulo, Brazil; nuneschris@yahoo.com (C.d.C.C.N.); eleni@ibilce.unesp.br (E.G.)
4   Laboratório de Fisiologia Ecológica (LAFIECO), Departamento de Botânica, Instituto de Biociências, Universidade de São Paulo. Matão Street, 277, Cidade Universitária, 05508-090 São Paulo, Brazil; msbuckeridge@gmail.com
*   Correspondence: polizeli@ffclrp.usp.br; Tel.: +55-16-3315-4680

Received: 15 August 2019; Accepted: 20 September 2019; Published: 8 October 2019

Abstract: Statistical evidence pointing to the very soft change in the ionic composition on the surface of the sugar cane bagasse is crucial to improve yields of sugars by hydrolytic saccharification. Removal of Li^+ by pretreatments exposing -OH sites was the most important factor related to the increase of saccharification yields using enzyme cocktails. Steam Explosion and Microwave:H_2SO_4 pretreatments produced unrelated structural changes, but similar ionic distribution patterns. Both increased the saccharification yield 1.74-fold. NaOH produced structural changes related to Steam Explosion, but released surface-bounded Li^+ obtaining 2.04-fold more reducing sugars than the control. In turn, the higher amounts in relative concentration and periodic structures of Li^+ on the surface observed in the control or after the pretreatment with Ethanol:DMSO:Ammonium Oxalate, blocked -OH and O^- available for ionic sputtering. These changes correlated to 1.90-fold decrease in saccharification yields. Li^+ was an activator in solution, but its presence and distribution pattern on the substrate was prejudicial to the saccharification. Apparently, it acts as a phase-dependent modulator of enzyme activity. Therefore, no correlations were found between structural changes and the efficiency of the enzymatic cocktail used. However, there were correlations between the Li^+ distribution patterns and the enzymatic activities that should to be shown.

Keywords: lithium; sugarcane bagasse; saccharification; glycosyl-hydrolase; ToF-SIMS; surface ion distribution; second-generation ethanol; pretreatment

---

1. Introduction

Demand for renewable fuels has considerably increased in recent years. Thus, there has been a significant increase of interest in sugarcane.

The most widespread crop in Brazil is sugarcane, with 391,767 thousand tons coming from the 2018/2019 harvest [1]. Sugarcane bagasse is composed of an elaborate arrangement of polysaccharides and proteins, combined with inorganic and organic ions acquired during different stages of culturing and processing [2]. Apart from these different sources of ions and the composition of the plant material, the pretreatment of sugarcane bagasse has become extremely important for determining the ionic composition of the substrate for the enzymatic saccharification. The cost and success of bioethanol production process from lignocellulosic biomass depends largely on the recalcitrant biomass itself, as well as on the repertoire of enzymes involved in the depolymerization of the constituent polysaccharides in the cell wall [3].

The pretreatment of sugarcane is crucial to ensure the conversion of polysaccharides into sugars for bioethanol production, since their chemical composition and physical structure are altered [4,5]. The different pretreatments can contribute to improving hydrolysis of the cell wall due to the rupture of the lignin structure and the connection with the rest of the biomass; removal of hemicellulose; and reduction of crystallinity and degree of polymerization of cellulose [6]. Several organic chemical changes have been researched using ToF-SIMS with great success [7]. However, pretreatments have the potential to alter the composition, relative concentration, and spatial distribution of alkaline and alkaline-earth metal ions on the substrate surface. This critical aspect possibly related to enzyme failure is still poorly researched. The mineral distribution on the surface of the substrate and the ionic properties of the surface area are key properties related to enzyme action because they are responsible for the surface charge, acidity, phase transfer, and the availability of binding and walking sites for enzymes [8,9]. Metals, alkaline metals, and alkaline earth metal ions are known to interfere positively or negatively in the enzymatic activity [8]. Therefore, it is necessary to map these changes and evaluate which ones interfere in enzyme function.

The importance of substrate as an ionic carrier is stressed in complex substrates such as sugarcane bagasse used in the second-generation ethanol industry.

The aim of this study was to analyze pretreatments of sugarcane bagasse to be used prior to enzymatic hydrolysis and to understand the anatomical factors related to our yield using ION-TOF. Furthermore, this work evaluated the neglected effect of current pretreatments on the composition, concentration and distribution of metallic ions on the surface of the sugarcane bagasse. Here, it was reported the extensive imaging analysis of different pretreatments of sugarcane bagasse used in second-generation ethanol production and their correlation with enzyme cocktail activity.

2. Results

2.1. Control-Milled Sugarcane Bagasse in Natura

Milled sugarcane bagasse *in natura* was chosen as control and subjected to direct saccharification. That material was also the initial material subjected to pretreatments to obtain a more suitable substrate for saccharification.

Control presented a very compact and overlapped structure containing aromatics remnant of lignin, $(C_xH_yO_z)_n$ from carbohydrates and an almost uniform distribution of C_xH_y from side groups. Sum images containing -OH and $(C_xH_yO_z)_n + C_xH_y$ are shown in Figure 1. ToF-SIMS revealed a large number of impregnated ions in the structure including Na^+, K^+ and Mg^{2+}, as well as Ca^{2+} covalently bounded $(Ca^-C_3H_4^+)$ (Table 1). From those ions, Li^+ covered 22.08% of -OH sites which were released together with Li^+ sputtering (Table 2, Figure 2). Na^+ and K^+ spots were widespread on the surface (Table 1). Na^+ presented clustered distribution forming small high-bulk density aggregates. Na^+ was co-located with Cl^- and PO^-, PO_2^- and PO_3^- sites. However, only 48.7% of PO^-, PO_2^- and PO_3^- sites, co-located with Na^+, were sputtered from the surface of the sugarcane bagasse. All other elements had a random distribution. The density of H^+ on the surface was very large. The negative mode sputter eroded ions at a mass range of 160–180 Da related to surface glucose, arabinose, and xylose units of the exposed cellulose and hemicellulose, respectively.

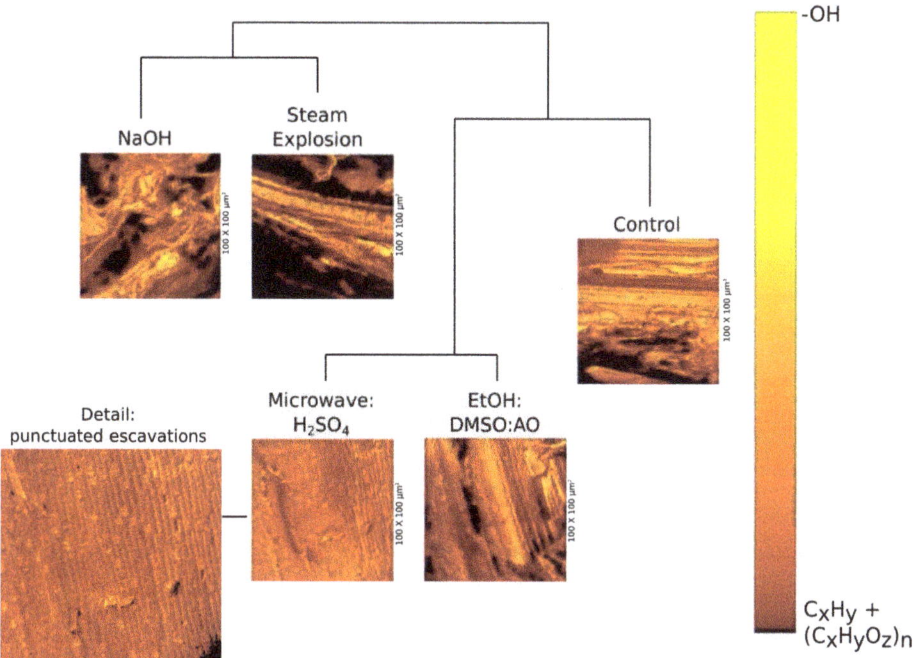

Figure 1. The overlapped signals of aromatics, $(C_xH_yO_z)_n$ residues and C_xH_y chains generated from the ION-TOF analysis of sugarcane bagasse submitted to different pre-treatments. Control was material *in natura*. The dendogram was obtained comparing aromatics, $(C_xH_yO_z)_n$ residues and the chains of C_xH_y distributions, total ion image surface entropy and roughness data of differently pretreated sugarcane bagasses at $p = 0.05$. NaOH and Steam Explosion pretreatments produced the most amorphous substrate because of the loss in periodicity of microfibril arrangements, while Microwave:H_2SO_4 essentially differed from Ethanol:Dimethyl Sulfoxide: Ammonium Oxalate (EtOH:DMSO:AO) pretreatment due to the production of slightly spherical excavations on the surface of the material, which can be observed in the chemical sputtering ion image. The images presented were obtained at negative mode and thus were dominated (88.3%) by overlapped signals of aromatics, $(C_xH_yO_z)_n$ residues and C_xH_y chains. Yellow stains are superposed -OH and O^- chemical images.

Table 1. Ion particle analysis after automated color threshold.

Ion/ Compost	Control *in Natura*			Steam Explosion Pretreatment			Microwave:H_2SO_4 Pretreatment			EtOH:DMSO:AO [a] Pretreatment			NaOH Pretreatment		
	Count (N)	Area (μm^2)	Average Size (μm^2)	Count (N)	Area (μm^2)	Average Size (μm^2)	Count (N)	Area (μm^2)	Average Size (μm^2)	Count (N)	Area (μm^2)	Average Size (μm^2)	Count (N)	Area (μm^2)	Average Size (μm^2)
Li^+	353	53.857	0.150	88	0.054	0.027	63	0.027	0.027	320	30.879	0.097	16	4.569	0.286
Na^+	71	349.574	6.221	35	185.866	5.310	121	376.032	3.108	61	383.557	6.288	23	374.151	16.267
K^+	19	432.026	26.729	50	254.450	5.089	94	364.368	3.876	38	387.346	10.193	15	376.596	25.106
Mg^{2+}	144	249.479	2.058	107	116.690	1.091	777	124.457	0.160	570	105.242	0.185	119	178.072	1.496
$Ca\text{-}C_3H^{4+}$	44	407.019	11.957	122	231.607	1.898	28	487.885	17.424	83	393.608	4.742	42	316.961	7.547
F^-	197	278.208	4.107	209	167.511	0.801	778	168.263	0.216	299	244.991	0.819	46	386.540	8.403
Cl^-	62	369.058	7.128	135	164.850	1.221	65	457.812	7.043	224	234.241	1.046	53	367.540	6.935
DDA_b	0	0.000	0.000	9	0.403	0.045	7	0.215	0.031	7	0.403	0.058	9	2.365	0.263

[a] Ethanol:Dimethyl Sulfoxide:Ammonium Oxalate. [b] Dimethyl Dialkyl Ammonium.

Table 2. Percentage of -OH sites uncovered by Li^+.

Pretreatment	Li^+-free-OH Area (%) *
Control (*in natura*)	77.92
EtOH:DMSO:AO [a]	78.50
Steam Explosion	84.98
Microwave:H_2SO_4	85.00
NaOH	84.16

[a] Ethanol:Dimethyl Sulfoxide:Ammonium Oxalate. * The percentage of lithium-free surface area observed after NaOH pretreatment was due to the presence of a heavy Li^+ aggregate located in only one site on the surface of the substrate. The Li^+ distribution for Steam Explosion and Microwave:H_2SO_4 were widespread on the surface of the substrate generated after both pretreatments (Figure 2).

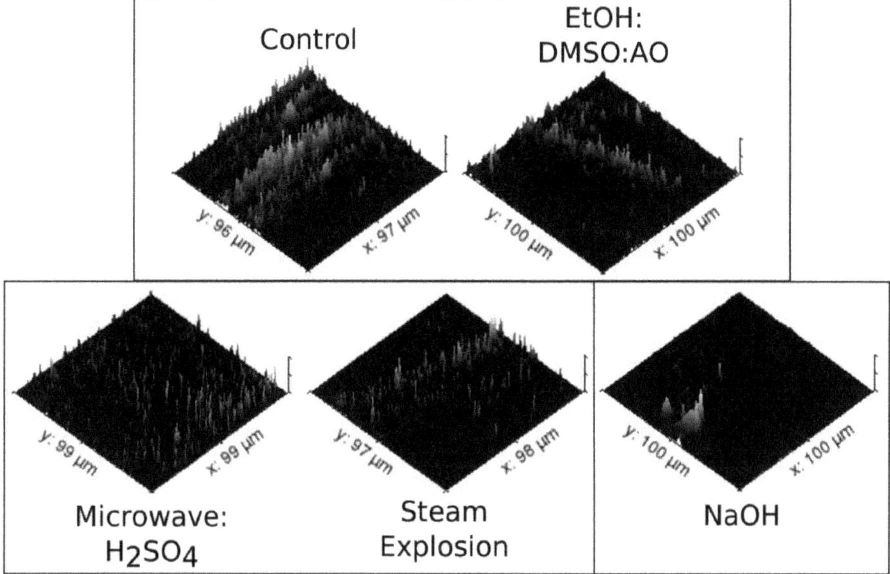

Figure 2. Mixture Discriminant Analysis and distribution patterns of Li^+ on the surface of sugarcane bagasse. Pretreatments are grouped into conjuncts. The normalized spatial concentration patterns for released Li^+, $Li^+[O^-]_n$ and $Li^+[OH^-]_n$ from the surface of pretreated sugarcane bagasse is presented. Parallel ridges of Li^+ records were observed on the surfaces of control materials, Steam Explosion and Ethanol: Dimethyl Sulfoxide: Ammonium Oxalate (EtOH:DMSO:AO) pretreatments. The periodic arrangements in fringe and lattice structures positively correlated with the positioning of fibrils only for control and EtOH:DMSO:AO pretreatments, once Steam Explosion destroyed that arrangement.

2.2. Steam Explosion Pretreatment

At positive mode, the mass spectrometric analysis of sugarcane bagasse subjected to steam explosion evidenced a decrease in the concentrations of metal and semi-metal ions (Table 1). The records for O^-, Cl^- and -OH were enriched 3.1-fold. The resulting chemical image of the surface of the substrate evidenced the increase in the exposure of compounds at 279–280 Da (benzoate) and at 417 Da (syringaresinol) from lignin. The pretreatment also increased 2.47-fold the $(C_xH_yO_z)_n$ at 160–180 Da (glucose and xylose from cellulose and hemicelluloses, respectively). The steam explosion produced 2.80-fold higher dimethyl dialkyl ammonium (DDA) amounts at random distribution than other pretreatments. In turn, Li^+ was dramatically reduced (2.20-fold) compared to control and presented random distribution (Figure 2). At negative mode, the exposure of anions and sites for an enzymatic attack such as -OH, O^-, $(C_xH_yO_z)_n$ and aromatics showed a 2.31-fold mean increase in signal intensity (Figure 1). Microscopic changes in fiber were evident by chemical image, presenting lattice deformation even at micrometer scale but without complete fibrils dismemberment and their periodic fringe ordering, whose remnants occupied 47% surface area.

2.3. Microwave:H2SO4 Pretreatment

Microwave:H_2SO_4 pretreatment decreased metal ions on the surface of the substrate (Table 1). At negative mode, the chemical image presented aromatics related to lignin superposed with C_xH_y signals. The image showed 1.22-fold lower records for $(C_xH_yO_z)_n$ on the surface than control, revealing higher hindering of celluloses and hemicelluloses. It is worth mentioning the low abundance of ions randomly distributed on the surface, 11.8% lower than control, but 43.2% higher than Steam Explosion. The surface of that pretreated substrate had 2.60-fold less DDA than the Steam Explosion and 28.8%

less Li^+ than the control. The periodic structure of fibrils was 97.2% preserved rather than prior to treatment. Microwave:H_2SO_4 produced a singular effect: the substrate pretreated surface showed 72 roughly circular excavations for 100×100 μm^2 area. Excavations presented an average radius of 0.72 μm and 2.45 nm depth, which were extremely rich in -OH and O^- (Figure 1). These excavations presented 68.4% of its surface covered by -OH and O^- from cellulose and hemicelluloses, and were devoid of Li^+, Ca-$C_3H_4^+$, Mg^{2+}, and -NH_3^+.

2.4. Ethanol:Dimethyl Sulfoxide:Ammonium Oxalate (EtOH:DMSO:AO) Pretreatment

The chemical image confirms the increase in the exposure of aromatic compounds and 1.21-fold increment in the exposal of $(C_xH_yO_z)_n$ at positive mode, followed by a 1.09-fold rise in the amounts of -NH_3^+, -CN and -CNH groups overlapped with $(C_xH_yO_z)_n$ on the substrate surface. These results pointed to overlaps between proteins and carbohydrates at the substrate surface. The pretreatment randomly added DDA and produced a low ionic average count (Table 1). At negative mode, the $(C_xH_yO_z)_n$ and aromatic exposure showed 1.30-fold higher signal intensity than the control because of the better exposure of fibrils' after that pretreatment (Figure 1).

2.5. NaOH Pretreatment

The chemical image revealed the complete deformation of the lattice structures observed in the control (Figure 1). Only 8.7% of the remnant parallel arrangements and lattice structures at the substrate surface were preserved in a 100×100 μm^2 area. That pretreatment also produced the biggest enrichment in Na^+ and F^- recorded, equivalent to a 1.73-fold mean increase relative to the control (Table 1). The increase in F^- and K^+ was routed to impurities in the NaOH solution (data not shown), while Na^+ enrichment can be a possible side effect of the pretreatment. In addition to the increased exposure of $(C_xH_yO_z)_n$ and C_xH_y, aromatics at 279—280 Da identified as benzoate and bi-phenolic residues from lignin. Li^+ was only observed in aggregates with DDA (Figure 2), but not on the remnant surface. Despite the higher relative concentrations of Na^+ and K^+, all other ion concentrations reduced (Table 1). Ions were distributed in lower numbers of aggregates occupying a reduced surface of the substrate. Li^+ presented a decrease of 1.41-fold in relation to the control (Table 1, Figure 2). That was observed with Microwave:H_2SO_4, but it was lower than those obtained with Steam Explosion and EtOH:DMSO:AO. Compared to the control, -OH and O^-, increased 1.65-fold (Figure 1). That was the biggest gain achieved with -OH exposition. Only the circular excavations verified in Microwave:H_2SO_4 pretreatment presented higher amounts of -OH exposed. However, circular excavations verified after Microwave:H_2SO_4 pretreatment occupied only 68.4% a 100×100 μm^2 area, while the area occupied by -OH after NaOH pretreatment represented 97.4% of the surface analyzed.

2.6. Clustering Pretreatments Using Discriminant Analysis with Machine Learning—Anatomical Parameters

Figure 1 shows the comparison of surface anatomical parameters ($p = 0.05$) of sugarcane bagasse differently pretreated. Aiming to compare all pretreatments, matrix discriminant analysis with machine learning was used, due to the non-linearity of data and the very complex nested matrix of parameters obtained. The surface structure properties analyzed were the percentage of the area occupied by C_xH_y, $(C_xH_yO_z)_n$ and aromatics (i.e., sugars and lignin) surface area covered by radicals (-OH and O^-), which are related to hydroxyl groups used by glycosyl-hydrolases during their action; the surface area occupied by periodic, parallel, and lattice structures containing C_xH_y and $(C_xH_yO_z)_n$; and surface entropy, waviness, and roughness. These parameters were automatically collected using the free and open-source softwares ImageJ and Gwyddion.

According to these parameters, NaOH and Steam Explosion pretreatments clustered together and produced the most amorphous substrate. It happened due to the loss of periodicity of the carbohydrate surface arrangements. On the other hand, Microwave:H_2SO_4 differed from the pretreatment with EtOH:DMSO:AO, because of the production of slightly spherical excavations on the surface of the material. The fibrils and microfibrils, observed in the control, presented intense cover-up by organic matter, visible in the images as amorphous deposits containing -OH and $(C_xH_yO_z)_n$ residues and chains of C_xH_y (Figure 1). The estimated differential entropy for the control surface was equal to -20.46 nats (natural units of information) with a deficit of 0.16266 nats, because of those roughly amorphous deposits.

2.7. Clustering Pretreatments Using Discriminant Analysis with Machine Learning – Ionic Parameters

The matrix discriminant analysis of the ionic parameters from the surface of the substrates clustered the following samples: EtOH:DMSO:AO pretreatment and Control; Steam Explosion and Microwave:H_2SO_4; and finally, isolated the NaOH pretreatment in a single group (Figure 2, Table 1). Using a jackknife procedure, deleting single ionic parameters before running the matrix discriminant analysis with machine learning, the ionic parameters related to Li^+ (Table 2) were identified as responsible by 37,6% of these results. Figure 2 showed the results of the matrix discriminant analysis with computerized images of Li^+ obtained using the Gwyddion software.

Steam Explosion and Microwave:H_2SO_4 pretreatments produced lower counting of aggregates, the lowest total area occupied by Li^+ and the lowest average size for Li^+ aggregates. However, both differed in DDA, only present after the Steam Explosion pretreatment (Table 1), and -OH-enriched excavations presented only after Microwave:H_2SO_4 pretreatment (Figure 1).

The NaOH pretreatment produced higher counts, percentage area and average size of aggregates of Li^+, but all Li^+ observed was not distributed accompanying fibrils of carbohydrates such as in Steam Explosion or control. In turn Li^+ presented random distribution and a very low abundance on the surface of the substrate, but one high bulk density aggregate of $13.79\ \mu m^2$ in a $100 \times 100\ \mu m^2$ chemical image area.

2.8. Performance of the Enzyme Cocktail Versus Pretreatment-the Enzymatic View Versus the Matrix Discriminant Analysis of Anatomical and Ionic Composition on the Surface of the Substrate

The best saccharification yield was obtained after NaOH pretreatment, which increased the release of reducing sugars by enzymatic saccharification in 2.04-fold more than the control at 10 h intervals (Figure 3). The right side of Figure 3 shows a tree made using the results from the MDA analysis of treatments used in each saccharification assay. The tree was placed next to the results of the saccharification assays to highlight the similarities between enzyme responses and MDA analysis results. Analysis was done at $p \le 0.05$. It was noted the relevant events in the dendrogram, such as the non-Michaelian branch related to milled material, the counts of lithium-ion clusters on the surface of the materials, the qualitative richness of Sodium observed in NaOH treatment, the total area free from metal ions less than $14,509\ \mu m^2$, and the qualitative changes in lithium-ion concentration on the surfaces analyzed. These raw data are in Table 1. Steam Explosion and Microwave:H_2SO_4 clustered in an intermediate subset during enzymatic hydrolysis increasing 1.74-fold the reducing sugars release at 10 h intervals compared to control. In turn, a 1.1-fold decrease in the production rates of reducing sugars was identified after EtOH:DMSO:AO pretreatment when compared to control. EtOH:DMSO:AO and control also presented time-dependent decay for the release of reducing sugars after 10 h intervals.

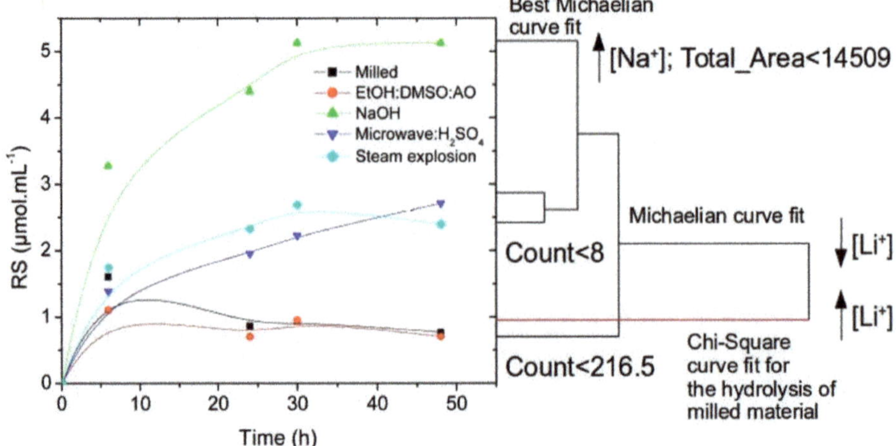

Figure 3. Time-dependent release of reducing sugars during the enzymatic hydrolysis of sugarcane bagasse. Generalized additive model (GAM) curves for reducing sugars (RS) released and the similarity analysis among enzyme cocktail activities were shown with the major traits related to the structural differences among each pretreatment using Mixture Discriminant Analysis (MDA) at $p = 0.05$.

2.9. Matrix Discriminant Analysis (MDA) of Anatomical and Ionic Parameters from the Surface of the Substrate Versus the Enzymatic View—Correlation Between Discriminant Analysis and Saccharification Yields

MDA for the anatomical parameters, including entropy, roughness, exposition of aromatics from lignin, $(C_xH_yO_z)_n$ residues and C_xH_y chains from carbohydrates, and microfibril and fibril arrangements (Figure 1), did not correlate with data obtained in this study about saccharification yields at $p = 0.05$. As enzymes are good topological recognizers, the information did not correlate with the topological view of the enzymes. These results clustered pretreatments as the least-squares method for the curves of time-dependent release of reducing sugars at $p = 0.05$ (Figure 3).

2.10. Enzyme Cocktail Response to Metal Ion Salts in Solution

As ionic parameters produced the same cluster profile for pretreatments as the saccharification and time-dependent releases of reducing sugars, the effects of the ions observed on the surface of the substrate were evaluated to check their impact on the saccharification when in solution. All ions tested activated glycosyl-hydrolases when in solution (Table 3), but decreased laccase activity up to 87% (for NH_4F). Therefore, it can be inferred that these ions negatively interfere in lignin degradation, but acted as non-essential activators for hemicellulases and cellulases when in solution. These results are opposed to the correlation observed between surface-bounded ions and saccharification.

Table 3. Effects of ion salt in solution on the enzyme activities.

Ion Salts	Laccase (%)	Xylanase (%)	Endoglucanase (%)	Cellobiohydrolase (%)	β-Glucosidase (%)
NH_4F	12.34	141.68	109.84	126.50	135.70
NaH_2PO_4	37.56	146.81	99.06	100.10	141.80
$MgCl_2 \cdot 6H_2O$	41.36	137.39	61.40	110.90	144.79
NH_4Cl	43.39	155.04	23.44	103.80	138.54
$CaCl_2$	43.07	147.14	120.15	119.30	144.23
KCl	42.72	148.24	140.31	109.50	146.57
LiCl	42.24	130.59	133.59	101.60	148.55
Na_2SO_4	29.41	181.60	110.78	137.30	152.23
$MnCl_2 \cdot 4H_2O$	30.01	216.97	164.68	129.00	143.83
NaCl	38.20	175.46	134.68	129.70	150.57
KH_2PO_4	33.56	162.18	124.22	148.30	143.17
BaCl	34.04	139.41	123.43	144.30	143.57
$Zn(NO_3)_2$	40.48	122.86	119.84	140.10	137.07

Control (without ions) corresponded to 100%.

3. Discussion

Glycosyl-hydrolases need to deal with a very complex surface area rich in C_xH_y and -OH sites for reaction. In addition to the ratio between hydrophobic areas and reaction sites containing -OH and O^-, the presence of several bounded metal ions on the surface of the substrate resulted in a new level of complexity to be considered for the development of a pipeline linking pretreatment and cocktail.

To be analyzed by ToF-SIMS, the material was deposited two-dimensionally on the disc for ion excavation. The specific surface area did not show visible relations with the enzyme hydrolysis. So, new aspects were evaluated for these materials. Interpreting the surface area is difficult, as some ions occupied the entire surface area evaluated for each material, such as Calcium and Chlorine ions, which can be seen in Table 1. Metal cation-free areas were rare, such as the 14,509 um^2 observed in NaOH treatment. No material had Anion-free areas at that micrometric scale. In addition, each material has its own surface area. The overlapped surface areas occupied by each metal ion and hydroxyl sites are more valuable, and the constant overlap between hydroxyl radicals and lithium ions were observed as exclusive for these materials. Perhaps the excess of other ions made it impossible to see overlaps among other cations and the hydroxyl-rich regions, spreading to areas containing C-H-rich hydrophobic regions. For example, the amounts of potassium, sodium, and the anions fluorine and chlorine were so high that they occupied the entire surface evaluated.

Contrary to expectations, the increased exposure of fibrils, Steam Explosion, and NaOH pretreatments did not answer equally during the enzymatic hydrolysis. On the other hand, Steam Explosion presented a saccharification statistically equivalent to that observed after Microwave:H_2SO_4 pretreatment. This pretreatment did not produce major changes in the arrangement of the microfibrils such as Steam Explosion, neither similar information about surface-sugars were observed, as the ones described above. However, both decreased the abundance and area occupied by metal-ions on the surface of the substrates compared to control. In particular, the areas containing exposed -OH and O^- free of Li^+, proved a recurring property in both. This property was shared by the substrates subjected to NaOH treatment.

The NaOH pretreatment increased surface-Na^+ while reducing surface-Li^+, reducing the surface acidity while increasing O^- and -OH on the surface area. The combination of amorphous structures, loss of periodicity in the spatial distribution of carbohydrates, and the reduced surface-bounded Li^+ together with the very large amounts of O^- and -OH sites were possibly good factors to develop productive E S complexes [10], because the strong increase in saccharification compared to other pretreatments and control.

The overall ionic cleaning of the surface increased the exposure of -OH and O^-, but the ion that most influenced the pattern of discrimination among substrates was lithium, which was sputter eroded with -OH, implying covering of OH by that ion. Although the changes in the concentration and distribution of nonessential ion activators were not limited to Li^+, the best results during the degradation of the substrate were obtained just after pretreatments capable of removing Li^+ from the -OH and O^- surfaces of the substrate, i.e., NaOH, Steam Explosion and Microwave:H_2SO_4 pretreatments. These pretreatments increased 1.23-fold mean the available Li^+-free -OH and O^- for ion sputtering than the control material.

Using only the main discriminating factors related to the ionic composition and distribution, i.e., the counting of Li^+ aggregates, the distance between Li^+ spots and the average size of Li^+ clusters, it was possible to identify relationships between Li^+ distribution and enzyme activity.

Steam Explosion pretreatment produced an almost uniform pattern of Li^+ distribution accompanying fibrils, while Microwave:H_2SO_4 produced random distribution patterns. Therefore, different distribution patterns, with statistically similar counts of aggregate and distancing of Li^+ on the surface of pretreated sugarcane bagasse, resulted in statistically similar production rates of reducing sugars by the presented enzyme cocktail. Once Steam Explosion and Microwave:H_2SO_4 do not have structural relationships related to fibril arrangement, the only aspect which linked both pretreatments, were the surface concentration and distribution of Li^+.

In turn, the pretreatment with NaOH produced the highest aggregation of Li^+, but the lowest concentration and the largest distribution of Li^+ on the substrate surface. It allowed the exposure of –OH, possibly, due to the reduced formation of coordination groups between Li^+ and -OH, which results in the highest sputtering of Li^+ free of OH and O^- radicals from the cane surface. NaOH also reduced the periodic arrangement of fibrils, $-OH^-$ and O^- enriched sites and reduced the surface occupied by Li^+, which positively correlated with the highest activity observed on pretreated substrates. Once Li^+ on the surface of the substrate can block -OH as revealed by the sputter erosion of Li^+ together with -OH, surface Li^+ can negatively interfere in the enzyme performance because glycosyl-hydrolases require free -OH to react [11]. This hypothesis could be risen using the results about the correlation between the increases of surface-bounded Li^+ and covering of -OH sites with the decrease in enzyme activity. Given the differences between the effects of ions in solution and the results from the matrix discriminant analysis, ions adhered to the substrate could affect the enzyme action differently from those in suspension. Thus, it could be hypothesized that the real substrate for the enzyme is an ionic polysaccharide coordination system.

An important feature found during the enzymatic hydrolysis of control materials was the great performance of the cocktail response during the first 10 h of the assay. Enzyme performance was second only to that shown after NaOH pretreatment of the substrate. Apparently, a depolymerase adaptation to control materials, i.e., the original substrate found in nature, allowed the rapid recognition of binding sites for a rapid phase transfer, resulting in large amounts of productive interactions, which did not exceed only those observed for the assays using NaOH-pretreated substrate. Therefore, the majority of pretreatments are only important during prolonged hydrolysis times.

On the other hand, the EtOH:DMSO:AO pretreatment produced an ionic surface very similar to that of control according to matrix discriminant analysis, especially regarding the pattern of surface distribution of Li^+. Nevertheless, the first 10 h of enzymatic attack recorded for this pretreatment was the least productive. As DDA was a substantial change in that substrate, it could be inferred a failure in the substrate recognition by enzymes. Although the solubility of the substrate could be improved after that pretreatment, caution is needed when adding methyl and amino groups in the carbohydrate structure, since its recognition as a substrate for enzymatic cocktails can be a problem.

EtOH:DMSO: AO pretreatment and control presented a strong decrease in reducing sugars after 10 h reaction times even without microbial activity. It was difficult to determine the cause for those decays. Once both control and EtOH:DMSO:AO pretreated biomass clustered together in every matrix discriminant analysis of chemical images, the observed decay can be related to ions trapped on the substrate surface and released into the reaction solution inducing reverse reactions or reducing sugar precipitation. The causes for those phenomena remain hindered by a very complex mixture produced during the saccharification.

It can be assumed that there is a complex causal mechanism controlled by phase-transfer of enzymes and ions affecting cooperative, competitive and deinhibitory processes, anchimeric assistance and [ion]$^-$ dependent induced-fit behavior of glycosyl hydrolases. Thus, it could be concluded that lithium was an activator in solution, but its pattern of presence and distribution in the substrate can act as an inhibitor. These results pointed to a phase-dependent action for alkali metal ions in the enzymatic activity.

4. Materials and Methods

4.1. Control—Milled Sugarcane Bagasse in Natura

Sugarcane bagasse was provided by Sugar and Alcohol Mills (Ribeirão Preto, São Paulo, Brazil). It was washed in tap water to remove reducing sugars, dried at 50 °C, and then milled in a knife mill SL 32 (SOLAB), 25 mesh. The material obtained was used as the control for the described pretreatments.

4.2. Sugarcane Bagasse Pretreatments

Sugarcane bagasse was submitted to four different pretreatment types. For all the pretreated sugarcane bagasse the resulting material was washed with deionized water until the complete removal of reducing sugars. The reducing sugars released were monitored using the 3,5-Dinitrosalicylic Acid (DNS) method [12]. After washing, the material was dried at 50 °C and stored at room temperature.

4.2.1. Steam Explosion Pretreatment

Milled sugarcane bagasse *in natura* was maintained in steam water at 14 kg.cm^{-2} for 8 min, followed by rapid steam water expansion.

4.2.2. Microwave:H_2SO_4 Pretreatment

Microwave pretreatment was made according to Moretti et al. [13] with minor changes. A sample of 10 g of sugarcane bagasse was immersed for 24 h in, a solution of 0.05 M H_2SO_4 and glycerol. After that, this sample was transferred to a 250-mL round-bottom flask into a microwave oven which was connected to a spinning reflux condenser. The released sample was irradiated at 2450 MHz for 5 min. An infrared thermometer was used to detect the temperature. Further, 30 mL of distilled water was added to the material, mixed, filtered and this suspension was used to determine the amount of reducing sugars.

4.2.3. Ethanol: Dimethyl Sulfoxide: Ammonium Oxalate Pretreatment

Cell wall components from sugarcane bagasse *in natura* were fractionated using the protocol described by Lima et al. [14]. For the removal of soluble sugars, samples of sugarcane bagasse (1 g) were incubated in 20 mL 80% ethanol at 80 °C for 20 min under constant stirring. The resulting material was centrifuged (11,000× *g*) for 15 min, and the supernatant was discarded. This step was repeated six times, and the resulting precipitate was washed with 20 mL of distilled water and dried overnight in an oven at 50 °C. Starch was removed incubating the dried material in 20 mL 90% Dimethyl Sulfoxide (DMSO) at 90 °C for 24 h, while pectins were extracted incubating the starch-free material in 20 mL ammonium oxalate solution, pH 7.0 at 80 °C for 3 h.

4.2.4. NaOH Pretreatment

50% dry mass of EtOH:DMSO:AO treated materials were hydrolyzed using NaOH to remove hemicelluloses [14]. The material was hydrolyzed at room temperature using a sequence of three steps: (1) 1-h hydrolysis time using 20 mL 0.1 M NaOH: 0.1 M sodium borohydride; (2) 1-h hydrolysis time at room temperature using 20 mL 1.0 M NaOH: 0.1 M sodium borohydride; and (3) 1-h hydrolysis time at room temperature using 20 mL 4.0 M NaOH: 0.1 M sodium borohydride.

4.3. Enzymatic Hydrolysis

The enzyme cocktail applied to achieve cell wall degradation used 0.122 U laccase (*Trametes versicolor*); 7 U xylanase (*Malbranchea pulchella* expressed in *Aspergillus nidulans*) [15]; 5 U endoglucanase (*Aspergillus terreus* expressed in *A. nidulans*) [16], 14 U cellobiohydrolase (*Aspergillus niveus* expressed in *A. nidulans*) [17], and 9 U β-glucosidase (*Aspergillus niger*) per gram of lignocellulosic biomass. Materials were suspended into 7 mL of 50 mM sodium citrate buffer, pH 5.0. Hydrolysis was conducted at 55 °C and 110 rpm during 48 h. Reducing sugars were determined using DNS method [12].

4.4. Effects of Dissolved Salts on Specific Enzyme Activities

The effects of the salts observed on the surface of sugarcane bagasse upon the enzyme cocktail activities were analyzed in 10 mM final concentration: NH_4F, NaH_2PO_4, $MgCl_2 \cdot 6H_2O$, NH_4Cl, $CaCl_2$, KCl, $LiCl$, Na_2SO_4, $MnCl_2 \cdot 4H_2O$, $NaCl$, KH_2PO_4, $BaCl$, and $Zn(NO_3)_2$. Endoglucanase and xylanase were measured using the substrates β-glucan and xylan beechwood, respectively. Determination of reducing sugars released used the DNS method [12] and glucose and xylose as controls for activity determination of endoglucanase and xylanase respectively. Cellobiohydrolases and β-D-glucosidases were determined by the cleavage of ρ-nitrophenyl-cellobioside (ρNPC) and ρ-nitrophenyl-β-D-glucopyranoside (ρNPG), respectively. ρ-Nitrophenol was used as standard. Laccase activity was determined using syringaldazine as substrate [18]. The oxidation of syringaldazine to quinone at room temperature was measured by the increase in the absorbance at 525 nm during 5-min of reaction. All assays used 50 mM sodium citrate buffer pH 5.0 at 55 °C. One unit of enzymatic activity was defined as the amount of enzyme that released 1 μmol min^{-1} of products. All experiments were performed in triplicate.

4.5. Chemical Image Analysis

Sugarcane bagasse was surfaced-analyzed using an ION-TOF TOF.SIMS 5 instrument at ION-TOF-TasconGmbh (Heisenbergstr, Münster, Germany) using Bi_3^+ as primary ion for the analysis of organic and inorganic materials. The primary ion energy was 30 keV, analysis current of 0.8 pA, analysis area from 25×25 μm^2 to 100×100 μm^2 and measurement time of 100 s. The measurement conditions used positive mode, suitable for metal ions and non-metallic salts and compounds containing amino groups, and the negative mode, used for ionization of carbohydrates (loss 1 H$^+$ or more protons). ION-TOF TOF.SIMS 5 instrument (ION-TOF GmbH, Münster, Germany) was controlled by the SurfaceLab software suite. This software was used for data acquisition and analysis using the included spectrum library for sample identification.

After a sum of the three-color channel images into a single channel, TIFF image files were converted to the standard 8-bit gray-scale file format using the ImageJ software [19]. Before analysis, each image for each ion and anatomical structure were standardized using the threshold method included in the ImageJ software. Standardization employed the IJ_IsoData algorithm and produced two-dimensional maps in red and black colors.

ImageJ software was used to count both, ionic spots numbers and intensities using densitometric analysis. The area recovered with ions, the diameter of ionic aggregates and the color intensities of ionic aggregates and areas covered with ions were also measured using the standardized images and the ImageJ software.

For anatomical analysis, the Gwyddion software [20] was used to measure the follow parameters: texture, roughness and waviness, the diameter of fibrils and fibril aggregates in lattice structures, and the diameter of excavations caused by microwave:acid treatments.

All anatomical and ionic data were summarized in matrices and analyzed using discriminant analysis.

4.6. Statistical Analysis

Data automatically collected using ImageJ and Gwyddion software were used to generate a numerical matrix for each treatment. Matrices summarizing ionic parameters at the surface of substrates and anatomical data included the ionic composition measured using ToF-SIMS and the relative concentration of each ion, ionic aggregate counts, total surface area occupied by ionic aggregates, the average area of ionic aggregates for each target ion, and the values measured for texture, roughness and waviness, the diameter of fibrils and fibril arrangements in lattice structures, and the diameter of punctuated excavations. Discriminant analysis with machine learning using R 3.3.1 [21] was employed to compare each matrix summarizing the ionic and anatomical parameters produced by each treatment. The discriminating analysis used the Mixture Discriminant Analysis method from the MDA [22] package at $p = 0.05$. This method is suitable to deal with difficult data sets. The discriminant analysis of the parameters in each pretreatment was made using function training, considering the levels = pretreatments. MDA greatly succeeded during the discrimination of all pretreatments because the non-normal distribution of data. The analysis produced a string of pretreatment names s when neighbor pretreatments were more similar than distant ones. A matrix was produced using the strings by assigning 1 to each neighbor pretreatment name in the resulting string and 0 to no neighbor. These binary matrices were used to generate a tree clustering pretreatment using R 3.3.1 and the tree package [21]. In order to estimate the best parameter to discriminate each pretreatment, data were jackknifed, which allowed the leaving of one column out at a time, and the discriminant analysis of the matrix was re-run. The parameter used to discriminate the major percentage of pretreatments in a string produced by matrix discriminant analysis was considered the best discrimination parameter to be used in the process.

Comparison among curves of enzymatic saccharification used the least-squares method implemented in R 3.3.1 [21]. All comparisons among time-dependent enzymatic saccharification were done at $p = 0.05$.

5. Conclusions

Anatomical parameters were not related to saccharification yields of sugarcane bagasse. However, a strong factor affecting the performance of the enzyme cocktail was lithium coordinated with the substrate. So, it can be concluded that Lithium was an activator in solution, but its presence and distribution pattern on the substrate can act as an inhibitor. These results pointed to a phase-dependent action for alkaline metal ions on enzyme activity.

Control and substrates produced by EtOH:DMSO:AO pretreatment also presented a time-dependent decay of reducing sugars in the solution possibly related to ion-dependent precipitation or reverse reactions. The cause for that phenomenon remains hindered by a very complex mixture produced during the saccharification process. Once all materials analyzed presented very similar ionic composition, but differed in relative concentration and distribution of lithium, it was necessary to focus the analysis and absolute quantification of ions and ionic rates related to these enzyme responses. It can be surmised that there is a complex causal mechanism controlled by phase-transfer of enzymes and ions affecting cooperative, competitive and deinhibitory processes, anchimeric assistance, and [ion]-dependent induced-fit behavior of glycosyl-hydrolases. So, it is necessary to develop a new integrative kinetics approach to deal with all these aspects related to ion-polysaccharide hydrolysis.

It is likely that these pretreatments will not generate such predicted structural responses as revealed by specific surface area data (Control: 5751.60 μm^2/ng; Steam explosion: 4375.16 μm^2/ng; Microwave:H_2SO_4: 6544.19 μm^2/ng; EtOH:DMSO:AO: 5235.62 μm^2/ng; NaOH: 5347.32 μm^2/ng). This problem prevented any expected correlation between pretreatment and saccharification response, but opened new possibilities for screening new characteristics. In this case, statistical indications of the importance of lithium distribution on enzymatic activity during the screening for correlations were found. It is not aware of investigations into the influence of cations adsorbed on the substrate. It only knows the effect of cations in solution. It is expected that these novelties open previously unexplored paths for the study of the influence of complexed or adsorbed cations on enzymatic activity assays, contributing to the development of more efficient enzymatic cocktails.

Author Contributions: A.S.d.A.S. conducted Ethanol:Dimethyl Sulfoxide:Ammonium Oxalate (EtOH:DMSO:AO) and NaOH pretreatments, saccharification and collaborated at redaction. A.F.S. analyzed, interpreted data and collaborated at redaction. C.d.C.C.N. and E.G. conducted the Microwave:H_2SO_4 pretreatment. A.C.V. contributed to design the study and revised the manuscript. M.S.B. and M.d.L.T.d.M.P. contributed to design the study, analyzed data and revised the manuscript. All authors read and approved the final manuscript.

Funding: This work was supported by the National Institute of Science and Technology of Bioetanol (Fundação de Amparo à Pesquisa do Estado de São Paulo, FAPESP, process n° 2008/57908-6 and 2014/50884-5) and Conselho de Desenvolvimento Científico e Tecnológico (CNPq, process 465319/2014-9).

Acknowledgments: Coordenação de Aperfeiçoamento de Pessoal de Nível Superior - Brasil (CAPES – Finance Code 001); Fundação de Amparo à Pesquisa do Estado de São Paulo (FAPESP processes 2010/12624-0; 2013/23385-5).Figures corresponding to 1A, B were obtained from TOF.SIMS 5 - 100 instrument operation. We thank Dr. Reinhard Kersting and Dr. Adam Sears at TASCON GmbH, Münster, Germany, and Fellipy Ferreira, at dp UNION Instrumentação Analítica e Científica Ltd.a., São Paulo, Brazil for the efforts. We thank Abilio Borghi for the technical assistance with the English language and Ricardo Alarcon, Mariana Cereia and Mauricio de Oliveira for technical assistance.

Conflicts of Interest: The authors declare no conflict of interest.

References

1. UNICA. Avaliação Quinzenal da Safra 2018/2019 da Região Centro-Sul. 2018. Available online: http://www.unica.com.br/documentos/documentos/ (accessed on 15 December 2018).

2. Rossetto, R.; Santiago, A.D. Cana-De-Açúcar. Available online: http://www.agencia.cnptia.embrapa.br/gestor/cana-de-acucar/arvore/CONTAG01_1_711200516715.html (accessed on 15 December 2018).

3. Mohanram, S.; Amat, D.; Choudhary, J.; Arora, A.; Nain, L. Novel perspectives for evolving enzyme cocktails for lignocellulose hydrolysis in biorefineries. *Sustan. Chem. Process.* **2013**, *1*, 1–15. [CrossRef]

4. Goldbeck, R.; Damásio, A.R.L.; Gonçalves, T.A.; Machado, C.B.; Paixão, D.A.A.; Wolf, L.D.; Mandelli, F.; Rocha, G.J.M.; Ruller, R.; Squina, F.M. Development of hemicellulolytic enzyme mixtures for plant biomass deconstruction on target biotechnological applications. *Appl. Microbiol. Biotechnol.* **2014**, *98*, 8513–8525. [CrossRef] [PubMed]

5. Buckeridge, M.S.; Grandis, A.; Tavares, E.Q. Disassembling the Glycomic Code of Sugarcane Cell Walls to Improve Second-Generation Bioethanol Production. In *Bioethanol Production from Food Crops*; Elsevier BV: Amsterdam, The Netherlands, 2019; pp. 31–43.

6. Van Dyk, J.; Pletschke, B.; Pletschke, B. A review of lignocellulose bioconversion using enzymatic hydrolysis and synergistic cooperation between enzymes—Factors affecting enzymes, conversion and synergy. *Biotechnol. Adv.* **2012**, *30*, 1458–1480. [CrossRef] [PubMed]

7. Mou, H.Y.; Wu, S.; Fardim, P. Applications of ToF-SIMS in surface chemistry analysis of lignocellulosic biomass: A review. *BioResources* **2016**, *11*, 5581–5599.

8. Purich, D.L. *Enzyme Kinetics: Catalysis & Control: A reference of theory and best-practice methods*, 1st ed.; Elsevier: London, UK, 2010; pp. 1–920.

9. Segel, I.H. *Enzyme Kinetics: Behavior and Analysis of Rapid Equilibrium and Steady-State Enzyme Systems*, 1st ed.; John Wiley & Sons INC: New York, NY, USA, 1993; pp. 1–957.

10. Shoseyov, O.; Shani, Z.; Levy, I. Carbohydrate Binding Modules: Biochemical Properties and Novel Applications. *Microbiol. Mol. Boil. Rev.* **2006**, *70*, 283–295. [CrossRef] [PubMed]

11. De Vries, R.P.; Visser, J. Aspergillus Enzymes Involved in Degradation of Plant Cell Wall Polysaccharides. *Microbiol. Mol. Boil. Rev.* **2001**, *65*, 497–522. [CrossRef] [PubMed]

12. Miller, G.L. Use of Dinitrosalicylic Acid Reagent for Determination of Reducing Sugar. *Anal. Chem.* **1959**, *31*, 426–428. [CrossRef]

13. Moretti, M.M.D.S.; Bocchini-Martins, D.A.; Nunes, C.D.C.C.; Villena, M.A.; Perrone, O.M.; Da Silva, R.; Boscolo, M.; Gomes, E. Pretreatment of sugarcane bagasse with microwaves irradiation and its effects on the structure and on enzymatic hydrolysis. *Appl. Energy* **2014**, *122*, 189–195. [CrossRef]

14. Lima, M.S.; Damasio, A.R.D.L.; Crnkovic, P.M.; Pinto, M.R.; Da Silva, A.M.; Da Silva, J.C.R.; Segato, F.; De Lucas, R.C.; Jorge, J.A.; Polizeli, M.D.L.T.D.M. Co-cultivation of Aspergillus nidulans Recombinant Strains Produces an Enzymatic Cocktail as Alternative to Alkaline Sugarcane Bagasse Pretreatment. *Front. Microbiol.* **2016**, *7*, 583. [CrossRef] [PubMed]

15. Ribeiro, L.F.; De Lucas, R.C.; Vitcosque, G.L.; Ribeiro, L.F.; Ward, R.J.; Rubio, M.V.; Damásio, A.R.; Squina, F.M.; Gregory, R.C.; Walton, P.H.; et al. A novel thermostable xylanase GH10 from Malbranchea pulchella expressed in Aspergillus nidulans with potential applications in biotechnology. *Biotechnol. Biofuels* **2014**, *7*, 115. [CrossRef] [PubMed]

16. Segato, F.; Berto, G.L.; De Araújo, E.A.; Muniz, J.R.; Polikarpov, I. Expression, purification, crystallization and preliminary X-ray diffraction analysis of Aspergillus terreus endo-β-1,4-glucanase from glycoside hydrolase family 12. *Acta Crystallogr. Sect. F Struct. Boil. Commun.* **2014**, *70*, 267–270. [CrossRef] [PubMed]

17. Segato, F.; Damasio, A.R.L.; Gonçalves, T.A.; Murakami, M.T.; Squina, F.M.; Polizeli, M.; Mort, A.J.; Prade, R.A. Two structurally discrete GH7-cellobiohydrolases compete for the same cellulosic substrate fiber. *Biotechnol. Biofuels* **2012**, *5*, 21. [CrossRef] [PubMed]

18. Szklarz, G.D.; Antibus, R.K.; Sinsabaugh, R.L.; Linkins, A.E. Production of Phenol Oxidases and Peroxidases by Wood-Rotting Fungi. *Mycologia* **1989**, *81*, 234–240. [CrossRef]

19. Schneider, C.A.; Rasband, W.S.; Eliceiri, K.W. NIH Image to ImageJ: 25 years of Image Analysis. *Nat. Methods* **2012**, *9*, 671–675. [CrossRef] [PubMed]

20. Nečas, D.; Klapetek, P. Gwyddion: An open-source software for SPM data analysis. *Open Phys.* **2012**, *10*, 181–188. [CrossRef]

21. RC Team. *R: A Language and Environment for Statistical Computing*; R Foundation for Statistical Computing: Vienna, Austria, 2008; Available online: https://www.R-project.org/ (accessed on 15 November 2018).

22. Leisch, F.; Hornik, K.; Ripley, B.D. *mda: Mixture and Flexible Discriminant Analysis*; R Package Version 0.4-9; Trevor Hastie & Robert Tibshirani: Stanford, CA, USA, 2016.

Sample Availability: Not available.

Article

Steam Explosion Conditions Highly Influence the Biogas Yield of Rice Straw

David Steinbach [1,2,*], Dominik Wüst [1], Simon Zielonka [3], Johannes Krümpel [3], Simon Munder [1], Matthias Pagel [2] and Andrea Kruse [1]

[1] Institute of Agricultural Engineering, University of Hohenheim, Garbenstrasse 9, 70599 Stuttgart, Germany; wuest.dominik@uni-hohenheim.de (D.W.); S_Munder@uni-hohenheim.de (S.M.); Andrea_Kruse@uni-hohenheim.de (A.K.)

[2] Karlsruhe Institute of Technology (KIT), Institute for Catalysis Research and Technology, Hermann-von-Helmholtz-Platz 1, 76344 Eggenstein-Leopoldshafen, Germany; matthias.pagel@kit.edu

[3] State Institute of Agricultural Engineering and Bioenergy, University of Hohenheim, Garbenstrasse 9, 70599 Stuttgart, Germany; simon.zielonka@uni-hohenheim.de (S.Z.); j.kruempel@uni-hohenheim.de (J.K.)

* Correspondence: David.Steinbach@kit.edu; Tel.: +49-7121-989724

Academic Editors: Ivet Ferrer, Cigdem Eskicioglu, Georgia Antonopoulou and Audrey Battimelli
Received: 1 September 2019; Accepted: 24 September 2019; Published: 26 September 2019

Abstract: Straws are agricultural residues that can be used to produce biomethane by anaerobic digestion. The methane yield of rice straw is lower than other straws. Steam explosion was investigated as a pretreatment to increase methane production. Pretreatment conditions with varying reaction times (12–30 min) and maximum temperatures (162–240 °C) were applied. The pretreated material was characterized for its composition and thermal and morphological properties. When the steam explosion was performed with a moderate severity parameter of $S_0 = 4.1$ min, the methane yield was increased by 32% compared to untreated rice straw. This study shows that a harsher pretreatment at $S_0 > 4.3$ min causes a drastic reduction of methane yield because inert condensation products are formed from hemicelluloses.

Keywords: steam treatment; pretreatment; lignocellulose; anaerobic digestion; biochemical methane potential; biomethane

---

1. Introduction

Rice straw is one of the most abundant lignocellulosic agricultural residues worldwide and is produced mostly in Asia as a byproduct of rice production. Rice was in the third place of crop production in 2013 with a world annual production of 746 million tons [1]. The production of rice straw as a byproduct can be estimated at about 1120 million tons using a straw-to-grain ratio of 1.5 [2]. A part of this agricultural residue is used, for example, as cattle feed [3]. Unfortunately, open-field burning of straw, which increases air pollution, is a common practice in Asia [4–6].

Possible energetic utilization of rice straw is limited by its low bulk density, which makes large-scale, centralized conversion technologies uneconomical. Therefore, decentralized conversion routs are of special interest. One such method is anaerobic digestion of lignocellulosic biomass, which is a most efficient conversion technology regarding the energy output-to-input ratio [7–9].

However, the structure of lignocellulosic biomass generally causes a low digestibility during anaerobic digestion because lignocellulose is a strongly connected composite of cellulose, hemicelluloses, and lignin. Cellulose is a linear polymer consisting of glucose building blocks linked together by glycosidic bonds. The intermolecular hydrogen bonds between adjacent cellulose chains result in a highly ordered, water-insoluble configuration that makes cellulose crystalline [10]. Hemicelluloses are a group of amorphous heteropolymers that are significantly shorter than cellulose macromolecules.

They are fixed in the lignocellulosic fiber structure because they provide a linkage between lignin and cellulose. Hemicelluloses are connected to lignin via covalent links and surround cellulose. Lignin is a complex, three-dimensional macromolecule constructed of phenylpropane units and is non-biodegradable via anaerobic digestion [8].

Hydrolysis is the rate-limiting step in the biogas process when a solid feedstock, such as lignocellulose, is used [8]. A pretreatment is required to increase the availability of cellulose and hemicellulose for the hydrolysis step. Pretreatment generally aims at the disintegration and separation of biomass to release the different components [11]. A variety of pretreatment methods have been under investigation for enhancing the biogas production of lignocellulose [9,12,13]. Among these methods, steam pretreatment was rated with a high potential [12]. It is a relatively inexpensive pretreatment as it does not require the addition of an external catalyst [14]. Also, it has energetic advantages because it can be carried out with approximately 1.5 kg of steam per kilogram of biomass, compared to 5–10 kg of hot water usually required for liquid phase pretreatments [15].

Steam explosion converts biomass at elevated pressures and temperatures in a steam atmosphere, followed by mechanical disruption of the biomass by discharging to atmospheric pressure. Steam hydrolysis is similar to the steam explosion process, except that it avoids the discharge of rapid pressure. Temperatures ranging from 140–240 °C have been applied over a wide range of residence times [16]. Ferreira et al. [17] named an optimum for steam explosion conditions of 150–220 °C and 5–20 min. They pointed out that pretreatment conditions that are too severe are unfavorable due to the formation of phenolic and heterocyclic compounds (e.g., furfural, hydroxymethylfurfural (HMF), and soluble phenolic compounds). These compounds could inhibit methane production, but the methane-producing microorganisms are, however, capable of adapting to such compounds at smaller concentrations [12].

Thermally labile acetyl groups in hemicelluloses are cleaved during steam explosion, and acetic acid is formed [11]. The liberated organic acids catalyze hydrolysis reactions of hemicelluloses and degradation reactions of a small part of lignin. The residue after steam explosion consists of cellulose, a chemically modified lignin, and residual hemicelluloses. The sum of hemicelluloses in the residue, as well as dissolved hemicellulose-derived sugars, declines with the severity of the pretreatment because of (1) furfural formation by dehydration of pentoses and (2) secondary reactions of dissolved compounds leading to solid pseudolignin by condensation reactions [18].

Steam explosion has been performed with various kinds of lignocellulosic biomasses prior to anaerobic digestion including triticale [19], corn [20], wheat straw [17,21–26], hay [27], sugarcane bagasse [28,29], sugarcane straw [29], rape straw [30], bulrush [31,32], miscanthus [33], birch [34], willow [35,36], and cedar [37]. Steam explosion of rice straw for anaerobic digestion was investigated in the study of Zhou et al. [38] who performed steam explosion under narrower parameter conditions (200–220 °C, 1–4 min) and focused on the microbial communities during anaerobic digestion. Other studies have focused on steam explosion of rice straw as pretreatment for enzymatic hydrolysis to obtain fermentable sugars [39–41].

The steam explosion pretreatment generally increases the specific methane yields while also increasing the speed of anaerobic degradation. However, the reported increase in methane yield ranged from 0%–5% [22,23] to 10%–30% [17,19–21,24,26–30,32] and up to 50%–345% [29,33,34,36–38] because of the differences in (1) steam explosion reactor setups and reaction conditions, (2) lignocellulosic plants species and (3) digestion procedures.

The aim of this work was to evaluate the influence of a previous steam explosion in a batch system on the anaerobic digestion of rice straw. Therefore, the steam explosion of rice straw was performed at different severities. The pretreated material was chemically characterized and, thereafter, subjected to batch anaerobic digestion tests to determine the methane yield.

2. Materials and Methods

2.1. Plant Material

Field-dried rice straw of the *Bahia* variety (*Oryza sativa* var. *Bahia*) was acquired from the Ebro Delta, Spain. The straw was cut in a Viking GE260 chaff cutter (Viking GmbH, Kufstein, Austria) to a length of less than 100 mm. Fines were removed by manual sifting using a perforated sieve with a hole diameter of 3.9 mm. The composition of the plant material is shown in Table 1. Elemental composition and ash content were determined according to the procedures listed in Section 2.3. Fiber analysis was performed via the Van Soest method.

Table 1. Composition of the rice straw based on dry matter. (If standard deviation is indicated, four repetitions were performed.)

Ash Content after 550 °C	Ash Content after 1000 °C	Neutral Detergent Fiber	Acid Detergent Fiber (wt.%)	C	H	N	S	O	Si	K	Ca	Mg	Na
12.0 ± 0.1	10.7 ± 0.2	70.8 ± 2.8	46.2 ± 1.7	42.2	6.0	0.8	0.3	38.7	3.6	1.4	0.5	0.3	0.1

2.2. Steam Explosion Pretreatment

The steam explosion was performed in a miniplant designed and constructed at the Karlsruhe Institute of Technology. The reaction vessel of the miniplant consisted of a stainless-steel reactor with a volume of 1 L and was constantly agitated with a cross-arm stirrer at 8 min^{-1}. Before each experiment, 26.8 ± 1.2 g of straw was inserted into the reactor. The reactor was electrically heated to a temperature of 110 °C. A constant flow of 5 g min^{-1} of steam was then introduced into the reactor. The explosion is technically a rapid equalization of pressure performed by pneumatically opening a ball valve. Volatile products and steam were discharged by the explosion step into a 240 L flash tank. The pretreated rice straw remained in the reactor and was manually removed, and stored at 4 °C. Figure 1 provides a piping and instrumentation diagram (P&ID) of the miniplant.

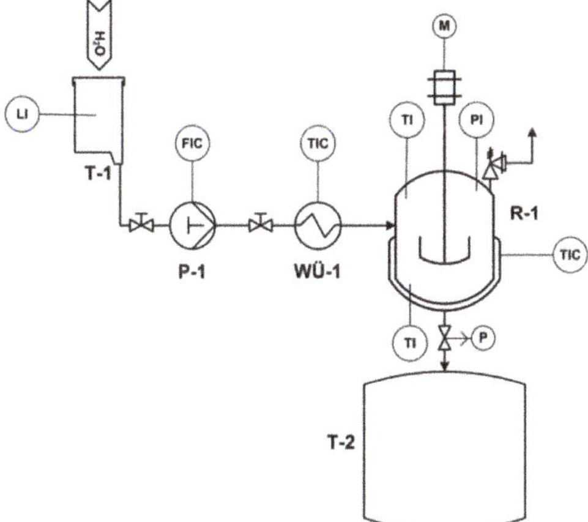

Figure 1. P&ID of the steam explosion miniplant. T-1: deionized water tank, P-1: HPLC pump, WÜ-1: electrical preheater for steam generation, R-1: steam explosion reactor with surrounding heating tape and insulation, and T-2: flash tank.

The reaction temperature was monitored by two type K thermocouples installed in the top and the bottom of the reactor. The arithmetic mean of the two temperatures measured was used for the calculation of the severity parameter S_0, combining time t and temperature T of the steam pretreatment in a single factor [18]. Equation 1 describes the time integral of S_0, which was used to describe the non-isothermal character of the heating process [42].

$$S_0 = log \int_0^t exp\left(\frac{T\,[°C] - 100\,°C}{14.75}\right) dt \tag{1}$$

Different pretreatment conditions with varying steam input times (12–30 min), steam input masses (60–150 g), and maximum temperatures (162–240 °C) were tested. Table 2 provides an overview of the experimental setup. The temperature profile during heat up is shown in the Supplementary Materials.

Table 2. Experimental setup for steam explosion of 27 g chipped rice straw.

Severity Parameter S_0 (min)	3.05	3.54	4.10	4.20	4.32	4.48	5.29
Steam input time (min)	30	30	30	16	12	30	32
Steam input mass (g)	150	150	150	78	60	150	60
Maximum reactor temperature (°C) [1]	162	174	206	222	229	222	240
Maximum pressure (bar) [1]	6.5	9.0	18.5	25.0	29.0	26.0	26.5

[1] before explosion step (°C)

2.3. Chemical Analysis of Untreated Rice Straw and Steam-Exploded Residue

The dry matter content of the untreated material was determined in triplicate by drying the rice straw at 105 °C for 16 h based on DIN EN 14774-1. The steam-exploded residue was dried at 40 °C for 87 h until its weight balanced to reduce evaporation of easily volatile reaction products of the steam explosion. The ash content of the untreated material was determined in triplicate via incineration in an electrically heated muffle oven at 550 °C for 4 h based on DIN EN 14775. The ash content of steam-exploded residues was measured in single runs. The organic dry matter content, which is also called volatile solids, was calculated from the difference between dry matter and ash content.

The stoichiometric carbon, hydrogen, nitrogen, and sulfur content (CHNS) was analyzed chromatically in a Vario EL cube (Elementar Analysesysteme GmbH, Langenselbold, Germany). The milled and dried samples of untreated straw and steam-exploded residue were analyzed in triplicate for CHNS. Sample milling to powder was performed in a freezer mill. The oxygen content of rice straw was estimated, closing the gap between CHNS and ash content to 100 wt.%. Other elements in the rice straw were analyzed after quantitative dissolution (HNO_3, HCl, HF 6:2:1 v v^{-1}) by ICP-OES type 725 (Agilent, Santa Clara, CA, USA).

Thermogravimetric analysis was performed with milled and dried samples in single runs at a STA 449 F5 (Netzsch-Gerätebau GmbH, Selb, Germany) with a heating rate of 10 K min^{-1} and nitrogen as the inerting agent. The thermogram obtained was differentiated to obtain the differential thermogravimetry curve (DTG).

The surface structures of dried untreated straw and steam-exploded residue were investigated via scanning electron microscopy (SEM) in a LEO 982 Gemini (Carl Zeiss AG, Jena, Germany) equipped with a Schottky-type thermal field emission cathode, secondary electron detectors (Everhart-Thornley, inlens), and a backscattered electron detector.

The acid-insoluble lignin content (Klason lignin) and acid-insoluble ash content were determined via an "ASTM protocol" [43] in triplicate for the untreated straw and in single runs for the steam-exploded residues.

Water-soluble components of rice straw and steam-exploded residues were extracted with hot water. Therefore, about 0.9 g milled and dried sample was introduced in an extraction thimble, which was placed in a boiling water bath for 3 h. Water extraction was performed in triplicate for untreated straw and in single runs for steam-exploded residues. After extraction, the liquid was analyzed via HPLC (Deutsche Metrohm GmbH, Filderstadt, Germany). Glucose and xylose were separated via HPLC at 35 °C in a Metrosep Carb 2 column (Deutsche Metrohm GmbH, Filderstadt, Germany) and quantified by an amperometric detector. An eluent with 0.1 mol L^{-1} sodium hydroxide and 0.01 mol L^{-1} sodium acetate was used with a flow rate of 0.5 mL min^{-1}. Hydroxymethylfurfural (HMF) and furfural were separated via HPLC at 20 °C using a Lichrospher 100 RP-18 column (Merck KGaA, Darmstadt, Germany) and quantified by a UV detector at 290 nm. A water–acetonitrile eluent (9:1 v v^{-1}) was used at a flow rate of 1.4 mL min^{-1}.

2.4. Specific Biogas Yield

The Hohenheim biogas yield test was used to assess the biochemical methane potential (BMP) of untreated and pretreated rice straw. The pretreated rice straw after steam explosion was directly used for the biogas yield test without any water washing. The Hohenheim biogas yield test is a feasible and well-established laboratory batch test developed at the University of Hohenheim and featured in the VDI-Guideline 4630-Digestion of organic materials [44]. It is used to evaluate and compare the methane production of different substrates as well as their biodegradability and gas production kinetics. Glass syringes (100 mL) were used as digesters and for gas storage. The syringes were fitted into a motor-driven rotor for mixing the sample, which was placed inside an incubator. Every run included a control variant and two standard substrates to ensure the correctness of the results and the comparability of different batches [45,46]. The methane percentage was measured by a gas transducer AGM 10 (Pronova Analysetechnik, Berlin, Germany) with a nondispersive infrared (NDIR) sensor, while the amount of produced biogas was recorded with an accuracy of 1 mL. Three replications were performed for each sample. The gas yield is expressed as m^3$_N$ kg $_{DM}$$^{-1}$, corrected for the gas production of the inoculum, and expressed for standard atmosphere (273.15 K, 1013.25 hPa).

2.5. Theoretical Methane and Biogas Yield

The theoretical maximum gas production from an organic substrate with the elemental composition $C_aH_bO_cN_dS_e$ can be determined stoichiometrically, see Equation (2) [47].

$$C_aH_bO_cN_dS_e + \left(a - \tfrac{b}{4} - \tfrac{c}{2} + \tfrac{3d}{4} + \tfrac{e}{2}\right)\cdot H_2O \rightarrow \left(\tfrac{a}{2} + \tfrac{b}{8} - \tfrac{c}{4} - \tfrac{3d}{8} - \tfrac{e}{4}\right)\cdot CH_4 + \left(\tfrac{a}{2} - \tfrac{b}{8} + \tfrac{c}{4} + \tfrac{3d}{8} + \tfrac{e}{4}\right)\cdot CO_2 + d\cdot NH_3 + e\cdot H_2S . \tag{2}$$

The maximal methane yield (Y_{CH4}), expressed as norm cubic meters per kilogram of organic dry matter (m^3$_N$ kg $_{DM}$$^{-1}$), can be obtained by Equation (3) using the molar volume of an ideal gas (22.4 mol L^{-1}). However, this theoretic value can never be achieved with lignocellulosic biomass in reality because lignocellulose contains compounds that are considered not biodegradable, such as lignin.

$$Y_{CH_4} = \frac{22.4\cdot\left(\tfrac{a}{2} + \tfrac{b}{8} - \tfrac{c}{4} - \tfrac{3d}{8} - \tfrac{e}{4}\right)}{12a + b + 16c + 14d + 32e}. \tag{3}$$

3. Results and Discussion

3.1. Characterization of the Solid Residue after Steam Explosion

3.1.1. Chemical Composition

The field-dried rice straw had a dry matter content of 91.6 wt.%, which was generally reduced during steam explosion (Figure 2a). However, the dry matter content among the residues increased with severity. An explanation might be that the residue becomes more hydrophobic at higher reaction severity, because of a decrease in the number of polar functional groups. The mass of steam introduced into the reactor might also influence the dry matter content. The experiments with the lowest steam inputs, namely S_0 = 4.20, 4.32 and 5.29 min, yielded the highest dry matter contents of 65, 66 and 68 wt.%, respectively.

The acid-insoluble lignin of water-free rice straw accounted for 14.7 wt.% and increased to 28–42 wt.% after steam explosion (Figure 2b). This relative increase has two causes: (1) hemicelluloses, and partly cellulose, are converted during steam explosion to water-soluble or volatile compounds. Thereby, the content of the more inert lignin is increased in the solid residue. (2) Hemicelluloses decompose and repolymerize to form a more stable solid that cannot be hydrolyzed to water-soluble components during the Klason method [11]. These repolymerization or condensation reactions involve not only hemicellulose-derived byproducts but also lignin [16]. These repolymerized pseudolignin compounds are more stable and increase the acid-insoluble lignin content.

The acid-insoluble ash content at low severity parameters was similar to the untreated material and accounts for 4.0 wt.%. The ash content increased at high severity (Figure 2c). This can also be attributed to the conversion of hemicelluloses, and partly cellulose, during steam explosion to water-soluble or volatile compounds. Thereby, the content of inert acid-insoluble ash increased in the solid residue. This parameter could also be used to roughly estimate the weight loss of organic material during steam explosion. This weight loss of organic material should be considered when the cascading of steam explosion and biogas production is evaluated.

Figure 2d shows the elemental composition of solid steam-exploded residues. The carbon content in the solid residue rose at high severity. By contrast, the hydrogen content decreased. This can be explained by the removal of hydrogen- and oxygen-rich volatile compounds during the steam explosion; thus, a carbon-rich solid residue remained.

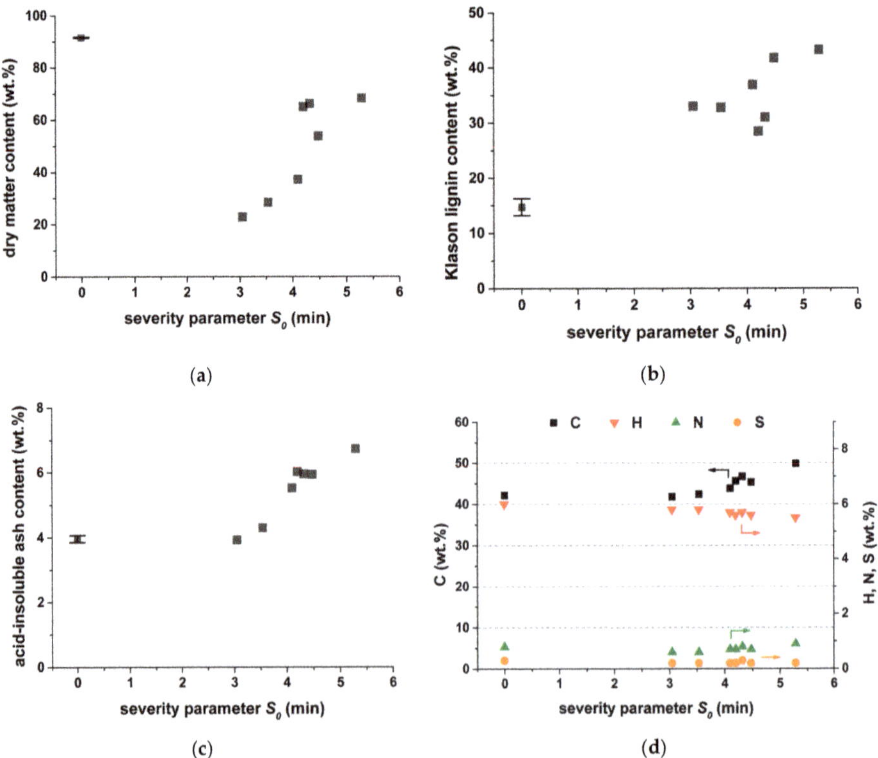

Figure 2. Chemical composition of untreated and pretreated rice straw depending on the severity parameter S_0 of the steam explosion, (**a**) dry matter content, (**b**) Klason lignin content, (**c**) acid-insoluble ash content, and (**d**) elemental composition.

3.1.2. Water-Extractable Components

After the hot water extraction of untreated rice straw, 80 mg g^{-1} glucose and 44 mg g^{-1} xylose were obtained in the liquid phase (Figure 3). Glucose was either present in the rice straw as free monosaccharide or originated from the degradation of polymers like starch or hemicelluloses [48]. In the latter case, the amorphous or rather labile hemicelluloses may partly hydrolyze during water extraction to yield glucose (but also xylose). No HMF was obtained after the water extraction of untreated rice straw; therefore, no dehydration of glucose occurred.

Compared to untreated rice straw, water-extractable xylose increased at a low severity of steam explosion. Xylose was no longer found in the extract at high severities ($S_0 > 4.1$ min). A possible explanation is that hemicelluloses are hydrolyzed to release additional xylose at low severities. At high severities, the released xylose is quickly decomposed to furfural [35]. Water-extractable HMF increases strongly with the severity factor. Thus, hexoses, such as glucose, are dehydrated at a higher severity to form HMF. The HMF is known as a fermentation inhibitor for anaerobic digestion [49]. However, reported values for inhibiting concentrations by HMF seem to be high, usually greater than 5 g L^{-1}. Moreover, the intensity of the inhibition is affected by the operational conditions and design of the anaerobic system, as well as the presence of other inhibitors, such as furfural or phenols [49,50]. Both inhibit the activity of acetate in utilizing methanogens.

Furfural was not detected after water-extraction of pretreated material. However, it was present in the gas phase of the steam-explosion reactor (data not shown). A sample of the gas phase was obtained via a relief valve, condensed afterwards, and analyzed for furfural. So, formed furfural was mainly removed as a volatile during steam explosion. A small part of furfural may still be present on the surface of the steam-exploded rice straw subjected to anaerobic digestion.

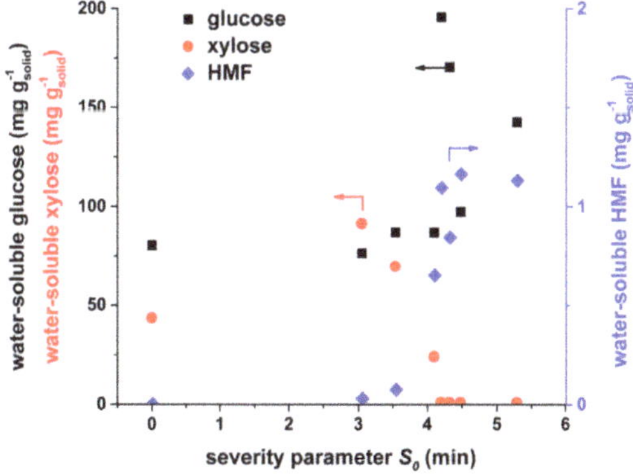

Figure 3. Water-soluble components (glucose, xylose, and hydroxymethylfurfural (HMF)) of untreated rice straw and pretreated residue depending on the severity parameter S_0 of the steam explosion. Extraction in boiling water for 3 h.

3.1.3. Thermogravimetric Analysis

The main components of lignocellulosic biomass differ in thermal stability. Hemicelluloses and cellulose consist of glycosidic linkages, which decompose within a narrow temperature region. The amorphous hemicelluloses decomposed at a lower temperature compared to the crystalline cellulose [51,52]. Lignin consists of phenylpropanoid units interconnected by different chemical linkages that have different binding energies [53]. Therefore, the decomposition of lignin proceeded over a broad temperature range.

Figure 4 shows the DTG curves of rice straw. The untreated material showed a peak at around 320 °C, which corresponds to cellulose. Hemicellulose showed a shoulder at 280–300 °C, which is still present for the lowest-severity experiments $S_0 = 3.05$ min and $S_0 = 3.54$ min. At a higher severity, the hemicellulose shoulder disappeared completely, indicating a destruction or alteration of hemicelluloses. An increased mass loss at 380–480 °C was detected for the high-severity experiments (S_0 = 4.49 min and S_0 = 5.29 min), which could be assigned to more temperature-stable repolymerization products of hemicellulose.

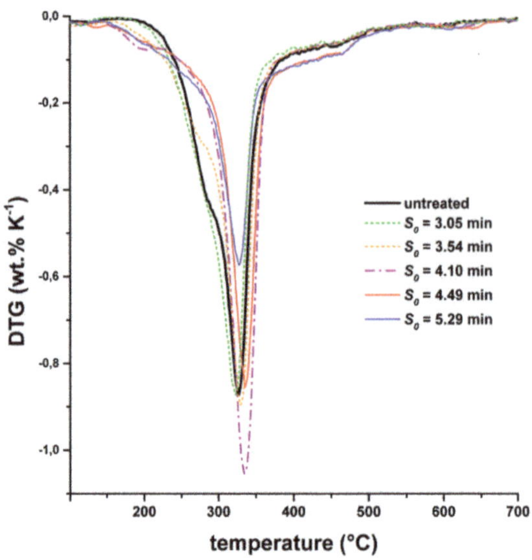

Figure 4. Differential thermogravimetric analysis (DTG) of untreated and pretreated rice straw depending on the severity parameter S_0 of the steam explosion.

3.1.4. Particle Morphology

The pretreated residue became darker and the particles more fragile with increasing severity. A difference in the biomass macrostructure was observed by SEM. After the steam explosion with a moderate severity parameter of S_0 = 4.10 min, a very porous structure was obtained (Figure 5b). However, at higher severities, no porous structure could be observed (Figure 5c). Photographs and SEM images of untreated rice straw and all steam-exploded residues are provided in Supplementary Materials. The surface structure for S_0 = 3.54 min indicated that parts of the solid melted and solidified later. Smaller fragments can be found, especially for high severities (e.g., S_0 = 4.20 min and S_0 = 4.48 min).

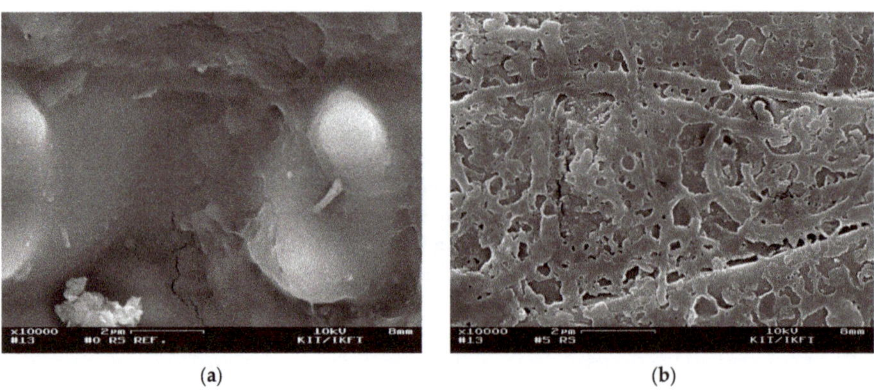

(a)   (b)

Figure 5. *Cont.*

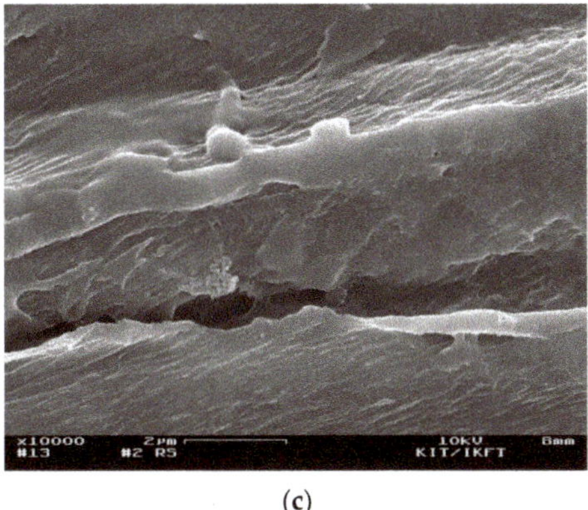

(**c**)

Figure 5. SEM images of (**a**) untreated rice straw and steam-exploded residue at a severity parameter of (**b**) S_0 = 4.10 min and (**c**) S_0 = 5.29 min.

3.2. Specific Biogas Yields

The theoretical maximum gas production of the untreated rice straw was calculated according to Equation (3). The elemental composition measured of rice straw $C_{1.000}H_{1.706}O_{0.688}N_{0.016}S_{0.003}$ resulted in a maximum methane yield of 0.478 m^3_N kg $_{DM}^{-1}$.

Figure 6 shows a comparison of the measured methane yield of rice straw with other common straws, such as wheat straw or maize straw, investigated in the Hohenheim biogas yield test. Rice straw showed a methane yield of 0.211 ± 0.009 m^3_N kg $_{DM}^{-1}$, which was lower compared to other common straws. The steam explosion variant of rice straw with the highest increase in methane yield S_0 = 4.10 min reached the level of untreated wheat straw.

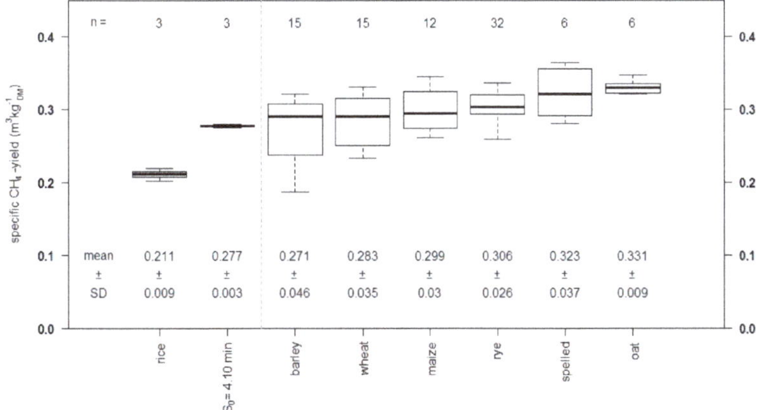

Figure 6. Specific methane yield of the untreated rice straw and the pretreated steam explosion variant S_0 = 4.10 min compared with other straws after 35 d (average yields of different samples regarding plant variety and particle size).

Figure 7 shows the specific methane yield of the steam exploded samples investigated over 35 d, which is also summarized in Table 3. The variants S_0 = 4.10 min and S_0 = 3.54 min reached a higher yield of 0.278 ± 0.003 m^3_N kg $_{DM}^{-1}$ (+32%) and 0.217 ± 0.007 m^3_N kg $_{DM}^{-1}$ (+3%) compared to untreated rice straw. These variants also showed faster gas production compared to the untreated straw. The other pretreatment conditions led to a reduction in specific methane yield of −76% ($S_0 =$ 5.29 min), −63% ($S_0 = 4.32$ min), and −10% ($S_0 = 3.05$ min).

When steam explosion was performed at a low severity ($S_0 = 3.05$ min and $S_0 = 3.54$ min), the methane yield was similar to the untreated rice straw. Thus, we concluded that the pretreatment conditions were not severe enough to open up the structure of the lignocellulosic biomass for anaerobic digestion (see SEM images in Supplementary Materials). On the other hand, high-severity parameters ($S_0 = 4.32$ min and $S_0 = 5.29$ min) caused a drastic decrease in the methane yield. An obvious increase in the methane yield was obtained only when the steam explosion pretreatment was performed at moderate severities ($S_0 = 4.10$ min). At this severity, the particle morphology had the most porous structure among all conditions investigated (Figure 5b). Therefore, the broken macromolecular structure as well as additional surface had a positive effect on further degradation during anaerobic digestion. Thermogravimetric analysis showed a change of hemicellulose at $S_0 = 4.10$ min, but there was no formation of temperature-stable repolymerization products. The latter could be an explanation for the drastic decrease in biogas yields of the high-severity experiments. We assumed that repolymerization products from hemicelluloses were poorly digestible in the biogas process because of their high resistance against (1) hydrolytic cleavage (increased Klason lignin content, Figure 3b) and (2) decomposition reactions (increased thermal stability shown by decomposition at high temperatures of 380–480 °C, Figure 4).

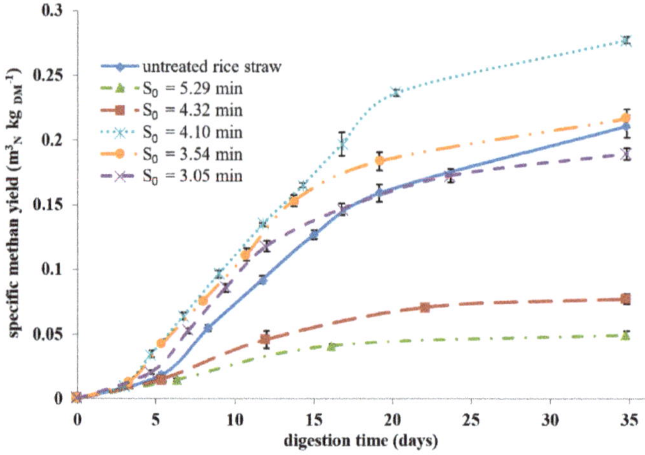

Figure 7. Cumulative specific methane production of rice straw and steam-exploded rice straw. Error bars indicate standard deviation for three repetitions.

Table 3. Biogas yield, methane content, and specific methane yield of the untreated rice straw and steam-exploded rice straw after 35 d (mean with standard deviation of three repetitions).

Severity Parameter	Biogas Yield (m^3_N kg $_{DM}$$^{-1}$)	CH$_4$ (vol.%)	Specific Methane Yield (m^3_N kg $_{DM}$$^{-1}$)
Untreated	0.368 ± 0.024	57.4 ± 1.5	0.211 ± 0.009
$S_0 = 3.05$ min	0.331 ± 0.009	57.3 ± 0.1	0.190 ± 0.005
$S_0 = 3.54$ min	0.393 ± 0.007	55.2 ± 0.9	0.217 ± 0.007
$S_0 = 4.10$ min	0.542 ± 0.011	51.3 ± 0.6	0.278 ± 0.003
$S_0 = 4.32$ min	0.128 ± 0.008	60.6 ± 2.0	0.078 ± 0.004
$S_0 = 5.29$ min	0.082 ± 0.002	60.8 ± 3.1	0.050 ± 0.003

4. Conclusions

Steam explosion of rice straw was performed as a pretreatment for anaerobic digestion to obtain a methane-rich biogas. Steam explosion was investigated at different reaction conditions, which resulted in severity parameters of S_0 = 3.05–5.29 min. The severity of steam explosion highly influences the methane yield:

- If the conditions of the steam explosion are too mild, the methane yield remains constant compared to untreated rice straw.
- If conditions are too severe, the methane yield drops dramatically. At these conditions, hemicelluloses are largely destroyed, and repolymerization leads to a more inert material.
- If conditions are moderate, the methane yield is increased, caused by a very porous structure and altered hemicellulose.

For process upscaling, the increase in methane yield at optimal steam explosion conditions has to be economically balanced with the mass loss of solid material during pretreatment and the effort of pretreatment itself.

Supplementary Materials: The following are available online at http://www.mdpi.com/1420-3049/24/19/3492/s1, Figure S1. Temperature profile of steam explosion experiments that lead to different severity parameters S_0; Table S1. Photographs and SEM images of untreated rice straw and steam-exploded residue.

Author Contributions: D.S., D.W., S.Z., and S.M. designed the experiments; D.S. and M.P. performed pretreatment experiments; D.S., D.W., S.Z., and J.K. acquired and analyzed the data. D.S., D.W., S.Z., S.M., and A.K. wrote the article.

Funding: This research received no external funding.

Acknowledgments: We thank Alexander Schäfer for the experimental work at the steam explosion miniplant. Thomas Tietz is acknowledged for his mechanical support. We thank Armin Lautenbach and Sonja Habicht for HPLC support. Wilhelm Habicht performed the SEM analysis.

Conflicts of Interest: The authors declare no conflict of interest.

References

1. Food and Agriculture Organization. *FAO Statistical Pocketbook 2015*; Food and Agriculture Organization: Rome, Italy, 2015; p. 236.
2. Lal, R. World crop residues production and implications of its use as a biofuel. *Environ. Int.* **2005**, *31*, 575–584. [CrossRef] [PubMed]
3. Binod, P.; Sindhu, R.; Singhania, R.R.; Vikram, S.; Devi, L.; Nagalakshmi, S.; Kurien, N.; Sukumaran, R.K.; Pandey, A. Bioethanol production from rice straw: An overview. *Bioresour. Technol.* **2010**, *101*, 4767–4774. [CrossRef] [PubMed]
4. Cao, G.; Zhang, X.; Wang, Y.; Zheng, F. Estimation of emissions from field burning of crop straw in China. *Chin. Sci. Bull.* **2008**, *53*, 784–790. [CrossRef]
5. Engling, G.; Lee, J.J.; Tsai, Y.-W.; Lung, S.-C.C.; Chou, C.C.K.; Chan, C.-Y. Size-resolved anhydrosugar composition in smoke aerosol from controlled field burning of rice straw. *Aerosol Sci. Technol.* **2009**, *43*, 662–672. [CrossRef]
6. Gadde, B.; Bonnet, S.; Menke, C.; Garivait, S. Air pollutant emissions from rice straw open field burning in India, Thailand and the Philippines. *Environ. Pollut.* **2009**, *157*, 1554–1558. [CrossRef] [PubMed]
7. Chandra, R.; Takeuchi, H.; Hasegawa, T. Methane production from lignocellulosic agricultural crop wastes: A review in context to second generation of biofuel production. *Renew. Sustain. Energy Rev.* **2012**, *16*, 1462–1476. [CrossRef]
8. Frigon, J.C.; Guiot, S.R. Biomethane production from starch and lignocellulosic crops: a comparative review. *Biofuel Bioprod. Bior.* **2010**, *4*, 447–458. [CrossRef]
9. Zhong, Y.; Zhao, J.; Xu, F.; Li, Y. Pretreatment of lignocellulosic biomass for enhanced biogas production. *Prog. Energy Combust. Sci.* **2014**, *42*, 35–53. [CrossRef]

10. Bobleter, O. Hydrothermal degradation of polymers derived from plants. *Prog. Polym. Sci.* **1994**, *19*, 797–841. [CrossRef]

11. Steinbach, D.; Kruse, A.; Sauer, J. Pretreatment technologies of lignocellulosic biomass in water in view of furfural and 5-hydroxymethylfurfural production—A review. *Biomass Convers. Biorefinery* **2017**, 1–28. [CrossRef]

12. Hendriks, A.T.W.M.; Zeeman, G. Pretreatments to enhance the digestibility of lignocellulosic biomass. *Bioresour. Technol.* **2009**, *100*, 10–18. [CrossRef] [PubMed]

13. Taherzadeh, J.M.; Karimi, K. Pretreatment of lignocellulosic wastes to improve ethanol and biogas production: a review. *Int. J. Mol. Sci.* **2008**, *9*. [CrossRef] [PubMed]

14. Schwald, W.; Breuil, C.; Brownell, H.H.; Chan, M.; Saddler, J.N. Assessment of pretreatment conditions to obtain fast complete hydrolysis on high substrate concentrations. *Appl. Biochem. Biotech.* **1989**, *20–21*, 29–44. [CrossRef]

15. Zimbardi, F.; Viola, E.; Nanna, F.; Larocca, E.; Cardinale, M.; Barisano, D. Acid impregnation and steam explosion of corn stover in batch processes. *Ind. Crop. Prod.* **2007**, *26*, 195–206. [CrossRef]

16. Ramos, L.P. The chemistry involved in the steam treatment of lignocellulosic materials. *Quim. Nova* **2003**, *26*, 863–871. [CrossRef]

17. Ferreira, L.C.; Nilsen, P.J.; Fdz-Polanco, F.; Pérez-Elvira, S.I. Biomethane potential of wheat straw: Influence of particle size, water impregnation and thermal hydrolysis. *Chem. Eng. J.* **2014**, *242*, 254–259. [CrossRef]

18. Overend, R.P.; Chornet, E. Fractionation of lignocellulosics by steam-aqueous pretreatments. *Philos. T. R. Soc. A* **1987**, *321*, 523–536. [CrossRef]

19. Schumacher, B.; Oechsner, H.; Senn, T.; Jungbluth, T. Thermo-mechanical pre-treatment of ripe triticale for biogas production. *Landtechnik* **2007**, *62*, 162–163, 190.

20. Schumacher, B. *Untersuchungen zur Aufbereitung und Umwandlung von Energiepflanzen in Biogas und Bioethanol*; Universität Hohenheim: Stuttgart, Germany, 2008.

21. Bauer, A.; Bösch, P.; Friedl, A.; Amon, T. Analysis of methane potentials of steam-exploded wheat straw and estimation of energy yields of combined ethanol and methane production. *J. Biotechnol.* **2009**, *142*, 50–55. [CrossRef]

22. Theuretzbacher, F.; Lizasoain, J.; Lefever, C.; Saylor, M.K.; Enguidanos, R.; Weran, N.; Gronauer, A.; Bauer, A. Steam explosion pretreatment of wheat straw to improve methane yields: Investigation of the degradation kinetics of structural compounds during anaerobic digestion. *Bioresour. Technol.* **2015**, *179*, 299–305. [CrossRef]

23. Risberg, K.; Sun, L.; Levén, L.; Horn, S.J.; Schnürer, A. Biogas production from wheat straw and manure— Impact of pretreatment and process operating parameters. *Bioresour. Technol.* **2013**, *149*, 232–237. [CrossRef] [PubMed]

24. Sapci, Z.; Morken, J.; Linjordet, R. An investigation of the enhancement of biogas yields from lignocellulosic material using two pretreatment methods: microwave irradiation and steam explosion. *Bioresources* **2013**, *8*, 1976–1985. [CrossRef]

25. Lindorfer, J.; Steinmuller, H.; Jager, A.; Eder, A.; Hofer, B.; Nidetzky, B.; Loncar, E.; Auer, W. Research into the production of bioethanol and biogas from Lignocellulose raw materials after pretreatment with steam explosion and cellulases. *Chem.-Ing.-Tech.* **2010**, *82*, 1197–1202. [CrossRef]

26. Ferreira, L.C.; Donoso-Bravo, A.; Nilsen, P.J.; Fdz-Polanco, F.; Perez-Elvira, S.I. Influence of thermal pretreatment on the biochemical methane potential of wheat straw. *Bioresour. Technol.* **2013**, *143*, 251–257. [CrossRef] [PubMed]

27. Bauer, A.; Lizasoain, J.; Theuretzbacher, F.; Agger, J.W.; Rincón, M.; Menardo, S.; Saylor, M.K.; Enguídanos, R.; Nielsen, P.J.; Potthast, A.; et al. Steam explosion pretreatment for enhancing biogas production of late harvested hay. *Bioresour. Technol.* **2014**, *166*, 403–410. [CrossRef] [PubMed]

28. Vivekanand, V.; Olsen, E.F.; Eijsink, V.G.H.; Horn, S.J. Methane potential and enzymatic saccharification of steam-exploded bagasse. *Bioresources* **2014**, *9*, 1311–1324. [CrossRef]

29. De Paoli, F.; Bauer, A.; Leonhartsberger, C.; Amon, B.; Amon, T. Utilization of by-products from ethanol production as substrate for biogas production. *Bioresour. Technol.* **2011**, *102*, 6621–6624. [CrossRef] [PubMed]

30. Vivekanand, V.; Ryden, P.; Horn, S.J.; Tapp, H.S.; Wellner, N.; Eijsink, V.G.H.; Waldron, K.W. Impact of steam explosion on biogas production from rape straw in relation to changes in chemical composition. *Bioresour. Technol.* **2012**, *123*, 608–615. [CrossRef]

31. Yue, Z.B.; Liu, R.H.; Yu, H.Q.; Chen, H.Z.; Yu, B.; Harada, H.; Li, Y.Y. Enhanced anaerobic ruminal degradation of bulrush through steam explosion pretreatment. *Ind. Eng. Chem. Res.* **2008**, *47*, 5899–5905. [CrossRef]

32. Wang, J.; Yue, Z.-B.; Chen, T.-H.; Peng, S.-C.; Yu, H.-Q.; Chen, H.-Z. Anaerobic digestibility and fiber composition of bulrush in response to steam explosion. *Bioresour. Technol.* **2010**, *101*, 6610–6614. [CrossRef]

33. Menardo, S.; Bauer, A.; Theuretzbacher, F.; Piringer, G.; Nilsen, P.J.; Balsari, P.; Pavliska, O.; Amon, T. Biogas production from steam-exploded miscanthus and utilization of biogas energy and CO2 in greenhouses. *Bioenerg. Res.* **2013**, *6*, 620–630. [CrossRef]

34. Vivekanand, V.; Olsen, E.F.; Eijsink, V.G.H.; Horn, S.J. Effect of different steam explosion conditions on methane potential and enzymatic saccharification of birch. *Bioresour. Technol.* **2013**, *127*, 343–349. [CrossRef] [PubMed]

35. Horn, S.J.; Estevez, M.M.; Nielsen, H.K.; Linjordet, R.; Eijsink, V.G.H. Biogas production and saccharification of Salix pretreated at different steam explosion conditions. *Bioresour. Technol.* **2011**, *102*, 7932–7936. [CrossRef] [PubMed]

36. Estevez, M.M.; Linjordet, R.; Morken, J. Effects of steam explosion and co-digestion in the methane production from Salix by mesophilic batch assays. *Bioresour. Technol.* **2012**, *104*, 749–756. [CrossRef]

37. Take, H.; Andou, Y.; Nakamura, Y.; Kobayashi, H.; Kurimoto, Y.; Kuwahara, M. Production of methane gas from Japanese cedar chips pretreated by various delignification methods. *Biochem. Eng. J.* **2006**, *28*, 30–35. [CrossRef]

38. Zhou, J.; Yan, B.H.; Wang, Y.; Yong, X.Y.; Yang, Z.H.; Jia, H.H.; Jiang, M.; Wei, P. Effect of steam explosion pretreatment on the anaerobic digestion of rice straw. *Rsc. Adv.* **2016**, *6*, 88417–88425. [CrossRef]

39. Wood, I.P.; Cao, H.-G.; Tran, L.; Cook, N.; Ryden, P.; Wilson, D.R.; Moates, G.K.; Collins, S.R.A.; Elliston, A.; Waldron, K.W. Comparison of saccharification and fermentation of steam exploded rice straw and rice husk. *Biotechnol. Biofuels* **2016**, *9*, 193. [CrossRef]

40. Moniruzzaman, M. Effect of steam explosion on the physicochemical properties and enzymatic saccharification of rice straw. *Appl. Biochem. Biotech.* **1996**, *59*, 283–297. [CrossRef]

41. Taniguchi, M.; Takahashi, D.; Watanabe, D.; Sakai, K.; Hoshino, K.; Kouya, T.; Tanaka, T. Effect of steam explosion pretreatment on treatment with Pleurotus ostreatus for the enzymatic hydrolysis of rice straw. *J. Biosci. Bioeng.* **2010**, *110*, 449–452. [CrossRef]

42. Montané, D.; Overend, R.P.; Chornet, E. Kinetic models for non-homogeneous complex systems with a time-dependent rate constant. *Can. J. Chem. Eng.* **1998**, *76*, 58–68. [CrossRef]

43. ASTM Standard. *Standard test method for acid-insoluble lignin in wood*; ASTM Standard: West Conshohocken, PA, USA, 2013; Vol. D1106 −96.

44. VDI. *Fermentation of organic materials*; VDI-Gesellschaft Energietechnik: Düsseldorf, Germany, 2006; Volume 4630.

45. Helffrich, D.; Oechsner, H. Hohenheimer Biogasertragstest-Vergleich verschiedener Laborverfahren zur Vergärung von Biomasse. *Agrartech. Forsch.* **2003**, *9*, 27–30.

46. Mittweg, G.; Oechsner, H.; Hahn, V.; Lemmer, A.; Reinhardt-Hanisch, A. Repeatability of a laboratory batch method to determine the specific biogas and methane yields. *Eng. Life Sci.* **2012**, *12*, 270–278. [CrossRef]

47. Boyle, W.C. Energy recovers from sanitary landfills – a review. In *Microbial Energy Conversion*; Schlegel, H., Barnea, J., Eds.; Elsevier: Pergamon, Turkey, 1977; pp. 119–138.

48. Park, J.Y.; Seyama, T.; Shiroma, R.; Ike, M.; Srichuwong, S.; Nagata, K.; Arai-Sanoh, Y.; Kondo, M.; Tokuyasu, K. Efficient recovery of glucose and fructose via enzymatic saccharification of rice straw with soft carbohydrates. *Biosci. Biotech. Bioch.* **2009**, *73*, 1072–1077. [CrossRef] [PubMed]

49. Monlau, F.; Sambusiti, C.; Barakat, A.; Quéméneur, M.; Trably, E.; Steyer, J.P.; Carrère, H. Do furanic and phenolic compounds of lignocellulosic and algae biomass hydrolyzate inhibit anaerobic mixed cultures? A comprehensive review. *Biotechnol. Adv.* **2014**, *32*, 934–951. [CrossRef] [PubMed]

50. Levén, L.; Nyberg, K.; Schnürer, A. Conversion of phenols during anaerobic digestion of organic solid waste—A review of important microorganisms and impact of temperature. *J. Environ. Manag.* **2012**, *95*, S99–S103. [CrossRef] [PubMed]

51. Xiao, L.-P.; Sun, Z.-J.; Shi, Z.-J.; Xu, F.; Sun, R.-C. Impact of hot compressed water pretreatment on the structural changes of woody biomass for bioethanol production. *BioResources* **2011**, *6*, 1576–1598.
52. Chen, W.H.; Ye, S.C.; Sheen, H.K. Hydrolysis characteristics of sugarcane bagasse pretreated by dilute acid solution in a microwave irradiation environment. *Appl. Energ.* **2012**, *93*, 237–244. [CrossRef]
53. Zakzeski, J.; Bruijnincx, P.C.A.; Jongerius, A.L.; Weckhuysen, B.M. The catalytic valorization of lignin for the production of renewable chemicals. *Chem. Rev.* **2010**, *110*, 3552–3599. [CrossRef]

Article

Enhancement of Volatile Fatty Acids Production from Food Waste by Mature Compost Addition

Yen-Keong Cheah, Joan Dosta and Joan Mata-Álvarez

1   Department of Chemical Engineering and Analytical Chemistry, University of Barcelona,
     08028 Barcelona, Spain
2   Water Research Institute, University of Barcelona, 08001 Barcelona, Spain
*   Correspondence: jdosta@ub.edu

Academic Editor: Derek J. McPhee
Received: 17 June 2019; Accepted: 8 August 2019; Published: 17 August 2019

Abstract: Food waste (FW) collected from a university canteen was treated in acidogenic fermenters to produce volatile fatty acids (VFA) under biological pretreatment with mature compost. Batch assays working at pH 6 revealed an increment of 9.0%, 7.9%, and 4.1% (on COD basis) of VFA concentration when adding 2.5%, 3.5%, and 4.5% *w/w* of mature compost, respectively, even though the volatile solids (VS) concentration of food waste was lower in the tests with increasing doses of mature compost. For batch tests at pH 7, this VFA generation improvement was lower, even though enhanced COD solubilization was recorded. Operating in semi-continuous conditions at 35 °C, pH of 6, and hydraulic retention time (HRT) of 3.5 days, the addition of 2.5% *w/w* of mature compost led to a VFA concentration up to 51.2 ± 12.3% more (on VS basis) when compared to a reference reactor without compost addition. Moreover, the percentage of butyric acid on VS basis in the fermentation broth working at a pH of 6 increased from up to 12.2 ± 1.9% (0% compost addition) to up to 23.5 ± 2.7% (2.5% compost addition). The VFA production was not improved when a higher percentage of mature compost was used (3.5% instead of 2.5% *w/w*), and it slightly decreased when mature compost addition was lowered to 1.5% *w/w*. When working at a pH of 7 in the semi-continuous fermenters with the addition of 2.5% *w/w* mature compost at an HRT of 3.5 days, an improvement of 79% and 104% of the VFA concentration (on VS basis) were recorded as compared to fermenters working at a pH of 6 with 2.5% and 0% *w/w* of mature compost addition, respectively. At a pH of 7, higher production of propionic and valeric acids was found with respect to the reactor working at a pH of 6. The effect of pH on VFA generation was estimated to have greater contribution than that of only biological pretreatment using mature compost. At a pH of 7, the VFA yield was higher for the fermenter working with 2.5% *w/w* mature compost but at a pH of 7 and HRT of 5 days, the effect of mature compost on VFA production improvement was lower than that obtained at a pH of 6. Moreover, higher solubilization in terms of soluble chemical oxygen demand and total ammonium was detected when biological pretreatment using mature compost was applied at both a pH of 6 and a pH of 7, which indicates enhanced hydrolysis in both conditions.

Keywords: acetic acid; acidogenic fermentation; biorefinery; butyric acid; HRT; pH; propionic acid

---

1. Introduction

The raising concern about climate change and sustainability have led to an increasing awareness of resource utilization [1] and, under this context, the demand for energy and materials is now a big challenge in this century [2]. Urban organic wastes are known to contain a great variety of fermentable and biodegradable materials, which include biodegradable organic compounds such as sewage sludge (primary and secondary), food waste (FW), and organic fraction of municipal solid waste (OFMSW), among others. The acidogenic fermentation of urban organic wastes is gaining attention due to their high accessibility and possible process improvement. In 2012, approximately 90 million tons of FW was generated in all European countries [3]. This number indicates that a huge quantity of volatile fatty acids (VFA) could have been recovered through anaerobic acidogenic fermentation. VFA is a key commodity to produce biomaterials, such as polyhydroxyalkanoates, and biodegradable bioplastics, which currently have a growing market [4–6]. Other than that, VFA could be used as external carbon sources for biological nutrient removal (nitrogen and phosphorus) in wastewater treatment plants or to produce bioenergy [7–9]. To improve the biomaterials utilization from FW, such as PHA, it is of the utmost importance to maximize the production of VFA and different strategies are possible [10].

Hydrolysis is usually the rate limiting step for VFA production [11–14]. Hence, development of treatment methods to improve hydrolysis and solubilization of complex organic compounds has been investigated and it is still under research. Currently, there are a number of pre-treatment methods [15–18] to enhance the hydrolysis step, which is usually classified as physical, chemical, and biological. Physical pretreatments (thermal and mechanical) increase the disintegration of cell membranes or the specific surface area, which can provide better contact between substrate and microorganisms [19–21]. In chemical pretreatments, the external addition of chemicals (acids, alkalis, ozonation, etc.) will somehow increase the solubilization of the substrate [17,18,22]. Biological pretreatments are getting more attention for acidogenic fermentation, since they do not require reagent additions and do not imply high-energy demands. Among biological pretreatments, bioaugmentation is a promising approach to enhance VFA production instead of other organic compounds [10]. The simplest use of bioaugmentation is to add to the fermentation medium, which is a pre-adapted mixed culture. An example is the rumen addition that has been successfully used in anaerobic fermentation of corn stover [23] or anaerobic digestion of high lignocellulose grass silage [24]. Rumen addition typically requires an immediate use of it, since these bacteria are very sensitive and could become inactive after a few hours. Even so, the net positive output has attracted many researchers' interest. Another approach that would overcome this problem is the use of mature compost, which is much more stable than rumen, but this biological pretreatment agent has been seldom reported in literature.

This paper focuses on the VFA production by adding mature compost to FW in acidogenic fermentation, to assess quantitatively the improvement yields and composition of VFA at different pHs and hydraulic retention time (HRT). To this end, batch and semi-continuous experiments were carried out for an operation period of 200 days.

2. Results and Discussion

2.1. Effect of Mature Compost Addition at pH of 6

Several batch tests were performed at a pH of 6.0 using mixtures 1:1 on the volatile solid (VS) basis of food waste and acidogenic fermentation inoculum of a digester treating FW [15] with different doses of mature compost (namely, 0%, 2.5%, 3.5%, and 4.5% *w/w*). Table 1 summarizes the main characteristics of these batch tests and Figure 1 shows the VFA production with different doses of mature compost. The results showed that, in a short-term period (10 days), an increment of 9.0%, 7.9%, and 4.1% of VFA concentration (on the COD basis) was observed when 2.5%, 3.5%, and 4.5% *w/w* of mature compost was added, respectively, as compared to the batch without compost addition (it should be noted that the quantity of food waste added in each batch decreased when the percentage of compost increased).

Table 1. Main characteristics of the batch tests of food waste using mature compost at a pH of 6.

Parameter	Units	0%	2.5%	3.5%	4.5%	FW Only	Compost Only
Food Waste weight	g	84.75	82.63	81.79	80.94	200	-
% VS of Food Waste in the mixture	%	50	44	41	39	100	-
Compost weight	g	-	5	7	9	-	27
% VS of Compost in the mixture	%	-	13	17	21	-	100
Inoculum weight	g	115.25	112.36	111.21	110.06	-	-
% VS of Inoculum in the mixture	%	50	44	41	39	-	-
Initial VS content	%	5.36	5.77	6.24	6.52	6.81	1.18
Initial soluble COD	g COD/L	36.95	34.64	34.56	35.14	41.22	5.21
Soluble COD at day 10	g COD/L	47.09	50.87	49.95	52.22	55.51	5.18
Initial NH_4^+-N	mg NH_4^+-N/L	355	345	353	339	47	1
NH_4^+-N at day 10	mg NH_4^+-N/L	1027	1058	1033	1058	458	16
VFA concentration and distribution at day 10							
VFA concentration	g COD/L	9.82	10.70	10.59	10.22	2.68	0.03
Acetic Acid	% COD	21.5	22.3	22.0	22.7	79.0	85.2
Propionic Acid	% COD	0.5	0.6	0.7	0.6	4.8	-
Isobutyric Acid	% COD	2.0	1.8	1.8	1.9	2.1	-
Butyric Acid	% COD	21.0	23.6	23.8	24.5	4.9	7.1
Isovaleric Acid	% COD	5.7	5.3	5.0	5.3	0.7	-
Valeric Acid	% COD	0.5	0.7	0.7	0.7	2.1	-
Isocaproic Acid	% COD	0.2	0.2	0.2	0.2	0.5	-
Hexanoic Acid	% COD	47.4	44.6	44.9	43.1	5.2	7.7
Heptanoic Acid	% COD	1.0	0.9	0.8	0.9	0.7	-

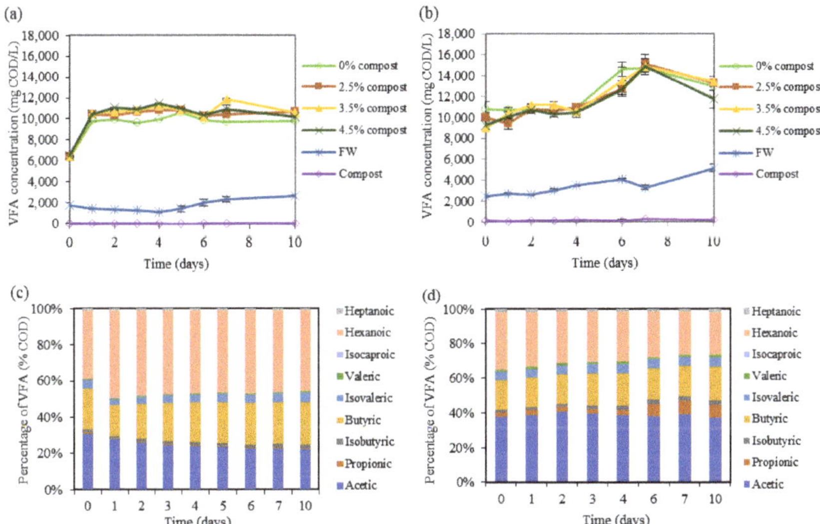

Figure 1. VFA concentration monitoring in a batch of acidogenic fermentation tests of FW at a pH of 6 (**a**) and a pH of 7 (**b**) with a different mature compost addition and evolution of the VFA distribution in the fermentation test with 2.5% *w/w* mature compost addition at a pH of 6 (**c**) and 7 (**d**).

Within one day, the batch containing inoculum and FW (with or without compost) increased to reach a stable VFA concentration. Only small changes were recorded afterwards. This could be related to the easily fermentable organic material of fresh FW and the high activity of the fermentation inoculum. Bottles containing only FW could have inherent microorganisms of these substrates, but this biomass did not contribute much to the formation of VFA in acidogenic fermentation in 10 days. For the batch containing only mature compost, VFA production remained below 26 mg VFA/L in this batch. When analyzing the VFA distribution between acidogenic fermentation of FW with and without addition of mature compost, the difference was not clear, since every individual VFA changed in a similar way. The biggest change in VFA composition could be seen from day 0 to day 1, which was the moment when most VFA were produced and reached stable conditions afterward. Table 1 summarizes the percentage of the three main VFA products obtained at the 10th day of these batch assays, which are mainly composed by acetic, butyric, and hexanoic acids. Figure 1c shows the evolution of the VFA profile during the batch test at a pH of 6 where 2.5% *w/w* was added. It was observed that butyric and hexanoic acids concentrations had some changes in the first two days. In contrast, acetic acid was approximately stable during the first five days and remained at 22% COD of total VFA concentration at day 10. When considering the soluble chemical oxygen demand (sCOD) analyzed at the end of the batch tests (Table 1), higher solubilization of organic compounds could be observed when mature compost was added. Without addition of mature compost, only 27.4% increase in sCOD was observed while 46.8%, 44.5%, and 48.6% increase in sCOD was monitored when 2.5%, 3.5%, and 4.5 % mature compost was added. These indicated that, with the addition of mature compost, hydrolytic enzymes were introduced into the fermentation broth with the aim to increase the hydrolytic rate [25,26]. These results were also consistent with the study by Fdez.-Güelfo et al. [27] who tested the biological pretreatment of OFMSW using mature compost and reported that 2.5% *v/v* was enough to increase the sCOD by roughly 50%, which was the indicator of the solubilization yield in that study. When compost was added, FW was microbiologically solubilized prior to acidogenic fermentation [28]. In terms of total ammonium nitrogen (TAN), the percentage of solubilization of ammonium also increased abundantly.

To study the effects in long-term conditions, three semi-continuous reactors (A, B, and C) were put into operation at a pH of 6 without compost addition working under a retention time of 3.5 days. After 15 days of operation with approximately the same specific VFA production in terms of g VFA/g VS fed, several doses of mature compost were added to the digesters. Table 2 summarizes the different periods of FW collection carried out where it is observed that fermenter A (fed with only FW) remained as a reference reactor, fermenter B worked always with 2.5% (on dry weight) of mature compost (mixed in the influent FW), and fermenter C worked under different conditions of mature compost addition. Figure 2 monitors the VFA concentration profiles (in terms of mg COD_{VFA}/L), the VFA yield (expressed in terms of g VFA/g VS_{fed}), and the Organic Loading Rate (OLR) applied (in terms of g VS/(L day)) during Phase 1. On the other hand, Table 3 shows the main characteristics of the effluent fermentation broth obtained in each period and Figure 3 shows the monitoring of the individual VFA during the whole experimental period.

Table 2. Summary of the working conditions for the three semi-continuous mesophilic fermenters in the periods tested.

Fermenter	Phase	1				2		
	Stage	1	2	3	4	1	2	3
	Period (Days)	0–14	15–34	35–84	85–114	115–146	147–168	168–198
A	pH	6	6	6	6	6	6	6
	HRT (days)	3.5	3.5	3.5	3.5	3.5	3.5	3.5
	Compost (% *w/w*)	-	-	-	-	-	-	-
B	pH	6	6	6	6	6	6	7
	HRT (days)	3.5	3.5	3.5	3.5	3.5	3.5	5
	Compost (% *w/w*)	-	2.5	2.5	2.5	2.5	2.5	-
C	pH	6	6	6	6	7	7	7
	HRT (days)	3.5	3.5	3.5	3.5	3.5	5	5
	Compost (% *w/w*)	-	-	3.5	1.5	2.5	2.5	2.5

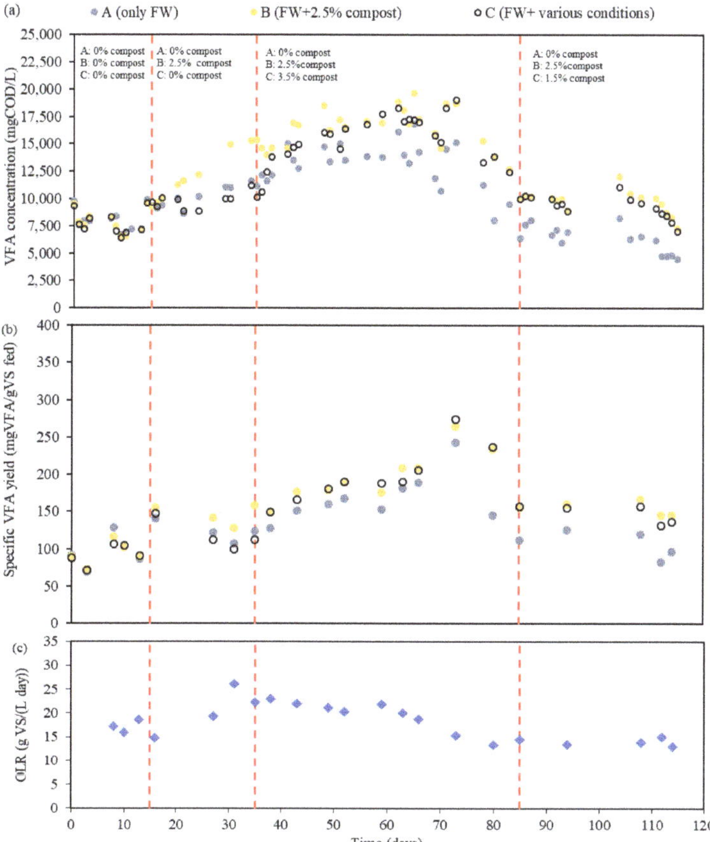

Figure 2. Evolution of (**a**) VFA concentration of the fermentation broth, (**b**) specific VFA production, and (**c**) applied organic loading rate (OLR) when treating food waste (FW) with several percentages of mature compost at a pH of 6 in fermenters A, B, and C.

Table 3. Summary of the main characteristics of the effluent fermentation broth for each fermenter (A, B, and C) depending on the food waste collection period.

	Parameters	Units	FW1	FW2	FW3	FW4	FW5	FW6	FW7	FW8	FW9	FW10	FW11
	Period	Days	1–20	21–40	41–57	58–72	73–96	97–114	115–132	133–145	146–155	156–180	181–198
FERMENTER A	pH	-	5.74 ± 0.33	5.83 ± 0.27	5.91 ± 0.16	5.95 ± 0.20	5.91 ± 0.17	5.92 ± 0.21	5.90 ± 0.35	5.95 ± 0.27	5.90 ± 0.28	5.99 ± 0.13	6.05 ± 0.17
	VFA	g/L	5.54 ± 0.68	7.34 ± 0.85	9.60 ± 0.57	9.42 ± 1.26	5.67 ± 1.74	4.15 ± 0.79	3.85 ± 0.78	5.39 ± 1.09	4.47 ± 1.03	6.18 ± 1.00	3.85 ± 0.64
	sCOD	g/L	46.3 ± 6.4	48.1 ± 3.7	49.9 ± 2.5	48.5 ± 1.6	43.4 ± 3.5	46.0 ± 2.4	47.3 ± 5.6	45.7 ± 1.6	28.8 ± 5.9	35.2 ± 6.8	27.2 ± 3.69
	COD_{VFA}/sCOD	%	20.4 ± 3.4	22.3 ± 3.1	26.8 ± 2.4	29.5 ± 1.8	21.6 ± 9.2	12.1 ± 2.7	12.2 ± 4.5	16.1 ± 3.7	25.6 ± 3.9	27.6 ± 3.5	19.6 ± 2.0
	NH_4^+-N	g NH_4^+-N/L	0.62 ± 0.27	0.52 ± 0.04	0.67 ± 0.04	0.39 ± 0.21	0.21 ± 0.06	0.25 ± 0.07	-	0.46 ± 0.07	0.59 ±0.02	0.81 ± 0.12	0.95 ± 0.15
	VFA yield	mgVFA_{eff}/gVS_{inf}	103.7 ± 29.9	120.5 ± 8.9	159.8 ± 8.1	174.8 ± 19.0	156.8 ± 59.4	100.4 ± 18.8	111.6 ± 14.8	78.7 ± 15.7	78.4 ± 12.8	107.4 ± 22.4	68.7 ± 12.2
	Acetic acid	% VS	53.7 ± 9.2	55.8 ± 4.7	60.3 ± 3.7	59.0 ± 2.8	58.3 ± 3.4	64.1 ± 3.6	68.9 ± 9.9	57.8 ± 5.6	51.0 ± 4.0	49.9 ± 4.1	49.6 ± 3.1
	Propionic acid	% VS	1.79 ± 0.25	1.31 ± 0.15	1.22 ± 0.11	0.96 ± 0.57	1.48 ± 0.54	1.64 ± 0.57	0.57 ± 0.55	0.77 ± 0.28	0.79 ± 0.26	0.79 ± 0.19	0.95 ± 0.20
	Isobutyric acid	% VS	0.82 ± 0.12	0.71 ± 0.06	0.72 ± 0.10	0.70 ± 0.09	0.58 ± 0.14	0.75 ± 0.06	0.38 ± 0.34	0.53 ± 0.24	1.05 ± 0.38	1.17 ± 0.29	1.59 ± 0.20
	Butyric acid	% VS	11.6 ± 3.0	11.2 ± 2.4	12.2 ± 1.9	10.7 ± 1.4	11.4 ± 1.0	7.4 ± 1.4	6.2 ± 2.6	11.2 ± 1.9	17.1 ± 0.9	17.2 ± 2.7	17.3 ± 1.8
	Isovaleric acid	% VS	1.82 ± 0.38	1.75 ± 0.28	1.86 ± 0.31	1.94 ± 0.32	1.27 ± 0.24	1.29 ± 0.29	0.91 ± 0.44	2.03 ± 0.72	3.36 ± 1.50	4.09 ± 1.17	5.38 ± 0.99
	Valeric acid	% VS	0.83 ± 0.12	0.58 ± 0.10	0.56 ± 0.05	0.55 ± 0.07	0.72 ± 0.20	0.91 ± 0.10	0.46 ± 0.45	0.49 ± 0.33	0.49 ± 0.25	0.63 ± 0.16	0.79 ± 0.31
FERMENTER B	pH	-	5.72 ± 0.33	6.01 ± 0.21	6.14 ± 0.16	6.02 ± 0.12	6.01 ± 0.10	6.14 ± 0.20	6.18 ± 0.18	6.05 ± 0.11	5.94 ± 0.15	6.47 ± 0.54	7.00 ± 0.21
	VFA	g/L	5.67 ± 0.83	8.74 ± 0.91	10.30 ± 0.81	10.60 ± 0.95	7.22 ± 1.51	6.21 ± 0.89	4.81 ± 0.70	6.15 ± 1.51	4.96 ± 0.42	7.88 ± 1.50	9.20 ± 1.09
	sCOD	g/L	46.1 ± 6.3	49.0 ± 0.7	49.1 ± 7.3	47.6 ± 7.4	49.1 ± 5.6	47.4 ± 3.2	46.3 ± 6.7	44.8 ± 1.3	30.3 ± 3.0	51.4 ± 20.3	40.3 ± 0.5
	COD_{VFA}/sCOD	%	20.8 ± 4.4	27.4 ± 3.8	34.6 ± 6.2	39.2 ± 7.1	25.7 ± 7.0	19.3 ± 1.4	18.8 ± 2.9	21.5 ± 2.6	29.4 ± 1.1	28.6 ± 10.1	34.4 ± 1.1
	NH_4^+-N	g NH_4^+-N/L	0.63 ± 0.28	0.54 ± 0.02	0.83 ± 0.18	0.55 ± 0.31	0.28 ± 0.07	0.36 ± 0.08	-	0.56 ± 0.22	0.56 ± 0.27	0.88 ± 0.24	1.31 ± 0.01
	VFA yield	mgVFA_{eff}/gVS_{inf}	104.4 ± 29.0	144.6 ± 12.8	181.5 ± 7.1	197.9 ± 19.0	203.7 ± 54.3	152.8 ± 11.9	140.0 ± 3.9	91.7 ± 8.2	88.5 ± 5.0	135.6 ± 28.1	140.4 ± 25.4
	Acetic acid	% VS	57.0 ± 6.3	44.2 ± 3.7	44.6 ± 2.3	39.4 ± 2.7	43.7 ± 5.9	45.7 ± 2.1	38.5 ± 6.9	38.4 ± 12.4	22.4 ± 4.8	25.7 ± 5.1	42.1 ± 4.0
	Propionic acid	% VS	1.71 ± 0.30	1.18 ± 0.08	1.98 ± 0.50	1.75 ± 0.28	1.45 ± 0.32	1.74 ± 0.07	0.37 ± 0.19	0.75 ± 0.16	0.63 ± 0.19	0.88 ± 0.26	5.16 ± 2.04
	Isobutyric acid	% VS	0.78 ± 0.12	0.60 ± 0.04	1.13 ± 0.30	1.08 ± 0.18	0.75 ± 0.12	0.93 ± 0.07	0.62 ± 0.14	1.04 ± 0.33	1.19 ± 0.41	1.61 ± 0.35	1.95 ± 0.19
	Butyric acid	% VS	10.6 ± 2.8	20.0 ± 2.3	21.7 ± 2.0	23.5 ± 2.7	23.5 ± 3.5	18.9 ± 0.8	20.5 ± 4.1	20.9 ± 5.4	38.2 ± 8.3	27.8 ± 3.8	17.9 ± 3.2
	Isovaleric acid	% VS	1.70 ± 0.45	1.46 ± 0.30	2.43 ± 0.43	2.30 ± 0.53	1.15 ± 0.29	1.69 ± 0.14	1.23 ± 0.22	2.62 ± 0.78	2.71 ± 1.22	4.03 ± 0.98	4.14 ± 0.42
	Valeric acid	% VS	0.78 ± 0.13	0.61 ± 0.05	1.21 ± 0.41	1.26 ± 0.19	0.89 ± 0.14	1.00 ± 0.05	0.45 ± 0.23	0.51 ± 0.16	0.76 ± 0.08	0.95 ± 0.27	3.97 ± 1.42
FERMENTER C	pH	-	5.66 ± 0.36	5.83 ± 0.30	6.19 ± 0.17	6.02 ± 0.12	6.00 ± 0.11	6.16 ± 0.20	6.81 ± 0.42	7.07 ± 0.06	6.82 ± 0.13	6.99 ± 0.11	6.96 ± 0.19
	VFA	g/L	5.54 ± 0.71	7.17 ± 1.13	10.2 ± 0.6	10.6 ± 0.7	7.25 ± 1.80	5.68 ± 0.65	6.40 ± 1.49	11.0 ± 1.0	11.6 ± 0.7	15.3 ± 2.1	10.7 ± 1.5
	sCOD	g/L	46.2 ± 6.0	48.4 ± 2.5	47.4 ± 5.4	51.7 ± 6.8	48.6 ± 8.2	48.5 ± 2.8	52.2 ± 12.1	60.0 ± 3.5	43.8 ± 1.4	59.3 ± 18.4	44.5 ± 1.2
	COD_{VFA}/sCOD	%	13.6 ± 2.2	13.4 ± 2.6	21.8 ± 3.1	21.0 ± 3.9	15.6 ± 3.4	10.8 ± 0.6	13.3 ± 4.5	19.2 ± 2.1	27.1 ± 1.7	27.1 ± 6.0	23.1 ± 2.2
	NH_4^+-N	g NH_4^+-N/L	0.63 ± 0.28	0.50 ± 0.03	0.77 ±0.19	0.55 ± 0.33	0.27 ± 0.06	0.33 ± 0.08	-	0.74 ± 0.29	1.12 ± 0.01	1.46 ± 0.19	1.56 ± 0.09
	VFA yield	mgVFA_{eff}/gVS_{inf}	101.6 ± 25.9	118.7 ± 21.5	179.0 ± 12.4	195.1 ± 9.8	205.8 ± 59.3	141.9 ± 13.2	190.0 ± 13.6	180.7 ± 15.4	202.9 ± 6.2	250.6 ± 41.6	199.4 ± 39.0
	Acetic acid	% VS	57.0 ± 7.6	58.6 ± 4.2	49.3 ± 4.4	41.9 ± 3.9	42.6 ± 3.4	43.1 ± 1.3	49.6 ± 4.7	53.0 ± 1.7	53.7 ± 1.2	50.2 ± 2.9	48.7 ± 3.0
	Propionic acid	% VS	1.79 ± 0.28	1.26 ± 0.17	1.77 ± 0.60	1.61 ± 0.31	1.52 ± 0.34	1.78 ± 0.06	0.79 ± 0.32	7.33 ± 1.72	7.52 ± 0.7	7.69 ± 1.48	4.62 ± 0.94
	Isobutyric acid	% VS	0.79 ± 0.13	0.73 ± 0.10	1.10 ± 0.37	1.05 ± 0.19	0.72 ± 0.11	0.87 ± 0.08	0.69 ± 0.05	1.12 ± 0.18	1.29 ± 0.19	1.71 ± 0.91	1.67 ± 0.19
	Butyric acid	% VS	10.1 ± 2.5	9.4 ± 1.8	19.5 ± 3.3	23.1 ± 3.3	24.8 ± 2.5	18.0 ± 2.1	20.4 ± 1.3	17.7 ± 1.3	17.1 ± 0.5	17.2 ± 1.6	18.4 ± 2.8
	Isovaleric acid	% VS	1.72 ± 0.45	1.77 ± 0.33	2.30 ± 0.45	2.12 ± 0.29	1.10 ± 0.26	1.69 ± 0.15	1.19 ± 0.31	1.66 ± 0.4	2.01 ± 0.32	2.63 ± 0.49	3.44 ± 0.65
	Valeric acid	% VS	0.81 ± 0.13	0.56 ± 0.08	1.03 ± 0.40	1.09 ± 0.21	0.93 ± 0.13	1.00 ± 0.04	0.64 ± 0.17	3.39 ± 2.02	5.32 ± 0.37	5.58 ± 0.67	4.63 ± 1.49

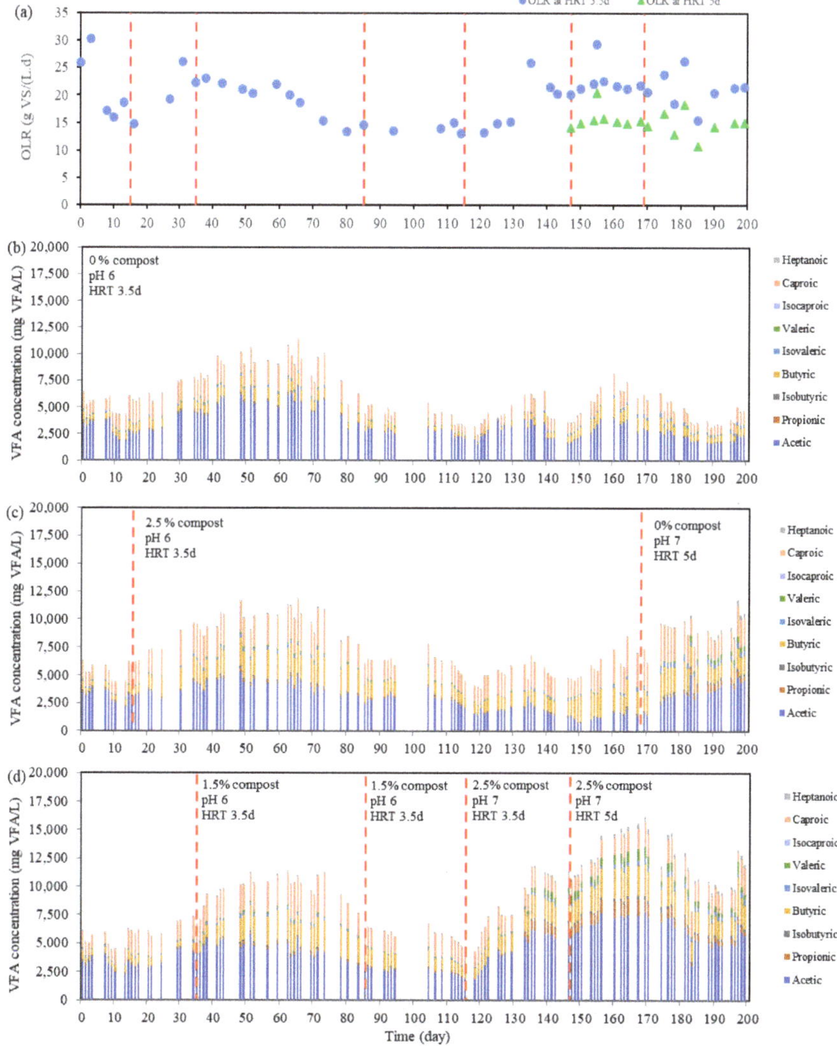

Figure 3. OLR (**a**) and individual VFA concentration in the whole experimental period of fermenters A (**b**), B (**c**), and C (**d**).

When 2.5% *w/w* of mature compost was mixed with FW as feeding of fermenter B, an increase of VFA production in comparison to the reference fermenter was observed. The VFA yield reached its maximum 0.27 gVFA/gVS on day 73 after the addition of 2.5% *w/w* of mature compost, with an average improvement in VFA yield of 30.6 ± 16.7% (VS basis) in that the FW collection period with respect to the reference reactor. When fermenter C was fed with FW and 3.5% *w/w* of mature compost (days 35–84), the production of VFA on average, was 29.8 ± 14.4% more (VS basis) than that of the reference fermenter. Therefore, it yielded similar results than the fermenter working with 2.5% *w/w* mature compost. To further understand the relationship between the percentage of mature compost and VFA production under these operating conditions, the percentage of mature compost in the feeding was decreased from 3.5 to 1.5% *w/w* in fermenter C. When the percentage of mature compost was lowered, the VFA yield slightly decreased with respect to fermenter with 2.5% *w/w* mature compost.

Therefore, in both long-term and short-term periods, with the presence of mature compost, VFA production could be limitedly boosted up in the beginning and maintained its dominance during the experiment, without considering the quantity of doses. Generally, higher solubilization expressed in terms of sCOD was detected in the fermenters working with mature compost addition (see Table 3), which is also consistent with a higher NH_4^+-N concentration in the fermentation broth. This might be due to the mature compost, which contains a variety of microorganisms, aerobic and anaerobic, with hydrolytic enzymes, which are capable of improving the solubilization of biodegradable organic matter [27]. Moreover, butyric acid percentage on a VS basis in the fermentation broth was enhanced due to mature compost addition at a pH of 6, which increases from up to 12.2 ± 1.9% (without compost addition) to up to 23.5 ± 2.7% (2.5% *w/w* compost addition).

Aside from the changes of mature compost percentage, it can be clearly seen that the VFA concentration of the reference fermenter was not constant, since the influent feedstock was not exactly the same in every collection and also the OLR varied from one period to the other. The mature compost addition resulted in a higher VFA yield with respect to the reference reactor in those periods when the OLR was lower. These results also reveal the current situation of VFA produced through acidogenic fermentation of real FW obtained directly from local restaurants, canteens, or households. Therefore, the fluctuations emerged during this experiment could be caused by the continuous changing of real FW every two to three weeks.

2.2. Effect of Mature Compost Addition at a pH of 7

Figure 1b shows the evolution of VFA concentration in batch assays treating FW under 35 °C and a pH of 7. Table 4 summarizes the main characteristics of these batch tests. From the results of batch assays, it could be observed that the VFA concentration steadily increased but did not show a large increment from day 0 to day 4. More VFA started to be produced from day 4 onward and reached the maximum observed VFA concentration on day 7. Batch tests with 2.5%, 3.5%, and 4.5% mature compost were able to reach 15.2 ± 0.9, 14.9 ± 0.8, and 14.9 ± 0.6 $gCOD_{VFA}$/L, respectively, at day 7. In this batch test, the bottles containing 0% mature compost followed the VFA production trend of those containing various percentages of mature compost ranging from 2.5% to 4.5% and their difference was relatively small (it should be noted that the quantity of food waste added in each batch decreased when the percentage of compost increased). A VFA concentration below 175 mg/L was detected in the control assay where only mature compost was added. The evolution of the distribution of individual VFA for the batch test with 2.5% compost (*w/w*) at a pH of 7 is shown in Figure 1d, where it is observed that the hexanoic acid percentage decreased with time while propionic and butyric acids increased. The day when the maximum VFA concentration was recorded (day 7) at a pH of 7 by adding 2.5% *w/w* compost. The main VFA constituent was acetic acid, accounting for approximately 5.99 gCOD/L, which was followed by hexanoic (3.79 gCOD/L), butyric (2.71 gCOD/L), propionic (1.16 gCOD/L), and isovaleric acid (0.78 gCOD/L). A similar proportion of VFA was reported in the experiment carried out by Kumar and Mohan [29] treating vegetable waste, considering only short chain carboxylic acids (C_2–C_5). Moreover, these authors [29] obtained 2.6% and 7.0% propionic acid in the fermentation broth at a pH of 6 and a pH of 7, respectively, which is in line with the results obtained in this study, where a significant increment of propionic acid production was observed from day 4 to day 6 when working under neutral pH and mature compost addition (see Figure 1d). These results brought to the conclusion that acidogenic fermentation at neutral pH changed the metabolic pathway, which leads to the production of propionic acid as compared to partially acidic pH. To define the state of acidogenic fermentation, some researchers used the degree of acidification [29–31], which represents the bioconversion of substrate into short chain carboxylic acids (VFA in this study). In a short-term period, an increase in sCOD and total ammonium nitrogen (TAN) can be observed (see Table 4). Higher sCOD after acidogenic fermentation proved the hydrolysis and solubilization occurred in this fermentation process. Without adding mature compost, sCOD could raise from 40.6 gCOD/L to 48.5 gCOD/L and the addition of 3.5% mature compost led to the highest increment, which was 25.4%, from 39.1 gCOD/L to 49.1gCOD/L. This value shrunk when the mature compost addition was increased from 3.5% to 4.5%. Fdez.-Güelfo et al. [27] compared the effect of solubilization with different doses of mature compost varying between 2.5% to 10% (*v/v*) and a big difference in the increment of sCOD was found between 2.5% and 5% mature compost, where 2.5% yielded the highest and 5% was only half of that. In their experiment, the lowest doses produced the highest increase of sCOD, which might be due to higher consumption of solubilized organic compounds when metabolic activity increases. On the other hand, TAN was directly proportional to the mature compost added without addition of mature compost. TAN was able to raise 60% respect to its initial value. This number was even higher when a different percentage of mature compost was added.

Table 4. Main characteristics of the batch tests of food waste using mature compost at a pH of 7.

	Units	0%	2.5%	3.5%	4.5%	FW Only	Compost Only
Food Waste weight	g	44.50	43.39	42.95	42.50	200	-
% VS of Food Waste in the mixture	%	50	43	41	39	100	-
Compost weight	g	-	5	7	9	-	9
% VS of Compost in the mixture	%	-	13	18	22	-	100
Inoculum weight	g	155.50	151.61	150.05	148.50	-	-
% VS of Inoculum in the mixture	%	50	43	41	39	-	-
Initial VS content	%	2.32	2.65	2.89	2.97	9.59	0.40
Initial soluble COD	g COD/L	40.56	39.54	39.14	38.73	35.61	6.53
Final soluble COD	g COD/L	48.47	47.88	49.06	42.93	n.a	6.86
Initial NH_4^+-N	mg NH_4^+-N/L	728	734	727	680	n.a	7
Final NH_4^+-N	mg NH_4^+-N/L	1164	1635	1724	1803	n.a.	195
VFA Concentration and Distribution at Day 10							
VFA concentration	g COD/L	12.94	13.28	13.43	11.74	5.18	0.23
Acetic Acid	% COD	37.5	37.4	38.2	35.9	76.2	53.0
Propionic Acid	% COD	6.8	7.3	7.5	7.3	13.5	-
Isobutyric Acid	% COD	2.6	2.5	2.6	2.6	2.9	-
Butyric Acid	% COD	18.4	19.6	19.4	19.4	1.2	22.0
Isovaleric Acid	% COD	5.7	5.3	5.2	5.4	1.1	-
Valeric Acid	% COD	1.3	1.4	1.4	1.5	1.6	
Isocaproic Acid	% COD	0.1	0.1	0.1	0.1	0.4	-
Hexanoic Acid	% COD	25.7	24.8	24.3	26.3	1.6	24.9
Heptanoic Acid	% COD	1.7	1.6	1.4	1.5	1.5	-

n.a.: Not analyzed.

Regarding the semi-continuous operation of the acidogenic fermenters, at the beginning of Phase 2, a change from pH 6 to pH 7 was applied to fermenter C. As shown in Figure 4, after changing to pH 7, the microorganisms responded quickly and the total VFA concentration of fermenter C was boosted five days later. A stepwise increment of VFA concentration from day 120 to 125 and from day 130 to 135 was related to a sudden increase in OLR, due to different FW collection periods. However, VFA production in fermenters A and B followed the same trends as in the previous period since they remained working at a pH of 6. FW is known by its heterogeneous composition in terms of carbohydrates, proteins, and lipids [32], and the randomly collected food waste from a University canteen leads to added complexity related to the different composition of this waste. Mixed microbial cultures (MMC) make VFA production possible through different metabolic pathways, which depend on the type of substrates (carbohydrates, proteins, and lipids) [33], among other factors. Nevertheless, there could exist some hindrances to retard or even stop the VFA generation by MMC, where one of them is steric hindrance from a residual extracellular polymeric substance (EPS) polymeric network [34]. EPS molecules result in the formation of a tight extracellular matrix (ECM) covering cells that constitute barriers, which will prevent the penetration of hydrolytic enzymes during digestion [35]. A large gap between fermenters working at a pH of 6 (A and B) and a pH of 7 (C) can be noted by the end of the first stage of Phase 2. In this phase, the biological pretreatment at a pH of 6 contributed to 11% to 25% improvement on VFA generation (on VS basis) with respect to the reference reactor at a pH of 6. The combined effect of regulating the pH to 7 while dosing mature compost yielded a net positive enhancement of VFA production. At a pH of 7 with the addition of 2.5% mature compost with an HRT of 3.5 days, an improvement of 79% and 104% of the VFA concentration (on VS basis) were recorded during this phase (days 133–145) as compared to fermenters B and A, respectively. From these comparisons, the effect of pH on VFA generation was estimated to have greater contribution than that of only biological pretreatment using mature compost. The regulation of pH to a value near 7 enhanced the solubilization of FW and resulted in having more sCOD and NH_4^+-N, as summarized in Table 3, and a concomitant higher VFA production. Moreover, the VFA profile at a pH of 7 with mature compost addition was different from that obtained at a pH of 6 (as stated in Table 3 and Figure 3). A rise in propionic, valeric, and isovaleric acid production was observed.

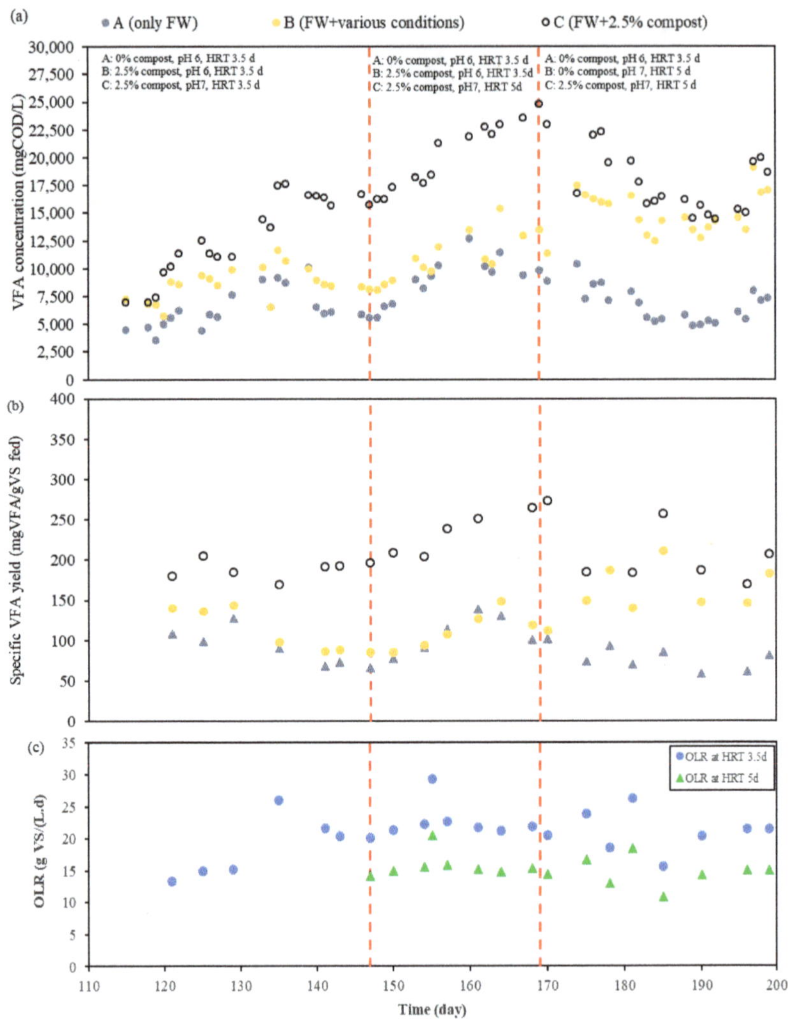

Figure 4. Evolution of (**a**) VFA concentration of the fermentation broth, (**b**) specific VFA production, and (**c**) applied OLR when treating FW with several percentages of mature compost at a pH of 6 and pH of 7 in fermenters A, B, and C.

In addition to working at a pH of 7, the HRT of Fermenter C was increased from 3.5 to 5 days. At longer HRT, more time was given to bacteria to undergo decomposition of long chain organic molecules, especially the rate-limiting hydrolysis step. Although it could be possible to achieve higher VFA production, risk of interference of methanogenic activities should also be taken into account [32]. Figures 3 and 4 show the VFA concentration of the semi-continuous operation of this fermenter. When HRT was increased to 5 days in fermenter C, VFA concentration in the fermentation broth increased linearly from 15.8 gCOD/L to 24.8 gCOD/L and remained within these limits in the third stage, with changes related mainly to OLR and the collected FW. In this case, it was unclear whether extending HRT from 3.5 days to 5 days had led to higher VFA yield, since, even though, the VFA yield increased during this period in Fermenter C, the VFA yield improvement with respect to the reference reactor (fermenter A) was in the same range. Kuruti et al. [36] performed some experiments to examine the optimal HRT for acidogenic fermentation and found out that, irrespective of the pH, batch fermentation broths were stabilized on day 4, 7, and 5 when working with glucose, cattle manure, and poultry litter, respectively. Moretto et al. [37] also suggested to consider HRT of 6 days to achieve best performance of process yield and maximum VFA level through the results of their experiments when treating a mixture of squeezed OFMSW and thickened waste activated sludge at a pH of 9 and mesophilic temperature (37 °C). In this study, the mesophilic fermenter containing 2.5% mature compost and operating at a pH of 7 produced the highest quantity of VFA when compared with the other two in operation.

Regarding the VFA profile (on VS basis) in this fermenter, acetic, propionic, butyric, isobutyric, valeric, and isovaleric acids accounted for the 48.7 ± 3.0%, 4.6 ± 0.9%, 18.4 ± 2.8%, 1.67 ± 0.19%, 4.63 ± 1.49%, and 3.44 ± 0.65%, respectively, of the total amount of VFA produced during the last stage of the experiment in accordance to the last FW collection period. Furthermore, in the last stage, the working pH of Fermenter B was regulated to 7, the retention time was extended to 5 days, and the mature compost addition was stopped. The net VFA production after making these modifications was positive, and a pH regulation to 7, together with increasing HRT, had greater effects on VFA generation. Comparing VFA production from fermenter B and C, VFA yield was higher for the fermenter working with mature compost but it was not possible to conclude that, at a pH of 7 and HRT of 5 days, the effect of mature compost on VFA production was as significant as when working at a pH of 6 (Phase 1). However, COD solubilization and NH_4^+-N release was higher when adding mature compost at a pH of 7 and HRT 5 days, which led to a lower ratio of VFA concentration with respect to soluble COD for the fermenter with mature compost addition, as stated in Table 3.

3. Materials and Methods

3.1. Substrate and Inoculum

The FW was collected from a university canteen every two weeks, approximately. Although the composition changed between collection periods, it mainly contained pasta, potato, meat, rice, noodles, bread, vegetables, and fruit peels. Once collected, FW was shredded (Bosch, Spain, MMB66G5M) and mixed with deionized water in a proportion of approximately 1:2 by volume, to obtain a concentrated feedstock of FW. This FW was stored in a refrigerator chamber at 4 °C until its usage. When the feedstock was needed for acidogenic fermentation, this substrate was diluted with deionized water to control the contents of total solids (TS, 4.45–7.57%) and volatile solids (VS, 3.78–6.43 %) so that their values were within a certain range during the experiments. The characterization of FW collected in different periods is shown in Table 5. The mature compost was collected from a full-scale mechanical-biological treatment (MBT) plant in the metropolitan area of Barcelona and came from tunnel composting of OFMSW and park and garden waste. The collected compost contained 52% of TS and 27% of VS. Once it was collected, compost was stored in a closed vessel to preserve its original moisture contents. Unlike FW, the mature compost was collected in one single time. Therefore, the characteristics of the mature compost were constant throughout the experiments. The inoculum used was effluent of acidogenic fermenter treating FW at a pH of 6 and hydraulic retention time (HRT) of 3.5 days from a previous work [15].

Table 5. Main characteristics of Food Waste from a university canteen in each collection period.

	Units	FW1	FW2	FW3	FW4	FW5	FW6	FW7	FW8	FW9	FW10	FW11	Range
Period	Days	1–20	21–40	41–57	58–72	73–96	97–114	115–132	133–145	146-155	156–180	181–198	-
pH	-	4.33 ± 0.61	6.54 ± 0.34	6.62 ± 0.75	6.46 ± 0.79	6.65 ± 0.23	6.67 ± 0.42	7.09 ± 0.40	6.11 ± 1.49	6.96 ± 1.15	5.73 ± 1.28	5.17 ± 0.85	4.33–7.09
Total Solids (TS)	% w/w	5.88 ± 1.63	7.31 ± 1.10	7.27 ± 0.35	6.57 ± 0.52	4.76 ± 0.38	4.45 ± 0.15	4.71 ± 0.42	7.03 ± 0.44	7.57 ± 0.96	6.39 ± 0.56	6.10 ± 1.23	4.45–7.57
Volatile Solids (VS)	% w/w	5.51 ± 1.68	6.12 ± 0.75	5.71 ± 0.24	5.46 ± 0.43	3.83 ± 0.24	3.78 ± 0.26	3.90 ± 0.28	6.40 ± 0.84	6.43 ± 1.03	5.79 ± 0.45	5.64 ± 1.18	3.78–6.40
VFA	g/L	1.08 ± 0.16	1.03 ± 0.08	1.21 ± 0.07	1.49 ± 0.18	0.98 ± 0.10	1.32 ± 0.13	0.93 ± 0.14	0.95 ± 0.28	0.75 ± 0.15	0.72 ± 0.19	0.83 ± 0.11	0.72–1.49
Soluble COD (sCOD)	g/L	40.6 ± 5.6	37.0 ± 2.9	38.7 ± 4.3	32.3 ± 14.7	31.1 ± 0.7	45.4 ± 1.6	51.0 ± 4.7	51.8 ± 4.7	14.9 ± 3.4	35.7 ± 21.8	31.9 ± 5.4	14.9–51.8
NH_4^+-N	mg NH_4^+-N/L	14.2 ± 3.9	34.1 ± 4.8	50.5 ± 1.3	26.7 ± 6.2	25.1 ± 7.1	13.0 ± 1.0	27.4	32.2 ± 12.0	26.6 ± 6.9	79.5 ± 23.8	107 ± 10	13–107

3.2. Experimental Set-Up

3.2.1. Batch Assays: FW with Mature Compost

Discontinuous assays were performed in serum bottles of a 200-mL effective volume with different mature compost doses (0, 2.5, 3.5, and 4.5% *w/w*) to study the effects of mature compost on VFA production and their individual VFA profile. Serum bottles were filled with inoculum and substrate according to their volatile solids content, to obtain an inoculum to substrate ratio (ISR) of 1:1 by weight. Corresponding mature compost doses (0, 2.5, 3.5, and 4.5% *w/w*) were then added. Blank samples were prepared as a control, including inoculum, substrate, and mature compost (three times the weight for 4.5% *w/w* and filling with deionized water). In different periods of time, two batches were carried out operating at a pH of 6 and a pH of 7 by using sodium bicarbonate ($NaHCO_3$) and hydrochloric acid (10 M HCl) for pH adjustment. Before the bottles were sealed with a stopper containing PTFE/Butyl septum, nitrogen gas was flushed through the headspace to remove residual air. These bottles were located inside an incubator (Memmert GmbH + Co. KG, Schwabach, Germany, Pass-through ovens UF750) working at a mesophilic temperature (35 °C). Furthermore, 2.5 mL of the sample were taken and centrifuged at 4000 rpm for 15 min. The supernatant was filtered by a 0.45-µm syringe filter. Part of this filtered sample was used for soluble chemical oxygen demand (sCOD) and total ammonium nitrogen (TAN) analysis in day 0 and day 7 while the other part of the filtered sample was prepared for VFA analysis where this operation was performed daily from day 0 to 7, and, on day 10, to check the stability of VFA production. The pH measurements were performed in the beginning (day 0) and the end of the batch test (day 10), and two selected days in between (day 3 and 7). In these days when the measurement was carried out, pH was adjusted to their initial pre-set values in order to avoid different behaviors on VFA production due to pH fluctuation. Analyses were performed in duplicate for the standard deviation calculation.

3.2.2. Semi Continuous Operation

For this study, three jacketed lab-scale reactors with an effective working volume of 4.5 L and mechanically stirred (using IKA-Werke, IKA Werke GmbH & Co. KG, Staufen, Germany, RW 16 basic functioning at approximately 150 rpm) were used as acidogenic fermenters at mesophilic conditions (35 °C) working with FW. These fermenters were operated at HRT of 3.5 days and the equivalent quantity of substrate (FW) was fed manually once per day (fed-batch culture). During feeding and draw-off operations, a minimum amount of nitrogen gas was flushed through the headspace of fermenters to avoid entering air.

At the beginning of the experiment, the three fermenters (A, B, and C) were fed with FW for 15 days. At this moment, all fermenters were working at HRT of 3.5 days and a pH of 6, and $NaHCO_3$ was used to increase alkalinity of the fermentation broth and to adjust pH to the pre-set value. The whole experiment was carried out under mesophilic conditions (35 °C). Fermenter A remained as the reference reactor and was always fed with FW without adding mature compost.

To analyze the long-term effect of biological pretreatment using mature compost on the acidogenic fermenter, fermenter B was fed with a mixture of FW and compost, which represented the 2.5% *w/w*. In fermenter C, different doses of mature compost were added to test the effect of the quantity of mature compost addition. Later, in addition, to study the effects of biological pretreatment, pH and HRT were taken into consideration. Table 2 summarizes the compost addition, working pH, and HRT applied in each acidogenic fermenter. As stated in this Table, to evaluate the fermenters' performance at a different pH and HRT conditions, the reference fermenter (A) worked at a pH of 6, HRT 3.5 days, and without mature compost addition and the other fermenters could work at a pH of 6 or 7, at different HRT of 3.5 or 5 days, and at different mature compost dosages (up to 3.5% *w/w*) depending on the parameter to be studied. Samples were taken daily on weekdays for continuous assessment of VFA and effluent pH, and other characterizations were analyzed two to three times per week.

3.3. Analytical Methods

The analysis of soluble COD, total solids (TS), and volatile solids (VS) were performed in accordance to the Standards Methods for the Examination of Water and Wastewater [38]. For the determination of the total ammonium nitrogen (TAN) concentration, the sample was centrifuged at 4000 rpm for 15 min, the supernatant was filtered through a 0.45 μm-pore size regenerated cellulose syringe filter. Samples were properly diluted to have the TAN range between 1 to 100 ppm and were analyzed with an ammonium ion selective electrode (Thermo Scientific, Beverly, MA, USA, Orion 9512HPBNWP). The pH was determined using a Crison pH electrode (pH series 52–04). Filtered samples were acidified with 85% phosphoric acid for VFA concentration analysis. Individual VFA including acetic, propionic, isobutyric, butyric, isovaleric, valeric, isocaproic, caproic, and heptanoic acids were determined by gas chromatography (Shimadzu GC 2010 plus) (Shimadzu Corporation, Kyoto, Japan) equipped with a capillary column (Nukol™, 15 m × 0.53 mm × 0.5 μm) and flame ionization detector (FID) (Shimadzu Corporation, Kyoto, Japan). The initial temperature of capillary column was 80 °C and increased at 10 °C/min to 110 °C, from 110 °C to 145 °C at 15 °C/min, which was followed by an increment of 20 °C/min to reach 190 °C. The temperature was 280 °C and 300 °C to the injector and detector, respectively. The carrier gas was helium, the fuel gas was hydrogen, and the oxidizing gas was synthetic air. Hence, individual VFA were detected by the programmed method for this gas chromatograph.

4. Conclusions

In this study, FW collected from a university canteen was treated in acidogenic fermenters to produce volatile fatty acids (VFA) under biological pretreatment with mature compost. Batch assays working at a pH of 6 revealed an increment of 9.0%, 7.9%, and 4.1% (on COD basis) of VFA concentration when adding 2.5%, 3.5%, and 4.5% *w/w* of mature compost, respectively, even though the VS concentration of food waste was lower in the tests with increasing doses of mature compost. When increasing the operating pH to 7, batch tests also revealed higher COD solubilization due to mature compost addition, even though it did not significantly impact enhanced VFA concentration in the fermentation effluent.

The addition of 2.5% *w/w* of mature compost to a semi-continuous acidogenic fermenter treating FW at mesophilic conditions (35 °C), pH of 6, and HRT of 3.5 days led to an improvement of the volatile fatty acids (VFA) yield up to 51.2 ± 12.3% (on VFA basis) when compared to the reference reactor working without mature compost addition. The mature compost dosage at a pH of 6 resulted in a higher percentage of butyric acid on a VS basis in the fermentation broth, which increases from up to 12.2 ± 1.9% (0% compost addition) to up to 23.5 ± 2.7% (2.5% *w/w* compost addition). Moreover, a higher sCOD content and a higher NH_4^+-N concentration was monitored in reactors working with mature compost, as a result of an enhanced hydrolytic activity. Considering the reactor working at a pH of 6 with 2.5% *w/w* addition, a higher compost addition of 3.5% *w/w* did not lead to higher VFA yield, while a slightly lower VFA production was recorded when working with 1.5% *w/w*.

When pH was changed from a pH of 6 to a pH of 7, VFA production was boosted and a higher production of propionic and valeric acids was recorded with respect to the reactor working at a pH of 6. In the semi-continuous fermenters with addition of 2.5% *w/w* mature compost at a pH of 7 with an HRT of 3.5 days, an improvement of 79% and 104% of the VFA concentration (on VS basis) were recorded as compared to fermenters working at a pH of 6 with 2.5 and 0% *w/w* of mature compost addition, respectively. Therefore, the effect of pH on VFA generation was estimated to have greater contribution than that of only biological pretreatment using mature compost. At a pH of 7, the VFA yield was higher for the fermenter working with mature compost but it was not possible to conclude that at a pH of 7 and HRT of 5 days, the effect of mature compost on VFA production was as significant as in a pH of 6. However, COD solubilization and NH_4^+-N release was higher when adding mature compost at a pH of 7 and HRT 5 days, which led to a lower ratio of VFA concentration with respect to soluble COD for the fermenter with mature compost addition.

Author Contributions: Conceptualization, Y.K.C., J.D. and J.M.A.; Methodology, Y.K.C., J.D. and J.M.A.; Validation, J.D. and J.M.A.; Formal analysis, Y.K.C. and J.D.; Investigation, Y.K.C.; Resources, Y.K.C., J.D. and J.M.A.; Data curation, Y.K.C.; Writing—original draft preparation, Y.K.C.; Writing—review and editing, Y.K.C., J.D. and J.M.A.; Visualization, Y.K.C. and J.D.; Supervision, J.D. and J.M.A.; Project administration, Y.K.C., J.D. and J.M.A.; Funding acquisition, J.M.A.

Funding: The European Union (Programme Horizon 2020) for RES URBIS project, grant agreement No. 730349 and the Spanish Ministerio de Economia y Competitividad (CTM2016-76275-R Project) funded this research.

Acknowledgments: The authors want to acknowledge the Institute of Water Research of University of Barcelona.

Conflicts of Interest: The authors declare no conflict of interest. The funders had no role in the design of the study; in the collection, analyses, or interpretation of data; in the writing of the manuscript, or in the decision to publish the results.

References

1. Fernández-Arévalo, T.; Lizarralde, I.; Fdz-Polanco, F.; Pérez-Elvira, S.; Garrido, J.; Puig, S.; Poch, M.; Grau, P.; Ayesa, E. Quantitative assessment of energy and resource recovery in wastewater treatment plants based on plant-wide simulations. *Water Res.* **2017**, *118*, 272–288. [CrossRef] [PubMed]

2. Yin, J.; Yu, X.; Wang, K.; Shen, D. Acidogenic fermentation of the main substrates of food waste to produce volatile fatty acids. *Int. J. Hydrog. Energy* **2016**, *41*, 21713–21720. [CrossRef]

3. Braguglia, C.M.; Gallipoli, A.; Gianico, A.; Pagliaccia, P. Anaerobic bioconversion of food waste into energy: A critical review. *Bioresour. Technol.* **2018**, *248*, 37–56. [CrossRef] [PubMed]

4. Bugnicourt, E.; Cinelli, P.; Lazzeri, A.; Alvarez, V. Polyhydroxyalkanoate (PHA): Review of synthesis, characteristics, processing and potential applications in packaging. *Express Polym. Lett.* **2014**, *8*, 791–808. [CrossRef]

5. Kourmentza, C.; Kornaros, M. Biotransformation of volatile fatty acids to polyhydroxyalkanoates by employing mixed microbial consortia: The effect of pH and carbon source. *Bioresour. Technol.* **2016**, *222*, 388–398. [CrossRef]

6. Singh, A.K.; Srivastava, J.K.; Chandel, A.K.; Sharma, L.; Mallick, N.; Singh, S.P. Biomedical applications of microbially engineered polyhydroxyalkanoates: An insight into recent advances, bottlenecks, and solutions. *Appl. Microbiol. Biotechnol.* **2019**, *103*, 2007–2032. [CrossRef]

7. Tong, J.; Chen, Y. Recovery of nitrogen and phosphorus from alkaline fermentation liquid of waste activated sludge and application of the fermentation liquid to promote biological municipal wastewater treatment. *Water Res.* **2009**, *43*, 2969–2976. [CrossRef]

8. Lim, S.-J.; Choi, D.W.; Lee, W.G.; Kwon, S.; Chang, H.N. Volatile fatty acids production from food wastes and its application to biological nutrient removal. *Bioprocess Eng.* **2000**, *22*, 543–545. [CrossRef]

9. Cavdar, P.; Yilmaz, E.; Tugtas, A.E.; Calli, B. Acidogenic fermentation of municipal solid waste and its application to bio-electricity production via microbial fuel cells (MFCs). *Water Sci. Technol.* **2011**, *64*, 789–795. [CrossRef]

10. Atasoy, M.; Owusu-Agyeman, I.; Plaza, E.; Cetecioglu, Z. Bio-based volatile fatty acid production and recovery from waste streams: Current status and future challenges. *Bioresour. Technol.* **2018**, *268*, 773–786. [CrossRef]

11. Lim, S.-J.; Kim, B.J.; Jeong, C.-M.; Choi, J.-D.-R.; Ahn, Y.H.; Chang, H.N. Anaerobic organic acid production of food waste in once-a-day feeding and drawing-off bioreactor. *Bioresour. Technol.* **2008**, *99*, 7866–7874. [CrossRef] [PubMed]

12. Lee, W.S.; Chua, S.M.A.; Yeoh, H.K.; Ngoh, G.C. A review of the production and applications of waste-derived volatile fatty acids. *Chem. Eng. J.* **2014**, *235*, 83–99. [CrossRef]

13. Yin, B.; Liu, H.; Wang, Y.; Bai, J.; Liu, H.; Fu, B. Improving volatile fatty acids production by exploiting the residual substrates in post-fermented sludge: Protease catalysis of refractory protein. *Bioresour. Technol.* **2016**, *203*, 124–131. [CrossRef] [PubMed]

14. Cesaro, A.; Belgiorno, V. Sonolysis and ozonation as pretreatment for anaerobic digestion of solid organic waste. *Ultrason. Sonochem.* **2013**, *20*, 931–936. [CrossRef] [PubMed]

15. Cheah, Y.K.; Vidal-Antich, C.; Dosta, J.; Mata-Alvarez, J. Volatile fatty acid production from mesophilic acidogenic fermentation of organic fraction of municipal solid waste and food waste under acidic and alkaline pH. *Environ. Sci. Pollut. Res.* **2019**, 1–14. [CrossRef] [PubMed]

16. Rughoonundun, H.; Granda, C.; Mohee, R.; Holtzapple, M.T. Effect of thermochemical pretreatment on sewage sludge and its impact on carboxylic acids production. *Waste Manag.* **2010**, *30*, 1614–1621. [CrossRef]
17. Hendriks, A.T.W.M.; Zeeman, G. Pretreatments to enhance the digestibility of lignocellulosic biomass. *Bioresour. Technol.* **2009**, *100*, 10–18. [CrossRef]
18. Carrère, H.; Dumas, C.; Battimelli, A.; Batstone, D.J.; Delgenès, J.P.; Steyer, J.P.; Ferrer, I. Pretreatment methods to improve sludge anaerobic degradability: A review. *J. Hazard. Mater.* **2010**, *183*, 1–15. [CrossRef]
19. Ariunbaatar, J.; Panico, A.; Esposito, G.; Pirozzi, F.; Lens, P.N.L. Pretreatment methods to enhance anaerobic digestion of organic solid waste. *Appl. Energy* **2014**, *123*, 143–156. [CrossRef]
20. Ma, J.; Duong, T.H.; Smits, M.; Verstraete, W.; Carballa, M. Enhanced biomethanation of kitchen waste by different pre-treatments. *Bioresour. Technol.* **2011**, *102*, 592–599. [CrossRef]
21. Bolzonella, D.; Battista, F.; Cavinato, C.; Gottardo, M.; Micolucci, F.; Lyberatos, G.; Pavan, P. Recent developments in biohythane production from household food wastes: A review. *Bioresour. Technol.* **2018**, *257*, 311–319. [CrossRef]
22. Strazzera, G.; Battista, F.; Garcia, N.H.; Frison, N.; Bolzonella, D. Volatile fatty acids production from food wastes for biorefinery platforms: A review. *J. Environ. Manag.* **2018**, *226*, 278–288. [CrossRef]
23. Hu, Z.H.; Yu, H.Q. Application of rumen microorganisms for enhanced anaerobic fermentation of corn stover. *Process Biochem.* **2005**, *40*, 2371–2377. [CrossRef]
24. Wall, D.M.; Straccialini, B.; Allen, E.; Nolan, P.; Herrmann, C.; O'Kiely, P.; Murphy, J.D. Investigation of effect of particle size and rumen fluid addition on specific methane yields of high lignocellulose grass silage. *Bioresour. Technol.* **2015**, *192*, 266–271. [CrossRef]
25. Fdez.-Güelfo, L.A.; Álvarez-Gallego, C.; Sales Márquez, D.; Romero García, L.I. Biological pretreatment applied to industrial organic fraction of municipal solid wastes (OFMSW): Effect on anaerobic digestion. *Chem. Eng. J.* **2011**, *172*, 321–325. [CrossRef]
26. Álvarez-Gallego, C.J.; Fdez-Güelfo, L.A.; de los Romero Aguilar, A.M.; García, L.I.R. Thermochemical pretreatments of organic fraction of municipal solid waste from a mechanical-biological treatment plant. *Int. J. Mol. Sci.* **2015**, *16*, 3769–3782. [CrossRef]
27. Fdez.-Güelfo, L.A.; Álvarez-Gallego, C.; Sales, D.; Romero, L.I. The use of thermochemical and biological pretreatments to enhance organic matter hydrolysis and solubilization from organic fraction of municipal solid waste (OFMSW). *Chem. Eng. J.* **2011**, *168*, 249–254. [CrossRef]
28. Gonzales, H.B.; Takyu, K.; Sakashita, H.; Nakano, Y.; Nishijima, W.; Okada, M. Biological solubilization and mineralization as novel approach for the pretreatment of food waste. *Chemosphere* **2005**, *58*, 57–63. [CrossRef]
29. Kumar, A.N.; Mohan, S.V. Acidogenic valorization of vegetable waste for short chain carboxylic acids and biohydrogen production: Influence of pretreatment and pH. *J. Clean. Prod.* **2018**, *203*, 1055–1066. [CrossRef]
30. Dahiya, S.; Sarkar, O.; Swamy, Y.V.; Venkata Mohan, S. Acidogenic fermentation of food waste for volatile fatty acid production with co-generation of biohydrogen. *Bioresour. Technol.* **2015**, *182*, 103–113. [CrossRef]
31. Naresh Kumar, A.; Venkata Mohan, S. Acidogenesis of waste activated sludge—Biohydrogen production with simultaneous short chain carboxylic acids. *J. Environ. Chem. Eng.* **2018**, *6*, 2983–2991. [CrossRef]
32. Jankowska, E.; Chwiałkowska, J.; Stodolny, M.; Oleskowicz-Popiel, P. Effect of pH and retention time on volatile fatty acids production during mixed culture fermentation. *Bioresour. Technol.* **2015**, *190*, 274–280. [CrossRef]
33. Garcia, N.H.; Strazzera, G.; Frison, N.; Bolzonella, D. Volatile fatty acids production from household food waste. *Chem. Eng. Trans.* **2018**, *64*, 103–108.
34. Ding, H.H.; Chang, S.; Liu, Y. Biological hydrolysis pretreatment on secondary sludge: Enhancement of anaerobic digestion and mechanism study. *Bioresour. Technol.* **2017**, *244*, 989–995. [CrossRef]
35. Sun, J.; Guo, L.; Li, Q.; Zhao, Y.; Gao, M.; She, Z.; Wang, G. Structural and functional properties of organic matters in extracellular polymeric substances (EPS) and dissolved organic matters (DOM) after heat pretreatment with waste sludge. *Bioresour. Technol. J.* **2016**, *219*, 614–623. [CrossRef]
36. Kuruti, K.; Nakkasunchi, S.; Begum, S.; Juntupally, S.; Arelli, V.; Anupoju, G.R. Rapid generation of volatile fatty acids (VFA) through anaerobic acidification of livestock organic waste at low hydraulic residence time (HRT). *Bioresour. Technol.* **2017**, *238*, 188–193. [CrossRef]
37. Moretto, G.; Valentino, F.; Pavan, P.; Majone, M.; Bolzonella, D. Optimization of urban waste fermentation for volatile fatty acids production. *Waste Manag.* **2019**, *92*, 21–29. [CrossRef]

38. APHA. *Standard Methods for the Examination of Water and Wastewater*, 22nd ed.; American Public Health Association, American Water Works Association. Water Environment Federation: Washington, DC, USA, 2012.

Sample Availability: Samples of the compounds in this article are not available from the authors.

Article

Comparison of Two Process Schemes Combining Hydrothermal Treatment and Acidogenic Fermentation of Source-Separated Organics

Long Lin [1], Ehssan Hosseini Koupaie [2], Armineh Azizi [2], Amir Abbas Bazyar Lakeh [2], Bipro R. Dhar [1], Hisham Hafez [2,3] and Elsayed Elbeshbishy [2,*

[1] Department of Civil and Environmental Engineering, University of Alberta, 116 Street NW, Edmonton, AB T6G 1H9, Canada; llin2@ualberta.ca (L.L.); bipro@ualberta.ca (B.R.D.)

[2] Environmental Research Group for Resource Recovery, Department of Civil Engineering, Faculty of Engineering, Architecture and Science, Ryerson University, 350 Victoria Street, Toronto, ON M5B 2K3, Canada; ehssan.hosseini.k@gmail.com (E.H.K.); armineh.azizi@ryerson.ca (A.A.); amir.bazyar@ryerson.ca (A.A.B.L.); Hisham.Hafez@greenfield.com (H.H.)

[3] Greenfield Global, 275 Bloomfield Road, Chatham, ON N7M 0N6, Canada

* Correspondence: elsayed.elbeshbishy@ryerson.ca; Tel.: +1-4169-79-5000 (ext. 7618); Fax: +1-4169-79-5122

Academic Editor: Ivet Ferrer

Received: 14 March 2019; Accepted: 6 April 2019; Published: 13 April 2019

Abstract: This study compares the effects of pre- and post-hydrothermal treatment of source-separated organics (SSO) on solubilization of particulate organics and acidogenic fermentation for volatile fatty acids (VFAs) production. The overall COD solubilization and solids removal efficiencies from both schemes were comparable. However, the pre-hydrolysis of SSO followed by acidogenic fermentation resulted in a relatively higher VFA yield of 433 mg/g VSS, which was 18% higher than that of a process scheme with a post-hydrolysis of dewatered solids from the fermentation process. Regarding the composition of VFA, the dominance of acetate and butyrate was comparable in both process schemes, while propionate concentration considerably increased in the process with pre-hydrolysis of SSO. The microbial community results showed that the relative abundance of *Firmicutes* increased substantially in the fermentation of pretreated SSO, indicating that there might be different metabolic pathways for production of VFAs in fermentation process operated with pre-treated SSO. The possible reason might be that the abundance of soluble organic matters due to pre-hydrolysis might stimulate the growth of more kinetically efficient fermentative bacteria as indicated by the increase in *Firmicutes* percentage.

Keywords: acidogenic fermentation; hydrothermal treatment; source separated organics; volatile fatty acids; particulate organics solubilization; microbial community analysis

---

1. Introduction

In the world, about 1.2 billion tonnes of food are wasted every year through the food supply chain, accounting for one-third of the food production [1]. In Canada, more than 27 million tonnes of food waste are disposed of annually [2]. If not adequately managed, the large amount of organic waste may cause significant social, environmental, and economic challenges. Source separated collection has been an effective management strategy to separate organic waste from other waste streams at the source, thereby minimizing the contamination of organic waste for the downstream process. Source-separated organics (SSO) is the term used to describe the organic waste stream collected at the source [3]. Besides food waste, SSO also contains other organic components, including yard trimmings (grass, leaves, etc.), fibres (paper towels, napkins, tea bags, etc.) and wood waste [4].

The bioconversion of SSO to various value-added products (e.g., CH_4, H_2, volatile fatty acids), and thereby reduce waste for ultimate disposal has attracted increasing attention. Specifically, waste-derived volatile fatty acids (VFA) production has been emerging due to the diverse application of VFAs [5]. The biological steps involved in VFA production are hydrolysis and acidogenesis. Hydrolysis, in which bacteria use enzymes to break down particulate matters or macromolecule into soluble compounds or monomers, is usually considered as the rate-limiting step. Pre-treatment of organic waste can be used to enhance the hydrolysis of particulate matters [6]. Among various pre-treatment methods, hydrothermal pre-treatment is considered environmentally friendly because no chemical addition is needed. The hydrothermal pre-treatment uses high temperature and pressure to increase the ionized products of water, which facilitate the hydrolysis of particulate matters into the soluble phase [7]. Also, the associated thermal energy can potentially be recovered at the end of the process, thereby reducing energy demand [8]. Moreover, various types of waste heat from boiler steam or flue gas can be utilized for the hydrothermal process at no additional cost [9].

Previous studies have reported positive impacts of hydrothermal pre-treatment of food waste on fermentative VFA production. For instance, a study showed hydrothermal pre-treatment of food waste at 160 °C for 30 min, and achieved 43% increase in SCOD and 55% increase in VFA after fermentation for 15 days [10]. Yin et al. [10] also reported that hydrothermal temperature higher than 170 °C led to the formation of some toxic and non-biodegradable products despite higher COD dissolution. Similarly, another study also examined the effect of hydrothermal pre-treatment of kitchen waste and obtained 1.3 times higher VFA production after fermentation for 21 days [11]. The compositions of SSO vary considerably from food waste, which could profoundly affect the hydrothermal pre-treatment and subsequent fermentation performance. However, most of these studies focused on food waste and did not identify detailed information on microbial composition during fermentation of pre-treated substrates. The differences between microbial communities involved in fermentation of SSO and pretreated SSO remained to be determined. Furthermore, as the hydrothermal process mainly enhances the solubilization of particulates, it may instead be used as post-treatment after fermentation to target only the solid fraction of fermented effluent, which has rarely been investigated. Thus, it is still unknown whether the order of applying the hydrothermal process affects solubilization of particulate organics and VFA production when combining with acidogenic fermentation.

Therefore, the objective of this study was to compare the performance of the hydrothermal process used as pre-treatment and post-treatment on solubilization of particulate organics as well as VFA production from SSO. The main parameters evaluated were solubilization of particulate matters, solids reduction, and VFAs production and composition. The microbial communities were also examined in the fermented effluent with/without hydrothermal pre-treatment to elucidate the possible VFA production pathways. Lastly, the implications of this study were also discussed comparing to literature. This study first compared the two process schemes combining hydrothermal treatment and acidogenic fermentation of SSO and provided new information into microbiome that drives fermentation with pretreated SSO. The findings may help advance understanding of fermentation process and operation integrating with hydrothermal treatment.

2. Results and Discussion

2.1. Solubilization of Particulate Matters

Figure 1 shows the concentrations and reduction efficiencies of suspended solids in raw, treated, and fermented SSO samples. The initial TSS and VSS concentrations in raw SSO were 62,000 mg/L and 45,400 mg/L, respectively. In system-1, raw SSO was fermented followed by post-hydrolysis of dewatered solids from the fermentation process. Conversely, in system-2, raw SSO was subjected to hydrothermal pre-treatment followed by fermentation. Both systems showed a step-wise decrease in TSS and VSS concentrations after each step of treatment. Moreover, each step in system-2 showed a superior performance in suspended solids reduction than the corresponding step in system-1. After fermentation (system-1), the TSS and VSS concentrations decreased to 52,500 mg/L and 35,700 mg/L respectively. The subsequent hydrothermal post-treatment further decreased the TSS and VSS concentrations to 39,000 mg/L and 27,000 mg/L, respectively. In system-2, the TSS and VSS concentrations first substantially reduced to 43,000 mg/L and 31,000 mg/L after thermal pre-hydrolysis; then, further reduced to 36,700 mg/L and 25,500 mg/L after fermentation, respectively. These results indicated that both pre- and post- thermal hydrolysis processes efficiently hydrolyzed particulate organics. The thermal hydrolysis process alone (31–32%) had a superior effect on the reduction of suspended solids of SSO than fermentation alone (16–21%). Notably, a major portion of the solid reduction in system-2 was contributed by the pre-hydrolysis step. Nonetheless, overall solids removal efficiencies from both schemes were comparable despite the slight advantage showed in system-2 (41–44% vs. 37–41%).

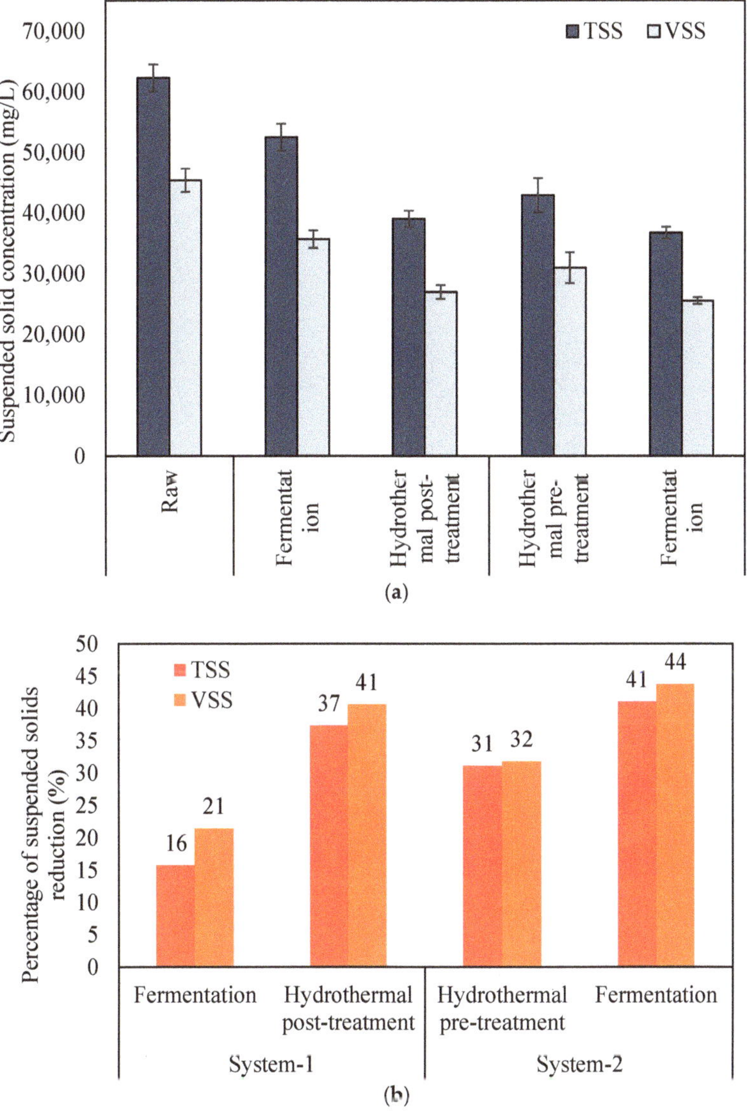

Figure 1. (**a**) The concentrations of suspended solids and (**b**) suspended solid reduction efficiency in raw, treated, and fermented SSO. (TSS: Total Suspended Solids, VSS: Volatile Suspended Solids)

Figure 2 shows concentrations of TCOD and SCOD, total and soluble carbohydrate, and total and soluble protein. The degrees of COD, carbohydrate, and protein solubilization are provided in the Supplementary Information (Figure S1). The initial concentrations of SCOD (32,500 mg/L), soluble carbohydrate (442 mg/L), and soluble protein (100 mg/L) in the raw SSO accounted for 32.8%, 3.5%, 6.3% of the TCOD, total carbohydrate, and total protein, respectively. Similar to solids reduction, both systems showed a step-wise increase in SCOD after each step of treatment, and each step in system-2 showed a better performance in terms of COD solubilization than the corresponding step in system-1. The fermentation in system-1 first increased the SCOD concentration from 32,500 mg/L to 45,600 mg/L, resulting in a 20% degree of solubilization. The subsequent post-hydrolysis further increased the SCOD to 54,200 mg/L, leading to an overall degree of solubilization of 33%. In comparison, the pre-hydrolysis in system-2 first increased the SCOD concentration to 49,000 mg/L, resulting in a 25% degree of solubilization. The subsequent fermentation process further increased the SCOD to 56,200 mg/L, corresponding to an overall degree of solubilization of 35%. Thus, the COD solubilization results were consistent with the trend observed for solids reduction. Furthermore, it is worth noting that the TCOD remained almost constant in both systems, indicating that negligible COD (<2%) were released as gaseous by-products (e.g., H_2, CO_2) from the fermentation or hydrolysis process, which was in line with the observation of negligible H_2 production in this study. As shown in Figure 2b,c, fermentation reduced the concentration of soluble carbohydrate and soluble protein in both process schemes, mainly due to the fermentation of these compounds to lower molecular organics, including various volatile organic acids.

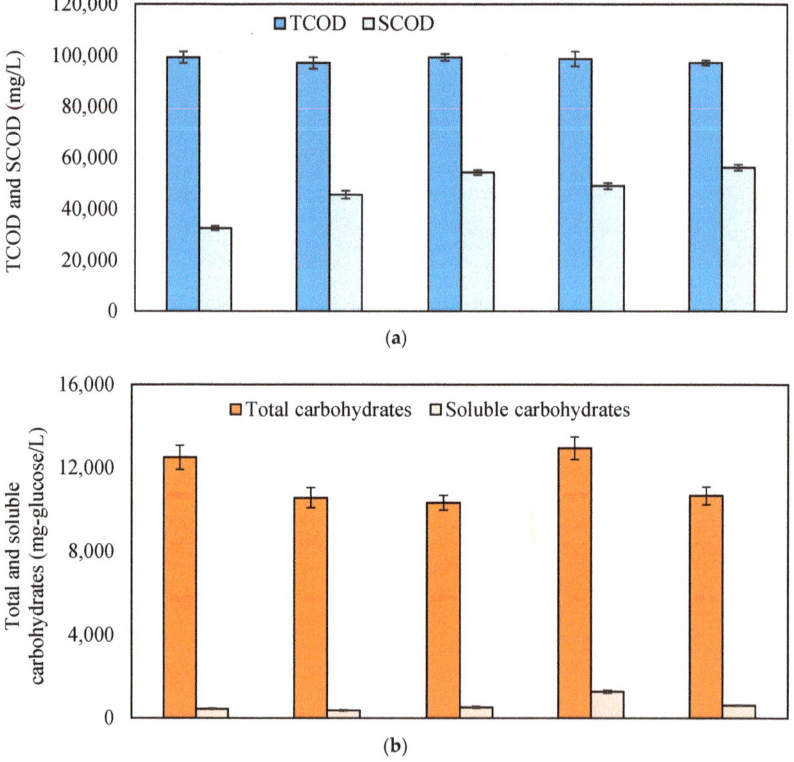

(a)

(b)

Figure 2. *Cont.*

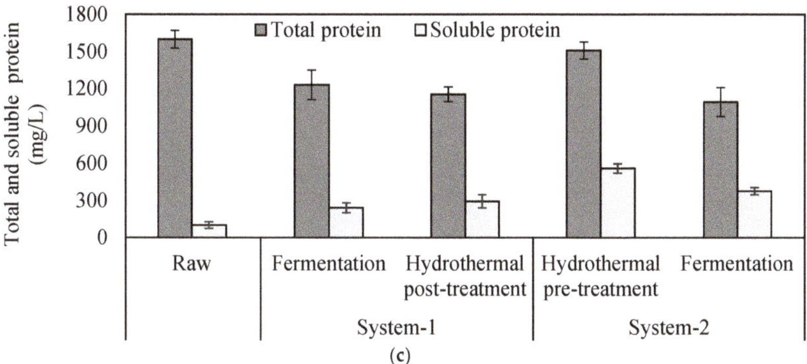

Figure 2. Concentrations of (**a**) TCOD and SCOD, (**b**) total and soluble carbohydrates, and (**c**) total and soluble protein in raw, treated, and fermented SSO. (TCOD: Total Chemical Oxygen Demand, SCOD: Soluble Chemical Oxygen Demand, SSO: Source Separated Organics)

The post-hydrolysis of dewatered solids from the fermentation process slightly increased the soluble carbohydrate and protein in system-1, while the pre-hydrolysis of SSO in system-2 considerably increased the concentrations of soluble carbohydrate and protein. According to the literature, the high temperature in the hydrothermal pre-treatment process hydrolyzed macromolecules (e.g., starch, cellulose, and hemicellulose, protein) into small molecular matter like oligosaccharide and monosaccharide (e.g., glucose and xylose), leading to the increase of soluble carbohydrates and protein [5,9]. Meanwhile, some soluble sugars and amino acids were further degraded into short-chain VFAs, such as acetic acid, thus decreasing the total carbohydrates/proteins. Lipids have been reported to be relatively stable under the hydrothermal process and fermentation [12]. Organic matters present in SSO are mostly carbohydrates, proteins, and lipids [13]. Therefore, the increase in the SCOD concentration of SSO during hydrothermal pre-treatment was most likely due to the solubilization of macromolecular organic compounds like carbohydrates and proteins. After fermentation, SCOD was increased while soluble carbohydrate and protein concentrations were decreased, which suggest that other intermediates (e.g., VFA) rather than soluble carbohydrate or protein were the main contributors of SCOD after fermentation.

2.2. Yields and Distribution of VFAs

Figure 3a shows the total VFA concentrations in raw, hydrolyzed, and fermented SSO samples. The corresponding VFA yields and SCOD yields (mg/g VSS) are provided in the Supplementary Information (Table S1). The initial VFA in raw SSO was 8400 mg/L. In system-1, the VFA was significantly increased to 15,000 mg/L after fermentation, which was 80% higher compared to raw SSO. The subsequent hydrothermal process only further increased the VFA concentration slightly to 16,700 mg/L. In contrast, pre-hydrolysis in system-2 only slightly increased the VFA concentration (18%) in SSO. The subsequent fermentation substantially increased the VFA concentration to 19,700 mg/L with an overall increase of 135% compared to the raw sample. Nonetheless, the VFA production primarily credited to the fermentation process in both systems. Accordingly, the highest VFA yields (433 mg/g VSS) and SCOD yields (1240 mg/g VSS) were obtained from fermentation of pre-treated SSO in system-2, which were 135% and 73% higher than those in raw SSO samples, respectively (Supplementary Information, Table S1). This result indicated that the order of applying the hydrothermal process affected the VFA production. While combining hydrothermal process and fermentation, applying hydrothermal process as pre-treatment showed a slight advantage over as post-treatment, resulting in 18% higher VFA production. Because the hydrothermal process solubilized particulate to soluble phase, providing soluble organics for fermentation, thereby enhancing the VFA production during the fermentation process. Similar to VFA production, ammonia concentration (Supplementary Information,

Figure S2) was substantially increased after fermentation, indicating the effective degradation of organic nitrogen (protein) during fermentation.

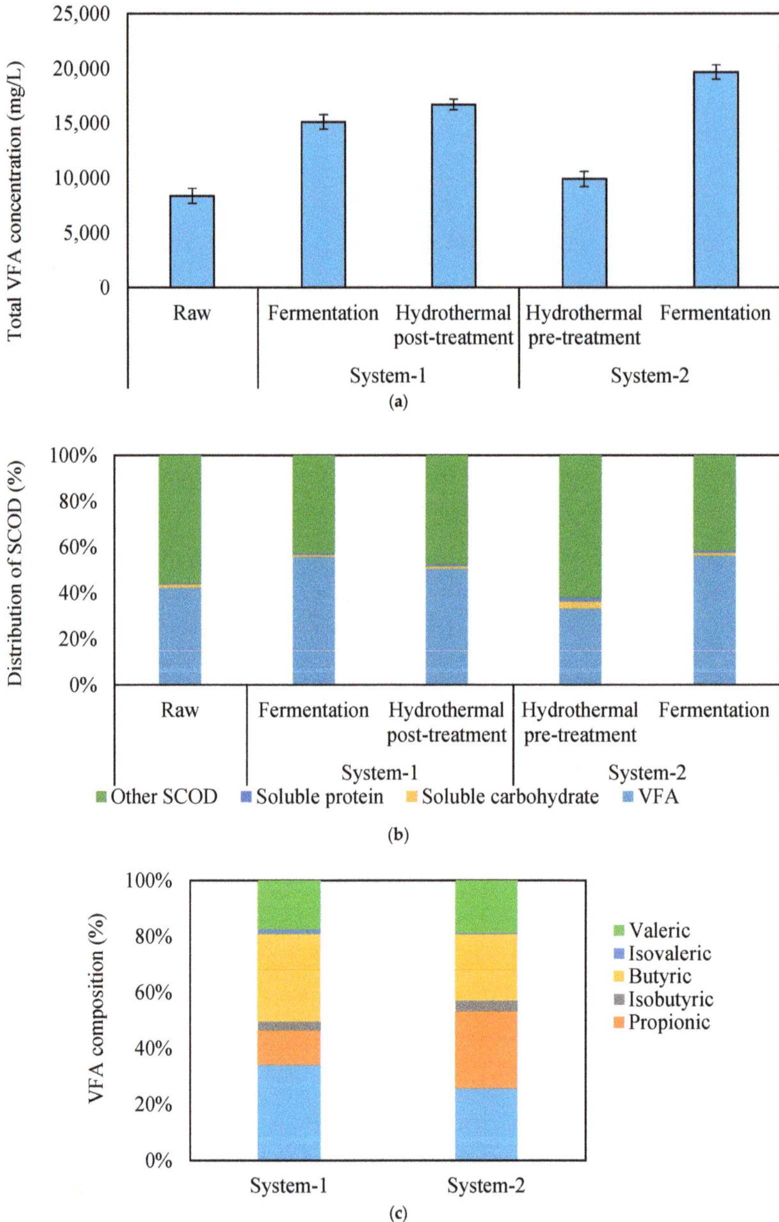

Figure 3. (**a**) The concentration of VFA, (**b**) distribution of SCOD component in raw, treated, and fermented SSO samples, and (**c**) composition of VFA in fermented SSO samples. (SCOD: Soluble Chemical Oxygen Demand, VFA: Volatile Fatty Acids, SSO: Source Separated Organics)

The highest VFA yield (433 mg/g VSS) attained in this study was further compared to those obtained in other studies. As shown in Table 1, VFA yields varied with operating conditions, such as hydrothermal temperature and holding time, and fermentation temperature and retention time. The substrate sources also had large impacts on VFA yields. Nonetheless, the VFA yield in this study is comparable to that reported by Ding et al., [12], but higher than that achieved by Li and Jin [14] and lower than those obtained by Yin et al., [10] and Yu et al., [11]. However, the VFA yield of SSO could be further improved by optimising other factors, such as HRT of fermentation, which should be investigated in future research.

Figure 3b shows the distribution of SCOD component in the raw, hydrolyzed, and fermented SSO samples based on the organic compounds, including VFA, soluble carbohydrate, and soluble protein. Among the soluble end products, VFA was the major contributor (56%) of SCOD after fermentation in both process schemes. In contrast, other SCOD rather than VFA was the main contributor (67%) of SCOD after pre-hydrolysis in system-2. In addition to soluble carbohydrates and soluble protein, other soluble SCOD may include alcohols, other organic acids, poly-phenols, furfural, etc. [9,15].

Figure 3c shows the composition of VFA in the fermented SSO samples from system-1 and system-2. In system-1, acetate (35%) and butyrate (30%) were the main products with a small production of propionate (12%). In system-2, besides the dominance of acetate (26%) and butyrate (22%), propionate (30%) was substantially increased compared to that in system-1. The difference in VFA compositions indicated that there might be different metabolic pathways for VFA production after pre-hydrolysis in system-2. According to the literature, acetate and butyrate were mainly produced from glucose fermentation [16]. Each glucose unit could be converted into either two acetate units and four hydrogen units via the acetate pathway or into one butyrate unit and two hydrogen via the butyrate pathway [17]. In contrast, propionate was mainly produced from the fermentation of amino acids [18] or various H_2-consumption pathways [19]. Thus, the higher propionate fraction in system-2 might be produced by amino acid fermentation bacteria or by H_2-consumption propionate producing bacteria. The dominance of propionate in VFAs has also been previously reported after the fermentation of thermally pretreated kitchen waste [14]. Meanwhile, previous studies have also shown that VFA composition in fermentation was highly dependent on the type of waste [12]. SSO characteristics markedly varied given the differences in origin, dietary customs, and collecting sites and seasons [20]. Therefore, the composition of VFAs in this study could be influenced by SSO characteristics.

Table 1. Comparison of VFA yield in this study to literature.

Substrate	Hydrothermal Process Condition	Hydrothermal Process Severity Index	Fermentation Condition	VFA Yield	Increase in VFA	Reference
Food waste from canteen	160 °C, 30 min	3.2	Batch, 30 °C, 15 days	910 mg/g $VS_{removed}$	55%	[10]
Restaurant, canteen, dining hall food waste	160 °C, 20 min	3.1	Batch, Fermentation: 35 °C, 48 h	450 mg/g vs.	240%	[12]
Food waste from canteen	160 °C, 30 min	3.2	Batch, 30 °C, 21 days	740 mg/g $VSS_{removed}$	130%	[11]
Kitchen waste from canteen	120 °C, 50 min	2.3	Batch, 35 °C, 21 days	32 mg/g vs.	No increase	[14]
Kitchen waste from canteen	160 °C, 120 min	3.8	-	280 mmol/L / 1.6 mmol/gVSS	120%	[21]
SSO	170 °C, 30 min	3.5	Semi-continuous, fed once a day, 38 °C, HRT = 3 days	433 mg/gVSS / 570 mg/gVSS$_{removed}$	135%	This study

Note: Severity Index = log {exp[(T − 100)/14.75] × t }, where T is the pretreatment temperature (°C), and t is the holding time (min) [13].

2.3. Microbial Community

Bacterial community at the phylum level and genus level is shown in Figure 4. At the phylum level, *Firmicutes* was the most dominant one in both samples, but was considerably higher in relative abundance in the sample from system-2 (54% vs. 70%) (Figure 4a). *Firmicutes* contains various fermentative bacteria. *Bacteroidetes* was the second most dominant one in both samples (~30%). In system-2, *Actinobacteria* was also slightly higher in relative abundance (1% vs. 3%), while the relative abundance of *Proteobacteria* dramatically decreased (12% vs. 1%).

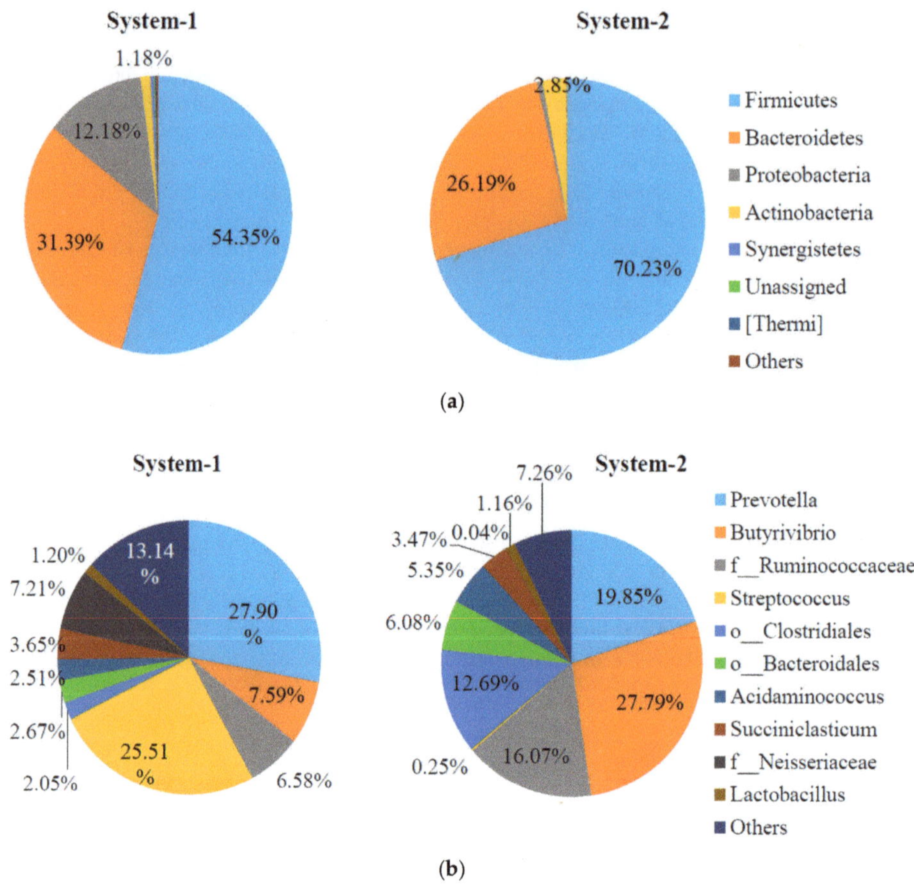

Figure 4. Relative abundance of bacterial composition at (**a**) phylum level and (**b**) genus level in fermentation process in system-1 and system-2. Note: those genera accounted for <1% of the total bacterial population were grouped into "others". Those unidentified genera were represented by the lowest taxonomic level assigned, such as family (f_) or order (o_) level.

At the genus level, *Prevotella* (*Bacteriodetes*) was dominant in both systems, and was relatively higher in relative abundance in system-1 (28% vs. 20%) (Figure 4b). Also, the percentage of *Streptococcus* (*Firmicutes*) was remarkably higher in system-1 (26% vs. 0.3%), which is well-known for lactic acid fermentation [22]. In contrast, *Butyrivibrio* (8% vs. 28%) and *Acidaminococcus* (2.5% vs. 5.4%) were higher in relative abundance in system-2. *Butyrivibrio* spp. contain butyric-acid-producing bacteria [23] and *Acidaminococcus* spp. contain amino-acids-degrading bacteria [24], both of which belong to the phylum *Firmicutes*. Previous studies have also reported the fermentation of model hemicelluloses by *Prevotella* strains and *Butyrivibrio fibrisolvens* [23]. Additionally, those unidentified genera sorted to the family *Ruminococcaceae* and the order *Clostridiales* (*Firmicutes*) were substantially higher in relative abundance in system-2, which contain a variety of fermentative bacteria [2,25]. These results indicated that there might be different metabolic pathways for VFA production in the fermentation process with pretreated SSO. One possible reason might be that the hydrothermal pre-treatment enhanced the soluble organic matters, thereby inducing the growth of various fermentative bacteria as indicated by the increased relative abundance in *Firmicutes*. Interestingly, the genus *Clostridium*, known as the main H_2-producer in the fermentation process [22], was not detected with high relative abundance in either system, which was in line with the observation of negligible H_2 production in this study.

3. Materials and Methods

3.1. SSO and Inoculum

In this study, the SSO was provided by the Disco Road Organics Processing Facility (Toronto, ON, Canada). This facility process SSO that consists of food waste, pet waste, houseplants, paper food packaging, diapers, and biodegradable plastics. At Disco facility, the raw materials first undergo visual inspection to remove the large items from the waste stream. After visual inspection, the waste streams enter a BTA® hydro-mechanical system (BTA International GmbH, Pfaffenhofen, Germany) that transforms the non-homogeneous mixture of the waste into a semi-homogeneous pulp. The BTA process starts with a screening followed by hydropulpers to remove the unwanted materials (e.g., plastic bags, sand, metals, and glass shards) from the produced pulp. It is worth noting that the SSO utilized in this study was collected from the pulp stream after the BTA process. The fermentation inoculum was collected from a full-scale mesophilic anaerobic digester located in Ashbridge Bay wastewater treatment plant (Toronto, ON, Canada). The digester was operated at a temperature of 35–37 °C and fed with a mixture of primary and secondary municipal sludge. Table 2 summarizes the characteristics of the SSO and inoculum used in this study.

Table 2. The characteristics of SSO and inoculum.

Parameters	SSO	Inoculum
pH	5.55 ± 0.1	7.9 ± 0.1
TS mg/L	65,300 ± 1900	15,700 ± 100
VS mg/L	47,900 ± 1304	9250 ± 150
VS/TS (%)	72 ± 0.3	59 ± 1
TSS mg/L	62,000 ± 3500	15,400 ± 300
VSS mg/L	45,400 ± 2000	10,000 ± 200
VSS/TSS (%)	72.9 ± 1.9	65.2 ± 0.3
TCOD mg/L	99,200 ± 3200	18,000 ± 200
SCOD mg/L	32,500 ± 900	500 ± 70
Total carbohydrates (mg/L glucose)	12,500 ± 450	NA
Soluble carbohydrates (mg/L glucose)	450 ± 30	NA
Total protein mg/L	1600 ± 150	NA
Soluble protein mg/L	100 ± 15	NA
Total nitrogen (mg/L N)	3100 ± 607	1300 ± 40
Total phosphate (mg/L PO_4^-)	1250 ± 354	1350 ± 30
Alkalinity (mg/L CaCO3)	5600 ± 300	4700 ± 15
Ammonia (NH_3-N)	900 ± 300	700 ± 50
VFA (mg/L HoAc)	8400 ± 680	680 ± 20

Total solids (TS), volatile solids (VS), total suspended solids (TSS), volatile suspended solids (VSS), total chemical oxygen demand (TCOD), soluble chemical oxygen demand (SCOD), volatile fatty acids (VFA), source separated organics (SSO).

3.2. Experimental Design and Set-Up

The hydrothermal treatment of SSO was carried out using a high temperature (up to 330 °C), and high pressure (up to 1900 psi) stirred reactor with a maximum sample volume of 2 L (Parr 4848, Parr Instrument Company, Moline, IL, USA). The Parr hydrothermal reactor was equipped with an automated controller with auto-tuning capabilities that allows for accurate monitoring of both the heating and cooling parameters including target temperature, holding time (soak) as well as the heating/cooling rate. The reactor content (SSO) was continuously mixed with the aid of a mechanical mixer connected to a speed controller. The hydrothermal treatment parameters such as temperature and pressure were monitored on a real-time basis via the SpecView software version 3.1.144 (SpecView Corporation, Gig Harbor, WA, USA). In this study, for each round of treatment, 2 L of SSO was delivered into the reactor vessel. After sealing the vessel, the mechanical mixer was set to the rotational speed of 150 rpm. The mixer was kept running until the end of the cooling cycle. The hydrothermal temperature of 170 °C with a holding time of 30 min was used in this study, which was the favorable condition obtained from the previous study [13]. For this purpose, the temperature of SSO increased gradually at a rate of 3 °C/min until it reached the target temperature of 170 °C (Cycle 1). Next, the temperature was kept constant at 170 °C for 30 min (Cycle 2). Afterwards, the cooling cycle was started and continued until the temperature reached below 50 °C (Cycle 3). The hydrothermal treatment condition (temperature, pressure, and retention time) was selected considering the optimum hydrothermal treatment condition obtained among 15 different tested scenarios in an earlier study performed by the same research group [13].

The experimental scheme of this study is shown in Figure 5. In system-1, the raw SSO was first fermented in a semi-continuous process. The fermenter effluent was then centrifuged for 30 min at 5000 rpm using a centrifuge (Sorvall Legend XT centrifuge, Fisher Scientific, Hampton, NH, USA) to separate the liquid and solid fraction of the fermented effluent. Then, the solid fraction of the fermented effluent was subjected to hydrothermal post-treatment. Afterwards, the liquid fraction (that was already separated through centrifugation) was mixed with the hydrothermally post-treated solid fraction to produce the final sample for analysis. In system-2, the raw SSO was first subjected to hydrothermal pre-treatment. Then, the pretreated SSO was used as the feedstock for a semi-continuous fermentation to produce a final sample for analysis.

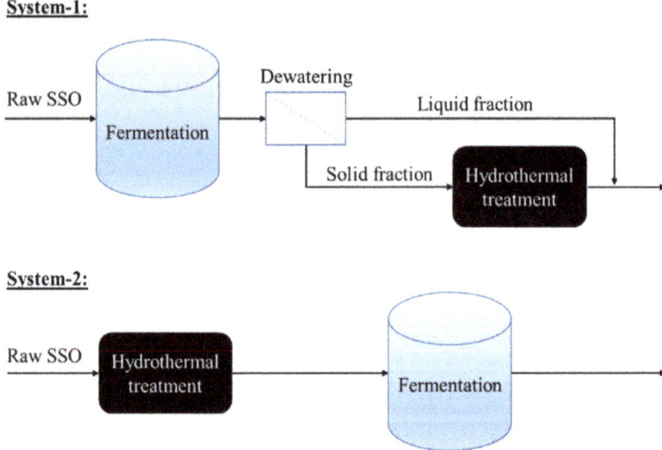

Figure 5. Schematic diagram showing the experimental set-up.

Feeding of the fermenters was done semi-continuously (once every 24 h). Before starting the fermentation process, the content of the fermenters was purged for a duration of 5 min with nitrogen gas to achieve a completely anaerobic condition. The pH of the fermenters was kept in the range of 5.4–5.6 by controlling the pH of the feed (raw and pre-treated SSO) throughout the operation period using sodium hydroxide (NaOH) and hydrochloric acid (HCl). The fermentation process was conducted in a water bath set at a temperature of 38 ± 1 °C. The fermenter content was continuously mixed throughout the operation period at a rate of 150 rpm using a sealed rotating plastic shaft. Every 24 h, 500 mL of the fermentation effluent was taken out, and the fermenters were fed again with 500 mL of either raw or thermally pretreated SSO, resulting in hydraulic retention time (HRT) of 3 d.

The feeding and decanting of the fermenters were done manually using a 500 mL aluminum syringe. Both the raw and thermally pretreated fermenters were operated for an initial period of 20 days to reach a steady-state condition and then the operation was continued for an additional 30 days. Eight samples (twice a week) were taken for analyses during the steady-state period.

3.3. Analytical Methods

The measurement of the samples solid content in terms of total solids (TS), volatile solids (VS), total suspended solids (TSS), and volatile suspended solids (VSS) was performed according to the Standard Methods [26]. The analysis of chemical oxygen demand (COD) was done through the closed reflux colorimetric method outlined by the Standard Methods. The carbohydrates and protein determination was conducted according to the colorimetric methods developed by Dubios et al. [27] and Frolund et al. [28]. The absorbance measurement of the samples was done using a spectrophotometer (DR 3900, HACH, Loveland, CO, USA) with the wavelengths of 600, 595, 490, 560, 650, 394, and 542 nm for the analysis of COD, protein, carbohydrates, ammonia, alkalinity, total nitrogen, and total phosphate, respectively. The COD, protein, and carbohydrate analyses were performed on both the total and soluble phases of the samples. To collect the soluble fraction for analysis, the samples were first centrifuged at 9000 rpm for 30 min using a Sorvall Legend XT centrifuge. Next, the liquid fraction (supernatant) of the centrifuged samples were filtered through 0.45 μm microfiber filters. The concentration of the total VFAs was determined calorimetrically using the TNT 872 kit (HACH). The VFAs composition was determined using a gas chromatograph (Varian 8500, Varian Inc., Toronto, ON, Canada) equipped with a fused silica column (30 m × 0.32 mm) and a flame ionization detector. Helium was used at a constant flow rate of 5 mL/min as the carrier gas. The temperatures of the detector and column were set at 250 and 110 °C, respectively.

3.4. Microbial Community Analysis

Microbial communities of the fermented effluent from both schemes were analyzed with high throughput 16S rDNA sequencing. The genomic DNA of the biomass samples were extracted using the PowerSoil® DNA Isolation Kit (MoBio Laboratories, Carlsbad, CA, USA). The extracted DNA concentrations were quantified using a spectrophotometer (NanoDrop 2000C, Thermo Fisher Scientific, Waltham, MA, USA). The extracted DNA samples were stored immediately at −70 °C prior to performing Illumina Miseq sequencing at the Research and Testing Laboratory (Lubbock, TX, USA) using the bacterial primer set 357Wf: CCTACGGGNGGCWGCAG and 785R: GACTACHVGGGTATCTAATCC to target the 16S rDNA. The demultiplexed sequencing data were processed and analyzed using the Quantitative Insights Into Microbial Ecology (QIIME v2) software [29]. The sequences were denoised (dada2 method) to remove and/or correct noisy reads, remove chimeric sequences and singletons, and join denoised paired-end reads [30]. The denoised sequences were assigned to species-equivalent operational taxonomic units (OTUs) at a 97% sequence similarity level using the open-reference OTU picking method (vsearch method against 2013-08 Greengenes database) [31].

3.5. Calculations

The degree of solubilization of carbohydrate, protein, and COD were calculated to assess the hydrolysis efficiency (Equation (1)):

$$\text{Degree of solubilization} = (S_t - S_0)/(T_0 - S_0) \times 100\% \tag{1}$$

where, S_t is the concentration of soluble organic materials (carbohydrate, protein, or COD) in the treated SSO (mg/L), S_0 is the initial concentration of soluble organic materials in the raw SSO (mg/L), and T_0 is the initial concentration of total organic materials in the raw SSO (mg/L).

The yields of SCOD and VFA were calculated (Equation (2)):

$$\text{Yield} = S_t/VSS_0, \text{ expressed as mg/g} - VSS \tag{2}$$

where, S_t is the final concentration of SCOD or VFA in the treated SSO (mg/L), and VSS_0 is the initial concentration of volatile suspended solids in the raw SSO (mg/L).

The conversion factors used for the calculation of COD of the carbohydrate (1.19 gCOD/g), protein (1.42 gCOD/g), acetate (1.07 gCOD/g), propionate (1.51 gCOD/g), butyrate (1.82 gCOD/g), and valerate (2.04 gCOD/g) were determined as previously described in the literature [32].

3.6. Statistical Analysis

The statistically significant effects of the independent variables (i.e., pretreatment temperature, pressure, and retention time) were tested by multi-factor analysis of variance (ANOVA) at a 95% confidence level ($\alpha = 0.05$) using Minitab Software 17 (Minitab, LLC, State College, PA, USA). The ANOVA analysis was conducted by considering 2-level interaction effects of the independent variables. The Fisher's least significant difference analysis was used to compare all pairs of means.

4. Conclusions

The process schemes of pre- and post-hydrothermal treatment of SSO were successfully performed and compared for solids removal and VFA yields. Results showed that solubilization of organic matters through pre-hydrolysis could boost subsequent acidogenic fermentation, resulting in an 18% increase in overall VFA yield over post-hydrolysis of fermented SSO. Both acetate and butyrate were dominant in both process schemes, while propionate was higher in the process with pre-hydrolysis. Meanwhile, various fermentative bacteria belonging to *Firmicutes* were higher in relative abundance in the process with pre-hydrolysis of SSO, suggesting different metabolic pathways for VFA productions in both process schemes.

Supplementary Materials: The following are available online. Table S1. Soluble chemical oxygen demand (SCOD) and volatile fatty acid (VFA) yields. Figure S1. (a) Suspended solid reduction efficiency and (b) degree of solubilization of COD, carbohydrate, and protein in system-1 and system-2. (TSS: Total Suspended Solids, VSS: Volatile Suspended Solids, COD: Chemical Oxygen Demand). Figure S2. Ammonia concentration of raw, treated, and fermented source separated organics (SSO) samples.

Author Contributions: Data curation, E.H.K.; Formal analysis, L.L., A.A. and A.A.B.L.; Funding acquisition, H.H.; Investigation, A.A. and A.A.B.L.; Resources, H.H.; Supervision, B.R.D. and E.E.; Validation, B.R.D. and E.E.; Writing—original draft, L.L.; Writing—review & editing, E.H.K.

Funding: This research was funded by Southern Ontario Water Consortium (SOWC), grant number SUB02523.

Acknowledgments: The authors would like to thank the Disco Road Organics Facility for providing the SSO samples.

Conflicts of Interest: The authors declare no conflict of interest. The funders had no role in the design of the study; in the collection, analyses, or interpretation of data; in the writing of the manuscript, or in the decision to publish the results.

References

1. Xu, F.; Li, Y.; Ge, X.; Yang, L.; Li, Y. Anaerobic digestion of food waste–challenges and opportunities. *Bioresour. Technol.* **2018**, *247*, 1047–1058. [CrossRef] [PubMed]
2. Paritosh, K.; Kushwaha, S.K.; Yadav, M.; Pareek, N.; Chawade, A.; Vivekanand, V. Food waste to energy: An overview of sustainable approaches for food waste management and nutrient recycling. *Biomed. Res. Int.* **2017**, *2017*, 1–19. [CrossRef] [PubMed]
3. Luk, G.K.; Bekmuradov, V. Energy products from source-separated organic waste. *WIT Trans. Ecol. Environ.* **2014**, *180*, 469–478.

4. Hosseini Koupaie, E.; Azizi, A.; Bazyar Lakeh, A.A.; Hafez, H.; Elbeshbishy, E. Comparison of liquid and dewatered digestate as inoculum for anaerobic digestion of organic solid wastes. *Waste Manag.* **2019**, *87*, 228–236.

5. Lee, W.S.; Chua, A.S.M.; Yeoh, H.K.; Ngoh, G.C. A review of the production and applications of waste-derived volatile fatty acids. *Chem. Eng. J.* **2014**, *235*, 83–99. [CrossRef]

6. Cesaro, A.; Belgiorno, V. Pretreatment methods to improve anaerobic biodegradability of organic municipal solid waste fractions. *Chem. Eng. J.* **2014**, *240*, 24–37. [CrossRef]

7. Zheng, Y.; Zhao, J.; Xu, F.; Li, Y. Pretreatment of lignocellulosic biomass for enhanced biogas production. *Prog. Energy Combust. Sci.* **2014**, *42*, 35–53. [CrossRef]

8. Dhar, B.R.; Nakhla, G.; Ray, M.B. Techno-economic evaluation of ultrasound and thermal pretreatments for enhanced anaerobic digestion of municipal waste activated sludge. *Waste Manag.* **2012**, *32*, 542–549. [CrossRef] [PubMed]

9. Ariunbaatar, J.; Panico, A.; Esposito, G.; Pirozzi, F.; Lens, P.N.L. Pretreatment methods to enhance anaerobic digestion of organic solid waste. *Appl. Energy* **2014**, *123*, 143–156. [CrossRef]

10. Yin, J.; Wang, K.; Yang, Y.; Shen, D.; Wang, M.; Mo, H. Improving production of volatile fatty acids from food waste fermentation by hydrothermal pretreatment. *Bioresour. Technol.* **2014**, *171*, 323–329. [CrossRef]

11. Yu, X.; Yin, J.; Wang, K.; Shen, D.; Long, Y.; Chen, T. Enhancing Food Waste Hydrolysis and the Production Rate of Volatile Fatty Acids by Prefermentation and Hydrothermal Pretreatments. *Energy Fuels* **2016**, *30*, 4002–4008.

12. Ding, L.; Cheng, J.; Qiao, D.; Yue, L.; Li, Y.Y.; Zhou, J.; Cen, K. Investigating hydrothermal pretreatment of food waste for two-stage fermentative hydrogen and methane co-production. *Bioresour. Technol.* **2017**, *241*, 491–499. [CrossRef] [PubMed]

13. Razavi, A.S.; Hosseini Koupaie, E.; Azizi, A.; Hafez, H.; Elbeshbishy, E. Hydrothermal pretreatment of source separated organics for enhanced solubilization and biomethane recovery. *Bioresour. Technol.* **2018**, *274*, 502–511. [CrossRef] [PubMed]

14. Li, Y.; Jin, Y. Effects of thermal pretreatment on acidification phase during two-phase batch anaerobic digestion of kitchen waste. *Renew. Energy* **2015**, *77*, 550–557. [CrossRef]

15. Dhar, B.R.; Elbeshbishy, E.; Hafez, H.; Lee, H.S. Hydrogen production from sugar beet juice using an integrated biohydrogen process of dark fermentation and microbial electrolysis cell. *Bioresour. Technol.* **2015**, *198*, 223–230. [CrossRef] [PubMed]

16. Xia, A.; Cheng, J.; Murphy, J.D. Innovation in biological production and upgrading of methane and hydrogen for use as gaseous transport biofuel. *Biotechnol. Adv.* **2016**, *34*, 451–472. [CrossRef] [PubMed]

17. Elbeshbishy, E.; Hafez, H.; Dhar, B.R.; Nakhla, G. Single and combined effect of various pretreatment methods for biohydrogen production from food waste. *Int. J. Hydrogen Energy* **2011**, *36*, 11379–11387. [CrossRef]

18. Cheng, J.; Ding, L.; Xia, A.; Lin, R.; Li, Y.; Zhou, J.; Cen, K. Hydrogen production using amino acids obtained by protein degradation in waste biomass by combined dark- and photo-fermentation. *Bioresour. Technol.* **2015**, *179*, 13–19. [CrossRef]

19. Gonzalez-garcia, R.A.; Mccubbin, T.; Navone, L.; Stowers, C.; Nielsen, L.K.; Marcellin, E. Microbial Propionic Acid Production. *Fermentation* **2017**, *3*, 21. [CrossRef]

20. Sundberg, C.; Franke-Whittle, I.H.; Kauppi, S.; Yu, D.; Romantschuk, M.; Insam, H.; Jönsson, H. Characterisation of source-separated household waste intended for composting. *Bioresour. Technol.* **2011**, *102*, 2859–2867. [CrossRef]

21. Jin, Y.; Li, Y.; Li, J. Influence of thermal pretreatment on physical and chemical properties of kitchen waste and the efficiency of anaerobic digestion. *J. Environ. Manag.* **2016**, *180*, 291–300. [CrossRef] [PubMed]

22. Lee, H.S.; Krajmalinik-Brown, R.; Zhang, H.; Rittmann, B.E. An electron-flow model can predict complex redox reactions in mixed-culture fermentative BioH2: Microbial ecology evidence. *Biotechnol. Bioeng.* **2009**, *104*, 687–697. [CrossRef] [PubMed]

23. Emerson, E.L.; Weimer, P.J. Fermentation of model hemicelluloses by Prevotella strains and Butyrivibrio fibrisolvens in pure culture and in ruminal enrichment cultures. *Appl. Microbiol. Biotechnol.* **2017**, *101*, 4269–4278. [CrossRef] [PubMed]

24. Rogosa, M. Acidaminococcus gen. n., Acidaminococcus fermentans sp. n., anaerobic gram-negative diplococci using amino acids as the sole energy source for growth. *J. Bacteriol.* **1969**, *98*, 756–766.

25. Cabrol, L.; Marone, A.; Tapia-Venegas, E.; Steyer, J.; Ruiz-Filippi, G.; Trably, E. Microbial ecology of fermentative hydrogen producing bioprocesses: useful insights for driving the ecosystem function. *FEMS Microbiol. Rev.* **2017**, *41*, 158–181. [CrossRef]

26. APHA. *Standard Methods for the Examination of Water and Wastewater*, 21st ed.; American Public Health Association: Washington, DC, USA, 2005.

27. DuBois, M.; Gilles, K.A.; Hamilton, J.K.; Rebers, P.A.; Smith, F. Colorimetric Method for Determination of Sugars and Related Substances. *Anal. Chem.* **1956**, *28*, 350–356. [CrossRef]

28. Frolund, B.; Griebe, T.; Nielsen, P.H. Enzymatic activity in the activated-sludge floc matrix. *Appl. Microbiol. Biotechnol.* **1995**, *43*, 755–761. [CrossRef]

29. Caporaso, J.G.; Kuczynski, J.; Stombaugh, J.; Bittinger, K.; Bushman, F.D.; Costello, E.K.; Fierer, N.; Peña, A.G.; Goodrich, J.K.; Gordon, J.I.; et al. QIIME allows analysis of high-throughput community sequencing data. *Nat. Methods* **2010**, *7*, 335–336. [CrossRef]

30. Callahan, B.J.; McMurdie, P.J.; Rosen, M.J.; Han, A.W.; Johnson, A.J.A.; Holmes, S.P. DADA2: high-resolution sample inference from Illumina amplicon data. *Nat. Methods* **2016**, *13*, 581–583. [CrossRef]

31. Rideout, J.R.; He, Y.; Navas-Molina, J.A.; Walters, W.A.; Ursell, L.K.; Gibbons, S.M.; Chase, J.; McDonald, D.; Gonzalez, A.; Robbins-Pianka, A.; et al. Subsampled open-reference clustering creates consistent, comprehensive OTU definitions and scales to billions of sequences. *PeerJ.* **2014**, *2*. [CrossRef]

32. Angelidaki, I.; Sanders, W. Assessment of the anaerobic biodegradability of macropollutants. *Rev. Environ. Sci. Bio/Technology* **2004**, *3*, 117–129. [CrossRef]

Sample Availability: Samples of the compounds are not available from the authors.

Article

On the Effect of Aqueous Ammonia Soaking Pre-Treatment on Continuous Anaerobic Digestion of Digested Swine Manure Fibers

Chrysoula Mirtsou-Xanthopoulou [1], Ioannis V. Skiadas [2] and Hariklia N. Gavala [2,*]

[1] Aalborg University Copenhagen, Department of Chemistry and Bioscience, A C Meyers Vænge 15, DK 2450 Copenhagen SV, Denmark

[2] Technical University of Denmark, Department of Chemical and Biochemical Engineering, Søltofts Plads 229, 2800 Kgs. Lyngby, Denmark

* Correspondence: hnga@kt.dtu.dk or hari_gavala@yahoo.co.uk; Tel.: +45-21171323

Academic Editors: Derek J. McPhee and Ivet Ferrer

Received: 4 June 2019; Accepted: 2 July 2019; Published: 5 July 2019

Abstract: (1) Background: The continuously increasing demand for renewable energy sources renders anaerobic digestion as one of the most promising technologies for renewable energy production. Due to the animal production intensification, manure is being used as the primary feedstock for most biogas plants. Their economical profitable operation, however, relies on increasing the methane yield from the solid fraction of manure, which is not so easily degradable. The solid fraction after anaerobic digestion, the so-called digested fibers, consists mainly of hardly biodegradable material and comes at a lower mass per unit volume of manure compared to the solid fraction before anaerobic digestion. Therefore, investigation on how to increase the biodegradability of digested fibers is very relevant. So far, Aqueous Ammonia Soaking (AAS), has been successfully applied on digested fibers separated from the effluent of a manure-fed, full-scale anaerobic digester to enhance their methane productivity in batch experiments. (2) Methods: In the present study, continuous experiments at a mesophilic (38 °C) CSTR-type anaerobic digester fed with swine manure first and a mixture of manure with AAS-treated digested fibers in the sequel, were performed. Anaerobic Digestion Model 1 (ADM1) previously fitted on manure fed digester was used in order to assess the effect of the addition of AAS-pre-treated digested manure fibers on the kinetics of anaerobic digestion process. (3) Results and Conclusions: The methane yield of AAS-treated digested fibers under continuous operation was 49–68% higher than that calculated in batch experiments in the past. It was found that AAS treatment had a profound effect mainly on the disintegration/hydrolysis rate of particulate carbohydrates. Comparison of the data obtained in the present study with the data obtained with AAS-pre-treated raw manure fibers in the past revealed that hydrolysis kinetics after AAS pre-treatment were similar for both types of biomasses.

Keywords: ADM1; anaerobic digestion; aqueous ammonia soaking pre-treatment; continuous; digested manure fibers; modelling

---

1. Introduction

Anaerobic digestion is one of the most promising renewable energy technologies, as it provides a solution to both environmental and energy considerations. Some of the benefits of this technology are that it reduces the odour, minimizes the size of the organic wastes, contributes to the reduction of the greenhouse gases emissions and produces a high value fertilizer as well as a renewable energy gas. Anaerobic digestion of swine manure has been a wide matter of discussion and it has proved a promising approach for renewable energy production in the form of methane. Nowadays, due to the intensification of animal production, manure seems to be the largest available substrate for biogas plants, in Europe. According to the European parliament [1] 55% of the available biomass for biogas production comes exclusively from animal manure. In fact, 22 large-scale biogas plants are currently under operation in Denmark using manure as primary feedstock, but their economical profitable operation relies on increasing the methane yield from manure [2].

Although manure is undoubtedly an excellent carrier for biogas plants, it contains an organic fraction which is not easily degraded. Many different strategies have been developed so far to increase the methane potential of the manure. New separation technologies that are being applied before anaerobic digestion have been tested over during the last few years [3]. The liquid fraction of the manure could be used as fertilizer in the farms while the solid fraction could be transported to centralized biogas plants for methane production [4]. Additionally, several different pre-treatment technologies have been applied to increase the methane yield of manure and manure fibers. Some representative results are those of Hartmann et al., [5], where mechanical maceration was applied on fibers with a 25% increase on the methane yield, and also the study of Bruni et al., [6] who applied hydrothermal, chemical, as well as enzymatic pretreatments on fibers, with a maximum increase of 66%. One chemical pre-treatment, which has been initially investigated for bioethanol production is Aqueous Ammonia Soaking (AAS). Ammonia is a proven delignification reagent, which is quite safe to handle, non-polluting and non-corrosive [7], and due to its high volatility, it can be easily recovered and recycled [7], thus avoiding the need of any further chemical consumption. More recently, AAS was also applied on the treatment of the fibrous fraction of raw and digested manure fibers, under room temperature [8,9]. In both studies, Aqueous Ammonia Soaking (AAS) treatment combined with subsequent removal of ammonia (which also gives the possibility for being recycled in a full-scale plant) was applied effectively, resulting in a high increase of the methane yield. According to Jurado et al. [8] an increase of up to 78% was observed compared to the non-treated raw manure fibers (control), during the first 16 days of digestion while optimisation of the pre-treatment conditions for raw manure fibers resulted in an impressive 244% increase of the methane yield [10]. Similar results were observed in the study of Mirtsou-Xanthopoulou et al. [9] where an increase of up to 110% was observed in the methane yield of digested manure fibers compared to the non-treated ones (control). Post-treatment of digested fibers focuses only on hardly biodegradable biomass in contrast to raw fibers where easily biodegradable material is also present. Therefore, the mass of fibers to be treated is expected to be significantly lower in case of digested fibers leading thus to a more economical process. The question that the present study addresses is whether models developed for digesters processing manure can actually be applied when the influent stream is supplemented with AAS-treated digested fibers and how disintegration/hydrolysis kinetics compare to those of AAS-treated raw fibers.

In general, mathematical models are used in order to efficiently describe and predict the efficiency of different biological processess. Different models for describing the anaerobic digestion process and predicting the methane production and effluent characteristics started to emerge as early as in 1970 [11]. The Anaerobic Digestion Model 1 (ADM1) is a generic mathematical model, which was developed specifically for the anaerobic digestion process in 2002 by the IWA anaerobic digestion task group [12]. ADM1 is a modeling platform that can be used for simulating the biogas process under different substrates and operating conditions, taking into account the most important biological and physicochemical processes of anaerobic digestion. Since ADM1's publication, numerous studies have been published applying, fitting and tuning ADM1 to different feedstocks, reactors and operating conditions [13]. Digesting manure with different substrates is a case that has also been lately applied in modeling with ADM1 [14–19], although to a lesser extent considering that there are around 300 ADM1 based studies and only 10% refer to manure as the main feedstock despite the fact that manure is one of the basic and most well-known feedstocks of anaerobic digesters.

In the present study, continuous experiments at a mesophilic (38 °C), CSTR-type, anaerobic digester fed with swine manure first and a mixture of manure with AAS-treated digested fibers in the sequel, were performed in order to verify the experimental data obtained from batch experiments. A modified

Anaerobic Digestion Model 1, accounting for co-digesting feedstocks of different composition [20] was used in order to assess the effect of the addition of AAS-pre-treated digested manure fibers on the kinetics of anaerobic digestion process.

2. Results and Discussion

2.1. Influent Characterization

The characteristics of manure and AAS-treated digested manure fibers are shown in Table 1. As anticipated, the AAS-treated fibers consisted mainly of particulate organic matter with carbohydrates being a substantial fraction while manure was rich in soluble organic matter.

Table 1. Characteristics of manure (A) and (B) and of Aqueous Ammonia Soaking (AAS)-pre-treated raw manure fibers.

Characteristics, Particulate Matter	Manure A [a] (Used in Phases 1 and 2)	AAS-Pretreated Digested Fibers (Used in Phases 2, 3 and 4)	Manure B [a] (Used in Phases 3 and 4)
COD, kg/100 kgTS	79.11	75.34	124.19
Carbohydrates, kg COD/100 kg TS	6.10	32.75	9.57
Proteins, kg COD/100 kg TS	28.00	21.41	43.95
Lipids, kg COD/100 kg TS	32.65	1.10	51.25
Inerts, kg COD/100 kg TS	12.37	20.08	19.42
Characteristics, soluble matter			
COD, kg/100 kgTS	91.83	23.7	55.90
Sugars, kg COD/100 kg TS	0	0	0.0
Aminoacids, kg COD/100 kg TS	18.92	1.02	11.52
Long chain fatty acids, kg COD/100 kg TS	21.25	19.49	12.94
Valeric acid, kg COD/100 kg TS	0	0.66	0.0
Butyric acid, kg COD/100 kg TS	3.98	0.00	2.42
Propionic acid, kg COD/100 kg TS	8.46	0.34	5.15
Acetic acid, kg COD/100 kg TS	34.75	2.07	21.15
Inerts, kg COD/100 kg TS	4.46	0	2.71
Inorganic carbon, kmole/100 kg TS	0.53	0.05	0.32
Inorganic phosphorus, kmole/100 kg TS	9.59×10^{-3}	6.52×10^{-3}	0.0058
Inorganic nitrogen (NH_3-N), kmole/100 kg TS	0.73	4.28×10^{-2}	0.44

[a] as reported in Jurado et al. [20].

2.2. Continuous Experiments and Modeling Results

Operating conditions and steady state characteristics of the anaerobic digester fed with manure A (first phase) and a mixture of manure A and AAS-pretreated digested manure fibers (second phase) are summarised in Table 2. Despite the fact that manure fibers were characterised by a much lower content in soluble organic material (Table 1) the biogas production of the second steady state (with manure and fibers as influent) was 16% higher than that of the first (with only manure as influent). The methane yield was calculated as 0.303l CH_4/g TS during the first steady state with manure as influent and 0.272 l CH_4/g TS during the second steady state when the digester was being fed with mixture of manure and AAS-treated digested fibers. Given that the TS ratio of manure:fibers in the influent of the digester was 0.52:0:48 and assuming that the methane yield due to the manure fraction was 0.303l CH_4/g TS also during the second steady state, then the methane yield of AAS-treated digested fibers was calculated to be 0.238 l CH_4/g TS. In previous studies, the methane yield from AAS-pretreated digested fibers was measured in the range of 141–160 mL CH_4/g TS in batch experiments (compared to a methane yield of 76 mL CH_4/g TS from non-treated digested fibers) [9]. Consequently, the continuous experiments verified what it was observed in batch experiments, namely that AAS pretreatment resulted to a significant increase of methane yield of digested fibers. The even higher methane yield obtained from the continuous experiment was most probably due to the adaptation of the mixed microbial culture on the different feedstock since the batch experiments were performed with inoculum which was adapted to manure only, where actually no hydrolysis had taken place [20].

Table 2. Operating conditions, experimental measurements and model predictions for the steady states on manure and mixture of manure and AAS-treated digested fibers.

Operating Conditions	Manure A (1st Phase)		Mixture of Manure A and AAS-Treated Digested Fibers (2nd Phase)	
HRT, d	24–25		24–25	
Flow rate, mL/d	114–134		117–125	
Organic loading rate (OLR), g COD/l/d	2.83–3.53		2.30–3.65	
Soluble OLR, g soluble COD/l/d	1.52–1.90		1.00–1.58	
Steady state characteristics	Manure A (1st phase)		Mixture of manure and AAS-treated digested fibers (2nd phase)	
	Experimental	Model	Experimental	Model
Methane, %	67.6 ± 2.1	64.2 ± 8.3	66.6 ± 1.8	67.0 ± 7.6
Biogas production, m^3/d	$2.32 \times 10^{-3} \pm 0.2 \times 10^{-3}$	$2.65 \times 10^{-3} \pm 0.1 \times 10^{-3}$	$2.70 \times 10^{-3} \pm 0.3 \times 10^{-3}$	$2.9 \times 10^{-3} \pm 0.2 \times 10^{-3}$
pH	8.2	8.1	8.1	8.0
Acetic acid, kgCOD/m^3	0.142 ± 0.025	0.087 ± 0.013	0.161 ± 0.084	0.081 ± 0.018
Propionic acid, kgCOD/m^3	0.023 ± 0.005	0.018 ± 0.002	0.012 ± 0.003	0.019 ± 0.002
Butyric acid, kgCOD/m^3	0.026 ± 0.002	0.011 ± 0.001	0.007 ± 0.0001	0.012 ± 0.001

ADM1 as developed in Jurado et al. [20] was used to simulate the first phase where only manure A was being added in the reactor. During the second phase, where the digester was fed with the mixture of manure A and AAS-pretreated digested fibers, fitting of the ADM1 to the experimental data was performed. The only kinetic parameters, which were mostly affected, were those of carbohydrates and proteins hydrolysis. Hydrolytic kinetic constants for carbohydrates and proteins in manure had been calculated as 0 and 2.8×10^{-4} d^{-1}, respectively [20], while for the mixture were increased to 6.8×10^{-2} and 7.0×10^{-3} d^{-1} and are in very good agreement with the hydrolysis constants estimated for AAS-pretreated raw manure fibers, 7.3×10^{-2} and 7.1×10^{-3} d^{-1}, respectively [20]. Hydrolysis constants were remarkably lower compared to all soluble substrate consumption rates, including volatile fatty acids and therefore it can be assumed that hydrolysis is the rate-limiting step for the overall anaerobic digestion process when adding manure fibers at the ratio tested hereby. Kinetics of volatile fatty acids uptake seemed to be unaffected as the behaviour of the digester under the pulse disturbances was satisfactorily simulated by the model without any further change in the kinetic parameters. The latter is very well depicted in Figures 1–3 where the experimental and predicted by the model VFA concentrations during the pulse experiments are presented.

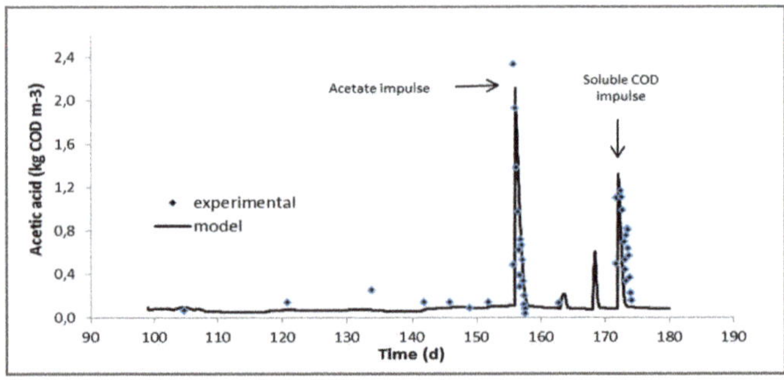

Figure 1. Experimental and predicted by the model acetic acid concentration during the experiment with mixture of manure and AAS-treated digested fibers as influent.

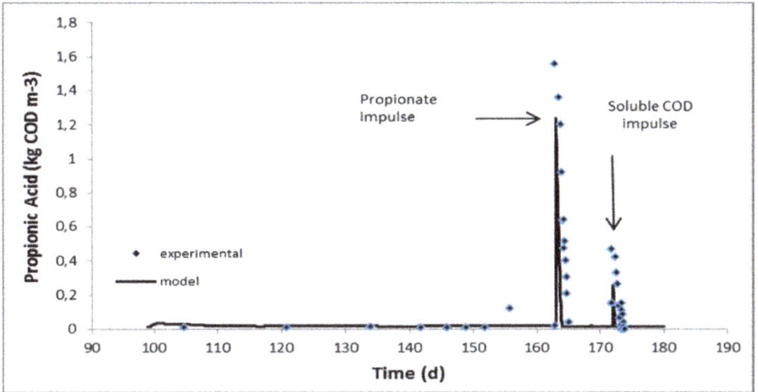

Figure 2. Experimental and predicted by the model propionic acid concentration during the experiment with mixture of manure and AAS-treated digested fibers as influent.

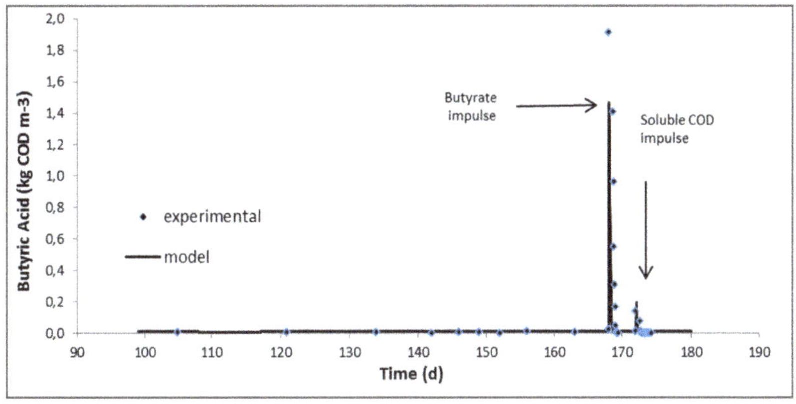

Figure 3. Experimental and predicted by the model butyric acid concentration during the experiment with mixture of manure and AAS-treated digested fibers as influent.

Model predictions for the steady state on manure A are also shown in Table 2. It is noticeable that the model, which was developed in another digester fed with manure, was able to accurately predict the steady state reached in the present study. Table 2 also includes model predictions for the second phase, after fitting of the model had been performed. It is obvious that the model was able to adequately predict the steady state reached on the mixture of manure A and AAS-pretreated digested fibers after alteration of just the hydrolytic constants.

Subsequently, ADM1, with the new values of the hydrolytic constants as obtained from the second phase, was used to simulate the performance of the digester during the third and fourth phases, where a mixture of manure B and AAS-treated digested fibers (at a ratio of 0.52:0.48 on Total Solids basis) was the influent at an HRT of 25 and 11.2 d., respectively.

The experimentally measured, as well as the calculated by the model, biogas production rate together with the organic (TS based) loading rate of the digester throughout the four phases of the continuous experiment are shown in Figure 4. The model exhibited a remarkable ability to predict the biogas production rate after significant change of the feeding characteristics (from manure A to manure B in the third phase) as well as of the HRT of the system (from 25 d to 11.2 d in the fourth phase). Last but not least, the model also satisfactorily predicted the methane concentration in the biogas, of the last two phases of the continuous experiments. Specifically, for the third phase, the model predicted an average percentage of methane in the biogas of around 64.6 ± 0.5 and experimentally it was measured as 63.2 ± 0.1. For the fourth phase, the methane percentages predicted from the model and measured experimentally were 68.2 ± 2.2 and 63.2 ± 0.4, respectively.

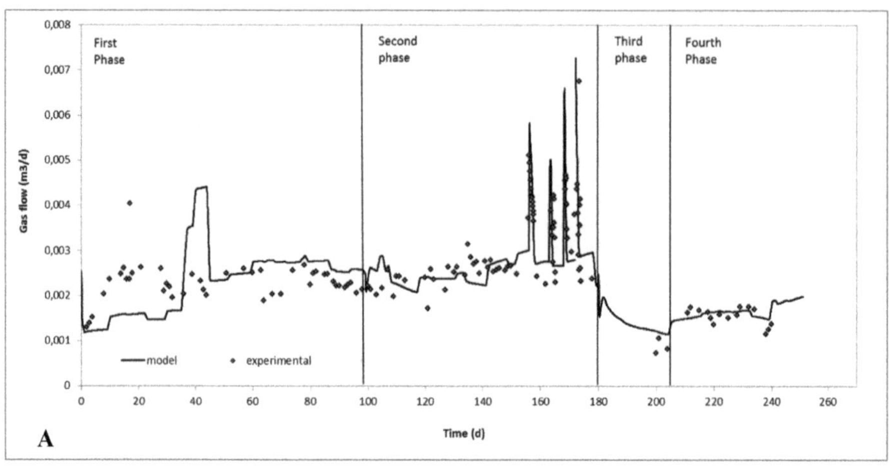

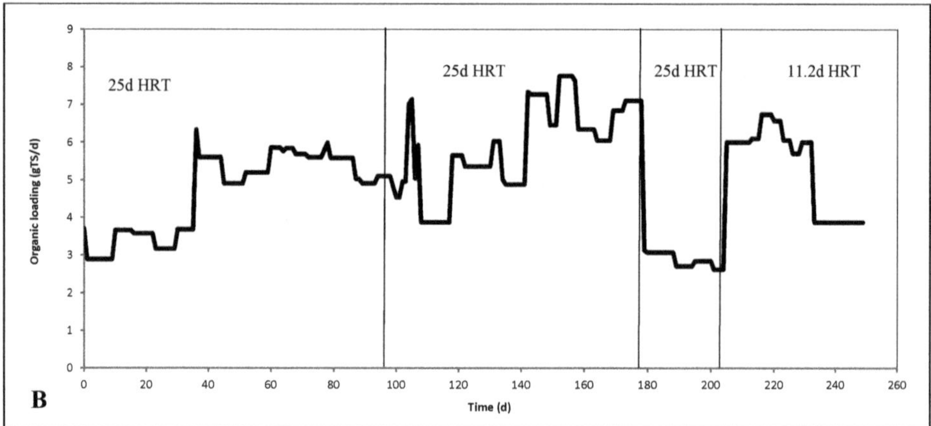

Figure 4. Experimental and theoretical biogas production rate in m^3/d (**A**) and organic loading rate in g TS/d (**B**) throughout the four phases.

3. Materials and Methods

3.1. Feedstock

Digested manure fibers (fibers collected after anaerobic digestion) and two batches of swine manure (A and B) were kindly provided by Morsø BioEnergi and stored at −20 °C until used. The two batches of manure were collected at a different period of the year, therefore their characteristics varied considerably—as shown in Table 1.

3.2. Analytical Methods

Characterization of the liquid fraction of manure and manure fibers included the determination of the soluble components and was done after centrifugation of the samples at 10,000 rpm for 10 min and filtration of the supernatant through 0.2 μm membrane filters.

Total (TS) and volatile (VS) solids were measured according to standard methods [21]. Total and soluble Chemical Oxygen Demand (COD) was determined using Hach Lange kits LCK_914 (5–60 g L^{-1} range) and LCK_514 (100–2000 mg L^{-1} range), respectively.

For total phosphorus and total nitrogen determination the material was dried at 42 °C overnight and powdered, while for the soluble and inorganic measurements, the material was centrifuged at 10,000 rpm for 10 min and the supernatant was passed through 0.2 μm membrane filters. For the determination of the non-soluble and soluble organic phosphorus persulphate digestion was applied to the solid and liquid fraction, respectively, followed by ascorbic acid photometric determination of phosphate ions [21]. For the determination of non-soluble and soluble organic nitrogen, digestion with the micro-kjeldahl apparatus was applied to the solid and liquid fraction, respectively, followed by distillation and titration (titrimetric method) for ammonium ions determination (APHA, 2005). For the inorganic forms of phosphorus (PO_4^{-3}-P) and nitrogen (NH_4-N) determination, analysis was carried out by applying ascorbic acid photometric determination [21] and by using the Hach Lange kit LCK_305 (1–12 mg L^{-1} range), respectively.

Two groups of carbohydrates were determined in the samples of manure and manure fibers: the first group was the total carbohydrates, including those bound in the lignocellulosic biomass and the second group was the simple sugars [22]. Analysis of the two groups of carbohydrates was carried out based on the NREL analytical procedures [23]. Detection and quantification of sugar monomers (glucose, xylose and arabinose) was made with HPLC-RI equipped with an Aminex HPX-87H column (BioRad, Hercules, CA, USA) at 60 °C. A solution of 4 mmol L^{-1} H_2SO_4 was used as eluent at a flow rate of 0.6 mL min^{-1}. Samples for HPLC analysis were acidified with a 10% *w/w* solution of H_2SO_4, centrifuged at 10,000 rpm for 10 min and finally filtered through a 0.45 μm membrane filter. Klason lignin was also determined according to NREL analytical procedures [23]. For the determination of both total carbohydrates and lignin the material was dried at 42 °C overnight and powdered, while for the determination of free sugars the material was centrifuged at 1000 rpm and the supernatant was passed through 0.2 μm size membrane filters.

For the quantification of Volatile Fatty Acids, VFA, 1 mL of sample was acidified with 17% H_3PO_4 and filtered through minisart high flow filter (pore size 0.5 μm). VFAs were analyzed on a gas chromatograph (PerkinElmer Clarus 400) with a flame ionization detector and a capillary column (Agilent HP-FFAP column, 30 m long and 0.53 mm inner diameter).

Biogas composition in methane was measured with a gas chromatograph (SRI GC model 310) equipped with a thermal conductivity detector and a packed column (Porapak-Q, length 6 ft and inner diameter 2.1 mm). The temperature for injector, column and detector was set to 80 °C. The volume of methane produced in sealed vials during methane potential tests was calculated by multiplying the biogas composition in methane with the headspace volume.

3.3. Experimental Set-Up

3.3.1. Ammonia Pre-Treatment

Samples of digested manure fibers were soaked in ammonia reagent (32% *w/w* in ammonia) at a ratio of 10 mL reagent per g TS, under room temperature. The treatment was performed in closed glass flasks to avoid ammonia evaporation. After the completion of the treatment, water was added at a ratio of 10 mL per g TS to facilitate the subsequent ammonia distillation step. Distillation was performed using a rotary evaporator (Buchi RII Rotavapor) with a vertical condenser.

3.3.2. Continuous Experiments

One mesophilic (38 °C) CSTR-type digester of 3 l useful volume was started-up using mixed anaerobic liquor from a swine manure treating digester and fed with swine manure A at a hydraulic retention time of approximately 25 d (first phase). After the digester reached steady-state, the influent was changed (second phase) to a mixture of swine manure A and AAS-treated digested manure fibers (at a ratio of 0.52:0.48 on Total Solids basis). The ratio has been chosen based on the maximum addition of fibers allowing for a smooth operation (no blockings of the flow of material in the pipes) of the lab-scale digesters and pumps. Subsequently and after the digester reached the second steady state, the manure provided by Morsø BioEnergi changed to a new batch of manure (B) with much lower TS content (1.63 gTS/100 g manure) than the one used for the first and second phase (4.4 gTS/100 g manure). The digester was allowed to reach steady state (third phase) with the new type of manure under the same operating parameters applied for the second phase (TS ratio of manure: AAS treated fibers and HRT). After the third steady state was reached, the HRT was reduced to 11.2 d (fourth phase).

Summarizing, the four phases involved are as following:

First Phase: manure A as influent at an HRT of 25 d.

Second Phase: mixture of manure A and AAS-treated digested fibers (at a ratio of 0.52:0.48 on Total Solids basis) as influent at an HRT of 25 d.

Third Phase: mixture of manure B and AAS-treated digested fibers (at a ratio of 0.52:0.48 on Total Solids basis) as influent at an HRT of 25 d.

Fourth Phase: mixture of manure B and AAS-treated digested fibers (at a ratio of 0.52:0.48 on Total Solids basis) as influent at an HRT of 11.2 d.

The feeding of the digester was intermittent and repeated once a day during all phases. The operation, however, can actually be considered as "continuous" due to the relatively long retention times applied. Daily monitoring of the digester included biogas production and composition in methane, pH, volatile fatty acids and soluble COD concentration. When the digester reached the second steady state while fed with the mixture of manure and fibers, complete characterisation was performed in terms of almost all measurable components of ADM1 (see paragraph 3.4.2). Subsequently, the digester was subjected to pulse disturbances of acetic, propionic and butyric acids and soluble influent fraction (obtained after centrifugation and filtration) in order to study the dynamics of the processes (those experiments are considered part of the second phase). Following this, 50 mL concentrated solutions of acetic, propionic and butyric acids and a 600 mL sample consisting of the soluble fraction of the mixture (manure and digested fibers), were consecutively injected into the digester in order to increase the reactor concentration of the injected substrate to around 1600, 1200 and 1600 mg COD/l for the acetic, propionic and butyric acids, respectively, and to around 400 mg COD/l for the soluble fraction of the mixture [24]. The response of the bioreactor to these disturbances was monitored mainly through measurements of biogas production and composition in methane, pH and VFA concentration until all components approach their levels before the pulse.

3.4. Modeling

Anaerobic Digestion Model No1 (ADM1) was used to simulate the anaerobic digestion process and to assess the effect that the addition of AAS pretreated digested fibers had on the kinetic parameters. In the study by Jurado et al. [20], IWA anaerobic digestion model (ADM1) had been fitted to the experimental data obtained from a swine manure-fed digester which was different than the one used in this study and hydrolysis constants of carbohydrates (khydr_ch), proteins (khydr_pr) and lipids (khydr_li) and maximum uptake rates of long chain fatty acids (km_fa) and volatile fatty acids (km_c4, km_pro and km_ac for butyric, propionic and acetic acid, respectively) had been calculated as shown in Table 3. That model was used to simulate the first steady state of the reactor (with only swine manure as influent). Subsequently, the model was fitted to the experimental data obtained throughout the second phase, where a mixture of manure and fibers was introduced as influent and new hydrolysis constants had been obtained. Finally, the experimental results obtained when the new batch of manure (B) was added as well as when the Hydraulic Retention time (HRT) was reduced to 11.2d (third and fourth phase, respectively) were used to validate the model fitted to the second phase. The software used was Aquasim 2.1 g and the secant method was applied for the parameters estimation.

Table 3. Kinetic parameters as calculated in the study of Jurado et al. for a manure fed anaerobic digester [20].

Kinetic Parameter	Units	Value
Carbohydrates hydrolysis constant, khydr_ch	d^{-1}	<<
Proteins hydrolysis constant, khydr_pr	d^{-1}	3.0×10^{-3}
Lipids hydrolysis constant, khydr_li	d^{-1}	2.8×10^{-4}
Maximum uptake rate of long chain fatty acids, km_fa	kg COD fa/kg COD x_fa/d	0.93
Maximum uptake rate of butyric acid, km_c4	kg COD c4/kg COD x_c4/d	13.1
Maximum uptake rate of propionic acid, km_pro	kg COD pro/kg COD x_pro/d	6.56
Maximum uptake rate of acetic acid, km_ac	kg COD ac/kg COD x_ac/d	45.02

3.4.1. Correction of the HRT in the Kinetic Model

Due to the structural characteristics of the digester, there was an accumulation of solids, which was quantified by measuring Total Suspended Solids (TSS) concentration in the interior (by taking sample directly from inside the digester) and the effluent of the reactor. The higher retention time of the solids was expressed by introducing a new variable into the model, called 'tres'. Tres was calculated by substracting the Hydraulic Retention Time (HRT) from the Solids Retention Time (SRT) (Equation (1)):

$$tres = SRT - HRT \tag{1}$$

In the model, it was considered that solids and microbial biomass were recycled in the reactor and their concentration was corrected according to Equation (2) [25].

$$x = x \cdot \left(Q_{out} - \frac{V_{reactor}}{tres + \frac{V_{reactor}}{Q_{out}}} \right) \tag{2}$$

where x is the solids and biomass recirculation flow (kg COD/d), $V_{reactor}$ is the reactor active volume (m^3) and Q_{out} the effluent flow (m^3/d).

3.4.2. Organic and Inorganic Inputs in the Kinetic Model

Swine manure characteristics were considered as measured and calculated in Jurado et al. [20]. AAS-treated digested manure fibers were characterized in terms of total and soluble COD, total carbohydrates and free sugars, total and soluble Kjendahl nitrogen and NH_3-N, volatile fatty acids (valeric, butyric, propionic and acetic acids), inorganic phosphorus and inorganic carbon. Particulate and soluble inerts were determined as the residual COD after three months of batch anaerobic digestion. Particulate carbohydrates and proteins were calculated as the difference between total and free carbohydrates/sugars and total and soluble Kjeldahl nitrogen, respectively. Aminoacids were calculated as the difference between soluble Kjeldahl nitrogen and NH_3-N. Calculation of particulate lipids was based on the difference between non-soluble COD (total − soluble COD) and the sum of particulate carbohydrates, proteins and inerts, while long chain fatty acids were calculated as the difference between soluble COD and all measured soluble components (sugars, aminoacids, volatile fatty acids and soluble inerts).

In the model, the influent concentration of all organic and inorganic constituents was initially expressed in kg COD/100 kg TS and kmole/100 kg TS, respectively, and subsequently was multiplied with the influent TS concentration and calculated in kg COD/m^3 and kmole/m^3, respectively. This was done in order to take into account the variations of the TS concentration in the feeding. The influent flow rate, Q_{in}, was given in m^3/d and the flow rate of each individual component (in kg COD/d and kmole/d for organic and inorganic substances, respectively) was calculated as the product of the volumetric concentration in the influent with the influent flow rate, Q_{in}.

4. Conclusions

Continuous experiments at a mesophilic (38 °C) CSTR-type anaerobic digester fed with swine manure first and a mixture of manure with AAS-treated digested fibers in the sequel, were performed. The methane yield of AAS-treated digested fibers under continuous operation was 49–68% higher than that calculated in batch experiments in the past. Anaerobic Digestion Model 1 (ADM1) previously fitted on manure fed digester was used and simulated very satisfactorily the first experimental phase, proving thus the robustness of the model and process. Fitting of the second experimental phase, where the influent stream was supplemented with AAS-treated digested fibers (digested fibers refer to manure fibers after anaerobic digestion), indicated that AAS treatment had a profound effect mainly on the hydrolysis rate of particulate carbohydrates and it is noticeable that the estimated disintegration/hydrolysis constants were comparable to those obtained with AAS-treated raw manure fibers (raw fibers refer to manure fibers before anaerobic digestion). Last but not least, pulse experiments with important intermediates of the process, i.e. acetic, propionic and butyric acids, indicated that the associated kinetic constants were not affected by the addition of the AAS-treated digested fibers in the influent.

Author Contributions: Conceptualization, H.N.G.; methodology, C.M.-X., I.V.S. and H.N.G.; validation, C.M.-X.; formal analysis, C.M.-X. and H.N.G.; investigation, C.M.-X.; writing—original draft preparation, C.M.-X.; writing—review and editing, H.N.G. and I.V.S.; visualization, C.M.-X. and H.N.G.; supervision, H.N.G. and I.V.S.; project administration, H.N.G. and I.V.S.; funding acquisition, I.V.S.

Funding: This research was funded by EUDP, Energistyrelsen, Copenhagen, grants RETROGAS and DEMONIAGAS.

Acknowledgments: The authors wish to thank the EUDP, Energistyrelsen, Copenhagen for the financial support of this work under RETROGAS and DEMONIAGAS project.

Conflicts of Interest: The authors declare no conflict of interest.

References

1. Birkmose, T.; Foged, H.L.; Hinge, J. *State of Biogas Plants in European Agriculture*; Danish Agricultural Advisory Service: Brussels, Belgium, 2007.

2. Angelidaki, I.; Ellegaard, L. Codigestion of manure and organic wastes in centralized biogas plants. Status and Future Trends. *Appl. Biochem. Biotechnol.* **2003**, *109*, 95–105.

3. Maibritt Hjorth, K.V.; Christensen, M.L.; Christensen, S.G.S. Solid-liquid Separation of Animal Slurry in Theory and Practice. In *Sustainable Agriculture Volume 2*; Springer: Dordrecht, The Netherlands, 2017; pp. 553–568.

4. Jurado, E.; Gavala, H.N.; Rohold, L.; Skiadas, I. Cost-effective production of biogas from manure—Retrogas project. In Proceedings of the WasteEng10, Beijing, China, 17–19 May 2010; p. 224.

5. Hartmann, H.; Angelidaki, I.; Ahring, B.K. Increase of anaerobic degradation of particulate organic matter in full-scale biogas plants by mechanical maceration. *Water Sci. Technol.* **2000**, *41*, 145–153.

6. Bruni, E.; Jensen, A.P.; Angelidaki, I. Comparative study of mechanical, hydrothermal, chemical and enzymatic treatments of digested biofibers to improve biogas production. *Bioresour. Technol.* **2010**, *101*, 8713–8717.

7. Kim, T.H.; Gupta, R.; Lee, Y. Pretreatment of Biomass by Aqueous Ammonia for bioethanol production. In *Biofuels: Methods and Protocols*; Springer: Dordrecht, The Netherlands, 2009; pp. 79–91.

8. Jurado, E.; Skiadas, I.V.; Gavala, H.N. Enhanced methane productivity from manure fibers by aqueous ammonia soaking pretreatment. *Appl. Energy* **2013**, *109*, 104–111.

9. Mirtsou-Xanthopoulou, C.; Jurado, E.; Skiadas, I.V.; Gavala, H.N. Effect of aqueous ammonia soaking on the methane yield and composition of digested manure fibers applying different ammonia concentrations and treatment durations. *Energies* **2014**, *7*, 4157–4168.

10. Lymperatou, A.; Gavala, H.N.; Skiadas, I.V. Optimization of Aqueous Ammonia Soaking of manure fibers by Response Surface Methodology for unlocking the methane potential of swine manure. *Bioresour. Technol.* **2017**, *244*, 509–516.

11. Gavala, H.; Angelidaki, I.; Ahring, B. Kinetics and modeling of anaerobic digestion process. In *Biomethanation I*; Springer: Dordrecht, The Netherlands, 2003; pp. 57–93.

12. Batstone, D.J.; Keller, J.; Angelidaki, I.; Kalyuzhnyi, S.V.; Pavlostathis, S.G.; Rozzi, A.; Sanders, W.T.M.; Siegrist, H.; Vavilin, V.A. The IWA Anaerobic Digestion Model No. 1 (ADM1). *Water Sci. Technol.* **2002**, *45*, 65–73.

13. Batstone, D.J.; Keller, J.; Steyer, J.P. A review of ADM1 extensions, applications, and analysis: 2002–2005. *Water Sci. Technol.* **2006**, *54*, 1–10.

14. Astals, S.; Ariso, M.; Gali, A.; Mata-Alvarez, J. Co-digestion of pig manure and glycerine: Experimental and modelling study. *J. Environ. Manage.* **2011**, *92*, 1091–1096.

15. Bulkowska, K.; Bialobrzewski, I.; Klimiuk, E.; Pokoj, T. Kinetic parameters of volatile fatty acids uptake in the ADM1 as key factors for modeling co-digestion of silages with pig manure, thin stillage and glycerine phase. *Renew. Energy* **2018**, *126*, 163–176.

16. Gali, A.; Benabdallah, T.; Astals, S.; Mata-Alvarez, J. Modified version of ADM1 model for agro-waste application. *Bioresour. Technol.* **2009**, *100*, 2783–2790.

17. Jablonski, S.J.; Biernacki, P.; Steinigeweg, S.; Lukaszewicz, M. Continuous mesophilic anaerobic digestion of manure and rape oilcake—Experimental and modelling study. *Waste Manag.* **2015**, *35*, 105–110.

18. Mata-Alvarez, J.; Dosta, J.; Mace, S.; Astals, S. Codigestion of solid wastes: A review of its uses and perspectives including modeling. *Crit. Rev. Biotechnol.* **2011**, *31*, 99–111.

19. Nordlander, E.; Thorin, E.; Yan, J. Investigating the possibility of applying an ADM1 based model to a full-scale co-digestion plant. *Biochem. Eng. J.* **2017**, *120*, 73–83.

20. Jurado, E.; Antonopoulou, G.; Lyberatos, G.; Gavala, H.N.; Skiadas, I.V. Continuous anaerobic digestion of swine manure: ADM1-based modelling and effect of addition of swine manure fibers pretreated with aqueous ammonia soaking. *Appl. Energy* **2016**, *172*, 190–198.

21. American Water Works Association; Water Environment Federation; American Public Health Association. *Standard Methods for Examination of Water and Wastewater*, 21st ed.; APHA: Washington, DC, USA, 2005.

22. Haagensen, F.; Skiadas, I.V.; Gavala, H.N.; Ahring, B.K. Pre-treatment and ethanol fermentation potential of olive pulp at different dry matter concentrations. *Biomass Bioenerg.* **2009**, *33*, 1643–1651.

23. Sluite, A.; Hames, B.; Ruiz, R.; Scarlata, C.; Sluiter, J.; Templeton, D.; Crocker, D. Determination of Structural Carbohydrates and Lignin in Biomass. National Renewable Energy Laboratory (NREL) Home Page. Available online: https://www.nrel.gov/ (accessed on 4 July 2019).

24. Kalfas, H.; Skiadas, I.V.; Gavala, H.N.; Stamatelatou, K.; Lyberatos, G. Application of ADM1 for the simulation of anaerobic digestion of olive pulp under mesophilic and thermophilic conditions. *Water Sci. Technol.* **2006**, *54*, 149–156.

25. Antonopoulou, G.; Gavala, H.N.; Skiadas, I.V.; Lyberatos, G. Modeling of fermentative hydrogen production from sweet sorghum extract based on modified ADM1. *Int. J. Hydrogen Energy* **2012**, *37*. [CrossRef]

Sample Availability: Not available.

Article

A Citrus Peel Waste Biorefinery for Ethanol and Methane Production

Maria Patsalou, Charis G. Samanides, Eleni Protopapa, Stella Stavrinou, Ioannis Vyrides and Michalis Koutinas *

Department of Environmental Science & Technology, Cyprus University of Technology,
30 Archbishop Kyprianou Str., 3036 Limassol, Cyprus
* Correspondence: michail.koutinas@cut.ac.cy; Tel.: +357-2500-2067

Academic Editor: Derek J. McPhee
Received: 1 June 2019; Accepted: 3 July 2019; Published: 4 July 2019

Abstract: This paper deals with the development of a citrus peel waste (CPW) biorefinery that employs low environmental impact technologies for production of ethanol and methane. Three major yeasts were compared for ethanol production in batch fermentations using CPW pretreated through acid hydrolysis and a combination of acid and enzyme hydrolysis. The most efficient conditions for production of CPW-based hydrolyzates included processing at 116 °C for 10 min. *Pichia kudriavzevii* KVMP10 achieved the highest ethanol production that reached 30.7 g L^{-1} in fermentations conducted at elevated temperatures (42 °C). A zero-waste biorefinery was introduced by using solid biorefinery residues in repeated batch anaerobic digestion fermentations achieving methane formation of 342 mL g$_{VS}$$^{-1}$ (volatile solids). Methane production applying untreated and dried CPW reached a similar level (339–356 mL g$_{VS}$$^{-1}$) to the use of the side stream, demonstrating that the developed bioprocess constitutes an advanced alternative to energy intensive methods for biofuel production.

Keywords: bioethanol; biomethane; citrus peel waste; biorefinery; biorefinery residues

---

1. Introduction

Vegetable and fruit waste account for 20–50% of household waste in various countries, while citrus peel waste (CPW) comprises a principal residue under the specific category [1]. Global citrus production (orange, lemon, lime, kinnow, sweet orange, etc.) constitutes over 121 × 106 t annually, while the industrial juice manufacturing sector generates about 25 × 10^6 t of CPW [2]. CPW formed during processing of the fruit consists mainly of peels and pressed pulp (seeds and segment membranes), accounting for 50% of the fruit's weight [3,4].

Current management options for CPW include burying in landfills and application as animal feed after drying. However, the thermal dehydration process of CPW to produce animal feeds is energy consuming and not always cost-effective [5], while the final product consists of rather poor animal feed, due to low protein content and high quantity of sugars [6]. Moreover, the waste includes elevated organic matter (approximately 95% of total solids) and water content (approximately 80–90%), as well as low pH (3–4), making CPW inappropriate for landfilling based on the latest EU Waste Framework Directive 2008/98/EC [7].

Although the disposal of CPW is opposed to EU regulations, the waste could serve as a valuable feedstock for the manufacture of biofuels and other commodities. Specifically, CPW comprises high levels of pectin, cellulose, hemicellulose, and soluble sugars, while 0.5% g g^{-1} of essential oils are included on a wet basis [8,9]. D-limonene, constituting the main component of essential oils in CPW, is a terpenic compound with antimicrobial properties and applicability in food, cosmetics, and pharmaceutical industries [10]. Thus, citrus processing industries usually recover essential oils from the waste streams produced as an added-value extractable component [11].

Global climate change associated with the extensive release of greenhouse gases has raised concerns about the application of fossilized hydrocarbons as the main energy source [12]. Thus, in recent years, exploitation of new renewable resources for the production of biofuels as a replacement for the use of the nonrenewable source of petroleum has received much global research interest [13]. The ability of microorganisms to use renewable resources for biofuel synthesis is exploited by the current industry manufacturing biofuels (e.g., bioethanol, biomethane, biobutanol), mainly from sugarcane, corn, and wheat [14]. However, various studies have investigated the application of CPW as a promising feedstock for biofuel production through bioprocessing, using different pretreatment approaches and microorganisms.

Due to the antimicrobial properties of D-limonene, CPW valorization for production of ethanol and methane requires the removal of essential oils to avoid biosystem inhibition [10,15]. Various pretreatment methods have been employed for extraction of essential oils, including steam explosion [10,16,17], hydrothermal sterilization [15,18], popping [19], and drying [20,21]. However, essential oils and pectin constitute high added-value commodities, which were not extracted in several existing studies. Other CPW-based valorization approaches produced extractables and fermentation products using energy intensive approaches, including elevated acid hydrolysis temperatures, low ethanol fermentation temperatures that contribute to additional cooling and separation costs, and anaerobic digestion under thermophilic conditions.

The current study explored alternative and low environmental impact technologies for CPW valorization through the development of a biorefinery strategy. Low temperature dilute acid hydrolysis as a single procedure and combined with enzyme hydrolysis was evaluated as a CPW pretreatment method, following extraction of essential oils (via distillation) and pectin. The hydrolyzate formed was tested for ethanol production employing the thermotolerant strain *Pichia kudriavzevii* KVMP10, which was compared with two industrial yeasts at elevated fermentation temperatures, aiming to reduce biofuel production costs. A zero-waste process was targeted, employing solid biorefinery residues (BR) for methane production, which was compared with the use of CPW and dried citrus peel waste (DCPW), constituting the first study to our knowledge comparing the specific citrus waste-based fractions in anaerobic digestion.

2. Results and Discussion

2.1. The CPW Biorefinery Strategy

The current biorefinery aimed for CPW valorization through the removal of essential oils and pectin, as well as fermentation of the hydrolyzate formed as fermentation feedstock for bioethanol production. A zero-waste concept was established by refinement of remaining solid fractions via anaerobic digestion for methane production. Specifically, the first step included extraction of essential oils using distillation, while the remaining solid residues of the process were dried and applied to dilute acid hydrolysis under varying conditions. Thus, pectin was extracted from the hydrolyzate generated through precipitation with the addition of ethanol, while the hydrolyzate was subsequently applied to distillation for ethanol removal. Following pectin isolation, the hydrolyzate was fermented to bioethanol, testing the effectiveness of three yeast strains (*P. kudriavzevii* KVMP10, *Kluyveromyces marxianus*, and *Saccharomyces cerevisiae*), while the remaining solid residues from dilute acid hydrolysis were anaerobically digested for methane generation (Figure 1).

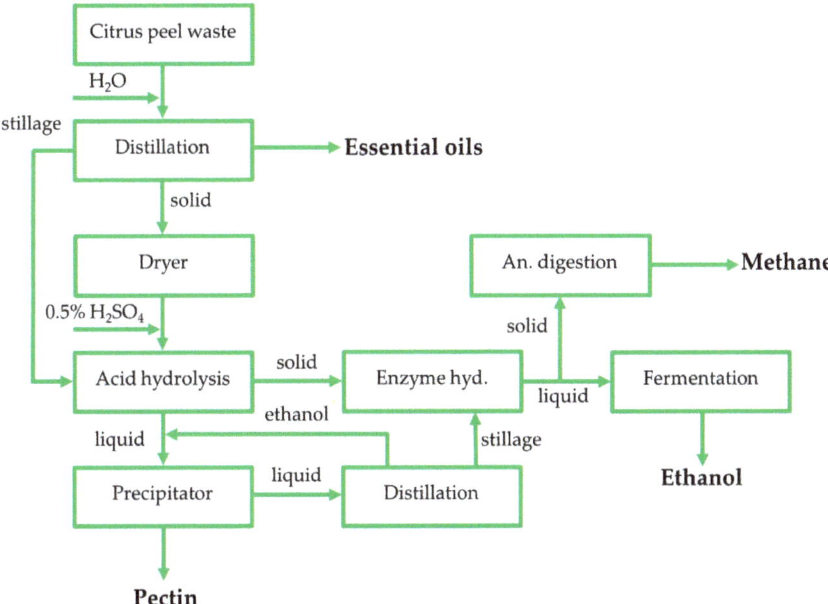

Figure 1. Process flow sheet of the zero-waste biorefinery used for citrus peel waste (CPW) valorization.

2.2. Ethanol Production

The industrial production of ethanol is currently performed by catalytic hydration of ethylene (chemical method) [22] and by fermenting agricultural feedstocks (biochemical method) [14]. Ethanol can serve as a green energy source, which is mainly produced using starch, sugar, and carbohydrates, such as corn, potato, molasses, sugarcane, and lignocellulosic biomass [23]. Sugars can be directly converted to ethanol, while starchy and cellulosic materials should be first pretreated mainly using enzymes or chemicals to hydrolyze the polymers into sugars [24]. Herein, CPW has been applied as a cellulosic material, which includes an additional significant content of soluble sugars that could be applied for production of a hydrolyzate rich in carbon sources for ethanol fermentations. The thermotolerant strain *P. kudriavzevii* KVMP10 was tested and compared with two industrial yeasts (*K. marxianus* and *S. cerevisiae*) for ethanol production from CPW hydrolyzates under technologically favourable fermentation conditions. Thus, an elevated bioprocess temperature (42 °C) was applied in an attempt to reduce operational costs associated with decreased energy use for cooling and lower contamination risk [25,26].

2.2.1. Ethanol Production Using CPW Hydrolyzates Obtained through Acid Hydrolysis

Dilute acid hydrolysis of CPW served the dual objective of breaking down polymers (cellulose, hemicellulose) into soluble sugars, while consisting of an essential processing step for pectin isolation [27]. Thus, six CPW hydrolyzates were produced using three hydrolysis temperatures (108 °C, 116 °C, 125 °C) for 10 min and 20 min, respectively. Optimal conditions for CPW saccharification through dilute acid hydrolysis were identified as 116 °C for 10 min, based on the concentration of ethanol produced and the final product yield during fermentations (Figure 2, Table 1). Thus, ethanol concentration and product yield reached 5.8 g L^{-1} and 0.48 $g_{ethanol\ (eth)}$ $g^{-1}_{total\ sugar\ consumed\ (TSC)}$, respectively, using *P. kudriavzevii* KVMP10, while *K. marxianus* and *S. cerevisiae* produced 4.6 g L^{-1} and 4.2 g L^{-1} of ethanol, respectively. Although, the highest ethanol concentration (6.7 g L^{-1}) was obtained using the hydrolyzate generated at 125 °C for 20 min with application of *P. kudriavzevii* KVMP10, the product yield decreased significantly to 0.32 g_{eth} g^{-1}_{TSC}. A *t*-test ($p < 0.05$) was performed to statistically assess differences between the mean values of ethanol concentration. Significant statistical difference was observed between ethanol titres obtained in *P. kudriavzevii* KVMP10 and *S. cerevisiae* fermentations, using both hydrolyzates exhibiting the highest product formation (116 °C for 10 min and 125 °C for 20 min). Nevertheless, the titre of the biofuel obtained in *P. kudriavzevii* KVMP10 fermentations fed with the aforementioned hydrolyzates was not statistically different. Overall, in nearly all experiments performed using *K. marxianus* and *S. cerevisiae*, the increase in hydrolysis duration reduced the final product titre and yield, indicating the potential formation of inhibitors at elevated preprocessing duration. This is in line with previous studies demonstrating the inhibitory effect that may occur in ethanol fermentations of *K. marxianus* [28] and *S. cerevisiae* [29] fed with feedstocks pretreated employing increased acid hydrolysis temperatures and duration. Specifically, hydrolyzates derived from acid pretreatment resulted in low product yield in *K. marxianus* fermentations [28], while the presence of acid hydrolysis products (furan derivatives, weak acids, phenolics) inhibited the growth of *S. cerevisiae* and caused a reduction in ethanol yield and productivity [29]. Therefore, considering that the overall performance of *K. marxianus* and *S. cerevisiae* was decreased, the lower product yield of *P. kudrivzevii* KVMP10 and the elevated energy demand expected with application of 125 °C for 20 min in hydrolysis, the use of 116 °C for 10 min was selected as suitable conditions for CPW hydrolysis. These results are in agreement with other studies demonstrating that the optimal conditions for dilute acid hydrolysis of CPW comprise application of 116 °C for 10–13 min with the use of 0.5% (*v/v*) H_2SO_4 and 5–6% (*w/v*) dry solids of the raw material (rm) [30,31].

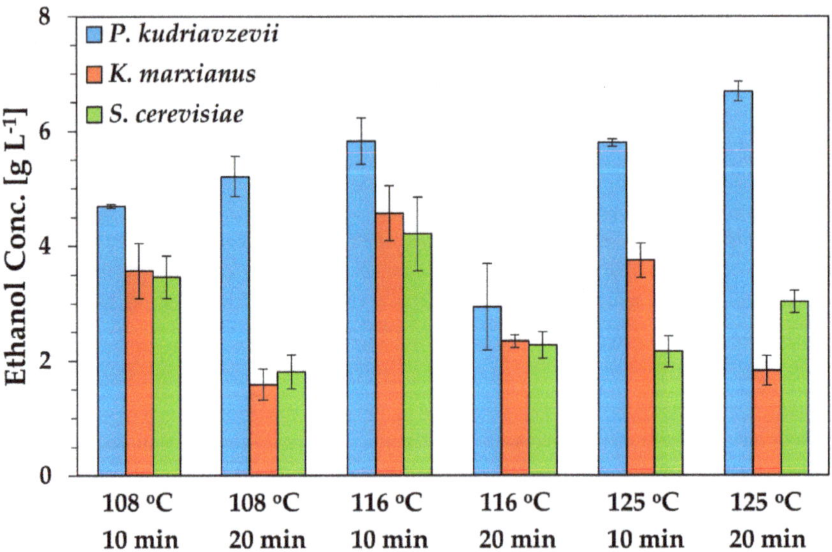

Figure 2. Ethanol titre achieved in *Pichia kudriavzevii* KVMP10, *Saccharomyces cerevisiae*, and *Kluyveromyces marxianus* fermentations of hydrolyzates obtained through dilute acid treatment of CPW.

A variety of CPW pretreatment approaches have been previously evaluated for ethanol production by different yeasts. The production of the biofuel was investigated in *S. cerevisiae* fermentations, using a CPW-derived hydrolyzate obtained through application of steam explosion as well as dilute-acid hydrolysis and pectin recovery, achieving a yield of 0.43 g_{eth} g^{-1}_{TSC} [16,32]. *S. cerevisiae* was also applied for bioethanol production using CPW hydrolyzates obtained through hydrothermal sterilization, steam explosion, and a combination of popping and enzyme hydrolysis, demonstrating ethanol production of 42 g L^{-1} [15], 60 L $t^{-1}_{raw\ material}$ [33] and 46.2 g L^{-1}, respectively [19]. Orange peel waste was pretreated by two-stage acid and enzyme hydrolysis, while the hydrolyzate formed was used as fermentation feedstock for bioethanol production by *S. cerevisiae* and *Mucor indicus*, achieving final product concentrations of 30.3 g L^{-1} and 15 g L^{-1} with a yield of 0.46 g$_{eth}$ g^{-1} $_{TSC}$ and 0.39 g$_{eth}$ g^{-1}$_{TSC}$, respectively [21,34].

Table 1. Consumption of reducing sugars and final titre of ethanol in fermentations of acid hydrolyzates.

Experiment	Reducing Sugars Consumed (g L^{-1})	Final Ethanol Titre (g L^{-1})
P. kudriavzevii **KVMP10**		
108 °C, 10 min	11.0	4.7
108 °C, 20 min	17.9	5.2
116 °C, 10 min	12.1	5.8
116 °C, 20 min	30.9	2.9
125 °C, 10 min	15.5	5.8
125 °C, 20 min	20.5	6.7
K. marxianus		
108 °C, 10 min	9.8	3.6
108 °C, 20 min	3.1	1.6
116 °C, 10 min	17.2	4.6
116 °C, 20 min	10.9	2.3
125 °C, 10 min	7.5	3.8
125 °C, 20 min	18.6	1.8
S. cerevisiae		
108 °C, 10 min	6.9	3.5
108 °C, 20 min	5.4	1.8
116 °C, 10 min	8.6	4.2
116 °C, 20 min	6.0	2.7
125 °C, 10 min	8.1	2.2
125 °C, 20 min	5.9	3.0

2.2.2. Optimization of the Fermentation Process

The final ethanol titre achieved using the hydrolyzates obtained was low compared to the relevant literature. Therefore, a number of parameters were evaluated to improve ethanol production through the biorefinery proposed. The effect of nitrogen source supplementation was initially tested to increase biofuel production. However, although nitrogenous compounds can substantially enhance the fermentation rate of ethanol [35,36], the use of yeast extract as a nitrogen source did not demonstrate any noticeable effect on ethanol formation in all yeast fermentations conducted.

Acid and enzyme hydrolysis were combined as sequential pretreatments to enhance the production of ethanol, due to potential increased release of fermentable sugars. The addition of the enzymatic pretreatment step in the biorefinery substantially enhanced the titre of bioethanol to 9.2 g L^{-1} in *P. kudriavzevii* KVMP10 fermentations, resulting in a product yield of 0.42 g$_{eth}$ g^{-1}$_{TSC}$. Pretreatment processes that involve two steps can be more efficient in lignocellulosic biomass saccharification, releasing increased concentrations of simple sugars that enhance ethanol generation [19,21]. Thus, sequentially applied acid and enzyme hydrolysis of CPW was capable of producing substantially elevated sugar yields that reached 0.58 g$_{total\ sugar}$ g^{-1}$_{dry\ raw\ material}$ [30].

Recycling of the remaining liquid stillage following essential oil extraction into the hydrolysis process was applied in an attempt to further increase the concentration of monosaccharides in the hydrolyzate and ethanol formation, as well as to reuse the process water generated for plant application. Stillage water recycling contributed to a three-fold increase of ethanol concentration that reached 30.7 g L^{-1} in *P. kudriavzevii* KVMP10 fermentations conducted at 42 °C, while biofuel production was also enhanced in *K. marxianus* cultures reaching 26.3 g L^{-1}. Nevertheless, ethanol production from *S. cerevisiae* remained at low levels with the reuse of the stillage, potentially due to the elevated process temperature employed. The present findings demonstrate that *P. kudriavzevii* KVMP10 constitutes a more efficient ethanol producer with application of CPW hydrolyzates as compared to the industrial yeasts tested (*K. marxianus* and *S. cerevisiae*). Moreover, the increased ethanol production was achieved at an elevated fermentation temperature, which is in agreement with other studies highlighting that *P. kudriavzevii* strains are robust ethanol producers exhibiting multiple tolerance at high temperatures and acidic environments [37]. Key novelties of the results obtained are that high ethanol production was achieved from CPW under mild pretreatment and harsh fermentation conditions, while saving of valuable resources was achieved through process water recycling. Thus, in accordance with the water reuse achieved, the remaining solid residue from CPW hydrolysis was applied in anaerobic digestion to produce biogas targeting the operation of a zero-waste biorefinery as described below.

2.3. Anaerobic Digestion of Biorefinery Residues

The biomethanization of CPW has been mainly studied following essential oil extraction, which serves the dual purpose of generating an added-value product and reducing the anaerobic digestion inhibition caused by the antimicrobial properties of the oil [5]. However, although essential oils were extracted from CPW in several studies prior to use in methane production [10,32], untreated CPW has also been used in anaerobic digestion of fresh or dried citrus fragments (peel, pulp, and seeds) under thermophilic [17,38] and mesophilic conditions [20,39]. Herein, methane production was tested using biorefinery solids BR, remaining as a side stream from acid hydrolysis of CPW, and their capacity to produce the biofuel was compared to the application of untreated CPW as well as DCPW under mesophilic conditions.

2.3.1. Methane Production Using BR, CPW, and DCPW

Biogas and methane production from BR as well as CPW and DCPW was evaluated with the addition of 6 g L^{-1} as initial content of volatile solids (VS) in anaerobic digestion (Figure 3). During the first batch, DCPW exhibited substantially higher production of methane at 35 d, as compared to the rest of the materials, generating 303 mL. The use of BR demonstrated prolonged biogas inhibition for the first 15 days of the process, accumulating high levels of acetate that reached 9.7 g L^{-1} (Figure 4), while CPW also caused inhibition of methane formation. However, the concentration of acetate was subsequently reduced in the experiment fed with BR, resulting in methane production that reached 264 mL at 35 days, similarly to biofuel production from CPW. Following 36 days, a subsequent batch was conducted through refeeding of the same content of each material in fermentations. The inhibitory effect was substantially reduced in the second batch, regarding the kinetics of methane production, indicating potential adaptation of the sludge to each material. Specifically, the production rate of methane was 11.3 mL d^{-1}, 15.9 mL d^{-1}, and 7.5 mL d^{-1} during the first batch experiment, while in the second batch, the rate increased to 14.9 mL d^{-1}, 25.6 mL d^{-1}, and 13.6 mL d^{-1} for CPW, DCPW, and BR, respectively. Following the second refeed of each material, the methane production rate was further increased only in the digestion of DCPW, reaching 27.2 mL d^{-1}. Moreover, cumulative methane formation was significantly increased with the use of BR, reaching 314 mL. Following 78 days, a second refeed of each material was applied to evaluate whether methane production was further increased. The production of methane was only slightly enhanced in the third batch, reaching 305–320 mL, for all materials tested. The data obtained demonstrate the capacity of BR to generate significant amounts of methane at a similar level to the use of dried or untreated CPW. Moreover, although methane production is consistent with other studies that reached similar biofuel formation from CPW [10,20,39], the production of the biofuel achieved here under mesophilic conditions was comparable to that of studies employing thermophilic anaerobic digestion temperatures [10,32]. Therefore, the approach followed not only enables the development of a zero-waste biorefinery, but also provides significant energy gains, due to the substantially reduced temperature used in anaerobic digestion.

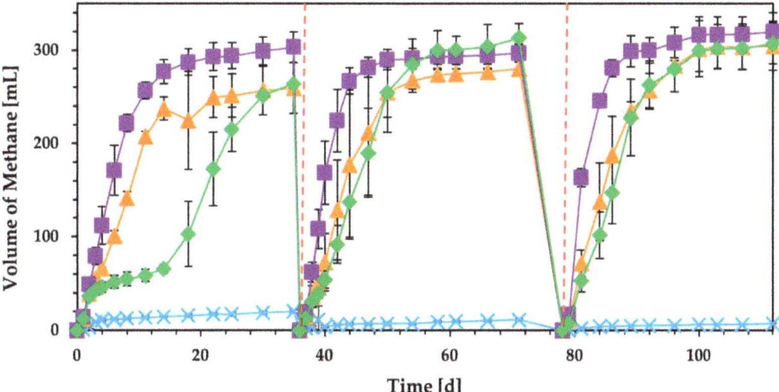

Figure 3. Cumulative methane production using 6 g_{VS} L^{-1} of untreated CPW (▲), dried citrus peel waste (DCPW) (■), and solid biorefinery residues (BR) (◆). A control fermentation was conducted without the addition of CPW (✕), while all experiments were conducted at 37 °C. Dashed lines represent the time where substrate refeed was applied.

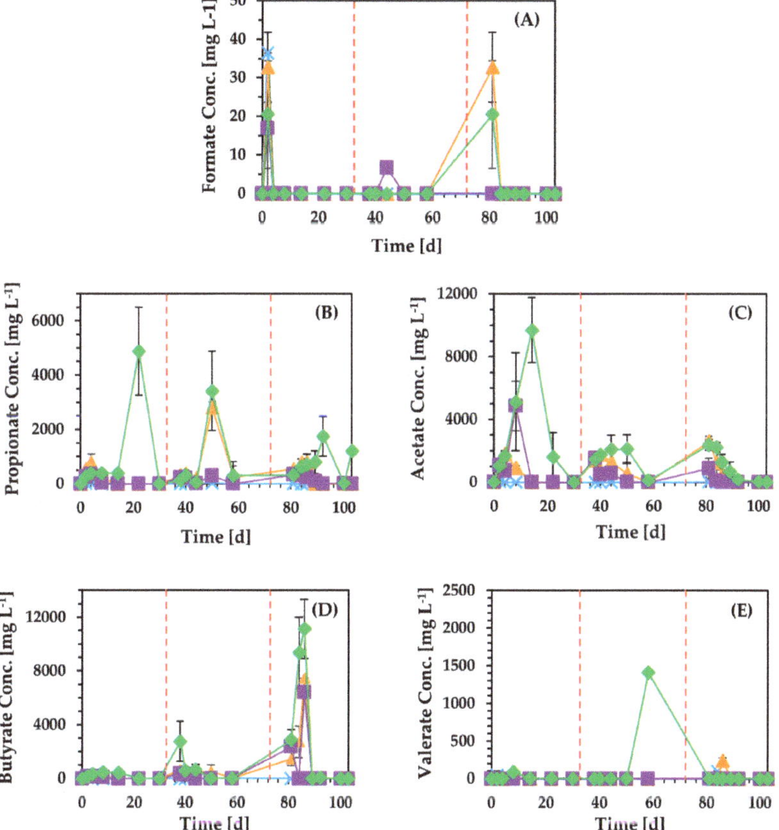

Figure 4. Concentration of volatile fatty acids, (**A**) formate, (**B**) propionate, (**C**) acetate, (**D**) butyrate, and (**E**) valerate, formed during anaerobic digestion of untreated CPW (▲), DCPW (■), solid BR (◆), and in the control experiment (✕).

Although the use of DCPW in anaerobic digestion resulted in the highest production of methane in the first batch as compared to the application of CPW and BR, calculation of bioprocess yield showed that employing CPW in the system enhanced the overall waste-to-energy conversion (Figure 5). The methane yield, defined as mL of methane produced per g of raw material (rm) used, was substantially higher, with the application of untreated CPW ranging between 72 and 84 $mL_{methane}$ g^{-1}_{rm}, respectively. However, employing DCPW and BR resulted in a similar and lower process yield that reached 42–51 $mL_{methane}$ g^{-1}_{rm} for both materials. The enhanced methane yield performed with the use of CPW was expected, given the higher organic content remaining in the specific material for treatment in anaerobic digestion as compared to DCPW and BR, where volatile compounds and/or hydrolyzed carbohydrates have been removed. This conclusion was clarified by the t-test ($p < 0.05$) performed to identify statistically significant differences in the mean values obtained for the final methane production (mL) per g of rm with each feedstock in the first two batches. Furthermore, a statistically significant increase ($p < 0.05$) was observed in the volume of biogas produced per g of rm between 35 days and 112 days with the use of BR, demonstrating the potential adaptation of the culture to the specific material following the first batch. Direct comparison of the three different citrus waste fractions demonstrated that even BR could produce high contents of methane, highlighting the applicability of biorefinery side streams for valorization through anaerobic digestion.

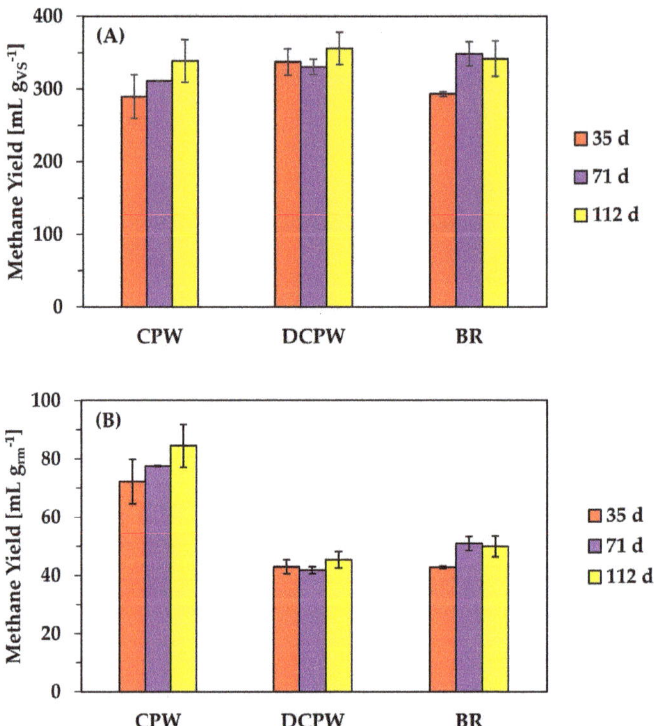

Figure 5. Yield of methane (**A**) as mL of product per g of initial volatile solids (VS) and (**B**) as mL of product per g of raw material (rm) generated from anaerobic digestion of CPW, DCPW, and BR following completion of each batch.

2.3.2. Biomethane Production Using Different Initial Quantities of CPW

Biomethanization of untreated CPW was previously evaluated by various studies. Forgacs et al. [17] achieved significantly low production of methane from untreated citrus waste that reached

102 mL g^{-1}_{VS}, while Koppar and Pullammanappallil [38] generated 644 mL g^{-1}_{VS} from untreated CPW under thermophilic conditions. The use of peel, seeds, and pulp of fresh ripe orange as distinct feedstocks in anaerobic digestion under mesophilic conditions resulted in methane production that reached 72 mL g^{-1}_{VS}, 581 mL g^{-1}_{VS}, and 288 mL g^{-1}_{VS}, respectively [39]. Herein, different initial quantities of untreated CPW volatile solids (ranging between 3–24 g L^{-1}) were applied in anaerobic digestion to evaluate the effect of increasing contents of organic material as well as essential oils in the bioprocess (Figure 6). The maximum cumulative production of methane in each experiment was monitored as 165 mL, 264 mL, 598 mL, and 1011 mL, employing 3 g_{VS} L^{-1}, 6 g_{VS} L^{-1}, 12 g_{VS} L^{-1}, and 24 g_{VS} L^{-1}, respectively. Methane production per g of VS applied was not significantly different in each experiment, ranging between 294 and 366 mL g^{-1}_{VS}. However, the increase in feedstock concentration resulted in a prolonged lag phase (Figure 6B). It has been previously shown that the increase in essential oil concentration can negatively impact biomethanization of orange peel waste, inhibiting the methanogenesis [40]. However, the results obtained here demonstrate that anaerobic digestion of CPW under mesophilic conditions can generate high methane yields, comparable to those obtained under thermophilic conditions. Thus, the maximum yield of methane achieved in the present study ranged between 294 and 366 mL g^{-1}_{VS}, which was three times higher than the yield achieved by Forgacs et al. [17], which reached 102 mL g^{-1}_{VS} under thermophilic conditions. A similar effect was reported by Lotito et al. [41], where the use of fresh and stored citrus peel waste achieved methane production that ranged between 333 and 471 mL g^{-1}_{VS} under mesophilic conditions. Nevertheless, significant quantities of α-terpineol and p-cymene were detected as final degradation products towards the end of digestion, while D-limonene was no longer present [40,41].

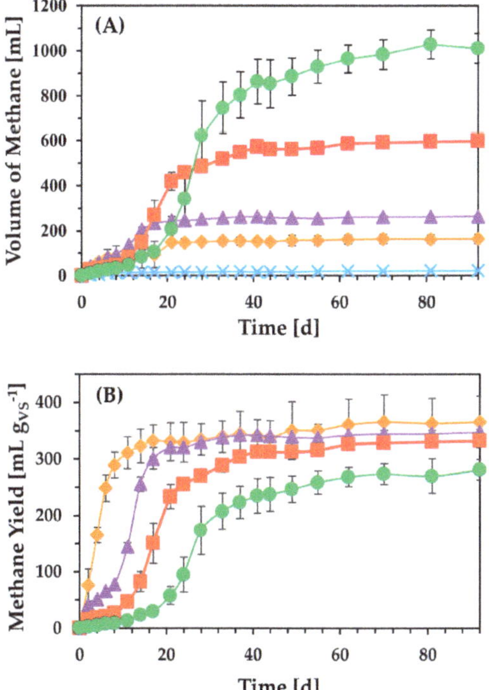

Figure 6. Cumulative methane production (A) and methane content per g of initial VS (B) using 3 g L^{-1} (━♦━), 6 g L^{-1} (━▲━), 12 g L^{-1} (━■━), and 24 g L^{-1} (━●━) initial volatile solids of CPW. A control fermentation was performed without the addition of CPW (━✕━), while all experiments were conducted at 37 °C.

3. Materials and Methods

3.1. Citrus Peel Waste

The feedstock applied in the experiments of the present work constituted citrus residues (seeds, pulp, and segment membranes), which were collected from a local citrus processing industry (KEAN Soft Drinks Ltd., Limassol, Cyprus) and maintained at −20 °C until further use. CPW was thawed and ground to solids of less than 2 mm in diameter using a laboratory blender (Waring Commercial, San Antonio, TX, USA). The chemical composition of CPW has been previously determined as follows: 22.00% cellulose, 11.09% hemicellulose, 2.19% lignin, 8.10% glucose, 12.00% fructose, 2.80% sucrose, 25.00% pectin, 6.07% protein, and 3.78% limonene [32].

3.2. Pretreatment of Citrus Peel Waste

3.2.1. Extraction of Essential Oils and Pectin

The first step of CPW pretreatment employed distillation for extraction and collection of essential oils as previously described [30], recovering 0.43% *w/w* of essential oils. Residues from essential oil distillation were dried at 70 °C for 24 h [42] and dilute acid hydrolysis was performed in an autoclave (SANYO MLS-3781L, Panasonic, Tottori, Japan) at 108 °C, 116 °C, and 125 °C for 10 min and 20 min, using 5% (*w/v*) of dry CPW with the addition of 0.5% (*v/v*) of sulfuric acid. Following dilute acid hydrolysis, centrifugation and filtration were employed to obtain the supernatant, which was mixed with an equal volume of ethanol (96% *v/v*) to precipitate pectin at room temperature for 4 h [32]. Subsequently, the mixture was centrifuged at 3000 rpm for 30 min. The precipitate was washed five times with ethanol (45% *v/v*), followed by drying at 50 °C to obtain pectin [43], isolating 23.25% *w/w* of the heteropolysaccharide, while the supernatant was applied in distillation at 80 °C three times to recover and recycle ethanol for pectin precipitation.

3.2.2. Enzyme Hydrolysis

Enzyme hydrolysis was applied as additional pretreatment prior to bioethanol fermentations. The procedure was performed with the addition of solid residues from acid treatment in the hydrolyzate obtained following ethanol recovery. The pH of the mixture was adjusted to 4.8 with the use of 1 M NaOH to ensure that process conditions were within the optimal pH range of 4.5–5.0 for the enzymes employed, while cellulases (Chem Cruz, Dallas, TX, USA) and β-glucosidases/pectinases (Oenozym FW, Lamothe-Abiet, Bordeaux, France) were added at 30 IU g^{-1}_{drm} and 25 BGL g^{-1}_{drm}, respectively [30]. Enzyme hydrolysis was performed at 50 °C for 48 h in shake flasks stirred at 100 rpm in a water bath, while upon hydrolysis completion, enzymes were inactivated via heating in an oven at 105 °C for 15 min [44].

3.3. Ethanol Fermentations

P. kudriavzevii KVMP10 was previously isolated as a thermotolerant ethanologenic yeast within our research group [45], while *K. marxianus* was obtained from the Leibniz Institute DSMZ—German Collection of Microorganisms and Cell Cultures (Braunschweig, Germany). Commercial pressed baker's yeast was used as a source of *S. cerevisiae*. All strains were maintained at −80 °C in glycerol stock cultures.

P. kudriavzevii KVMP10 and *S. cerevisiae* were precultured in liquid media, simulating a Valencia orange peel waste hydrolyzate [4], while the inoculum of *K. marxianus* was prepared using universal media for yeast strains (containing (g L^{-1}) yeast extract 3, malt extract 3, peptone 5, and glucose 10) incubated at 30 °C in shake flasks stirred at 100 rpm. Ethanol fermentations were performed in batch experiments using 100 mL flasks with a working volume of 60 mL at 42 °C and 100 rpm. The mixture of reducing sugars obtained from CPW pretreatment (dilute acid hydrolysis only or combined acid and enzyme hydrolysis) was used as a carbon source for the experiments. The feedstock was supplemented with 10 g L^{-1} of yeast extract as a nitrogen source. Moreover, recycling of the remaining liquid stillage following essential oil extraction into the hydrolysis process was tested to enhance the soluble sugar content in the hydrolyzate. All experiments were performed in duplicate, while three samples were analyzed for each replicate, constituting analyses of 6 samples at each time point.

3.4. Anaerobic Digestion Experiments

CPW, DCPW, and BR (remaining solids following acid hydrolysis of CPW) were used as feedstock in anaerobic digestion. Equal quantities of volatile solids (VS, 6 g L^{-1}) were employed to produce biogas under mesophilic conditions (37 °C), while anaerobic digestion was performed in batch experiments using 250 mL flasks with a working volume of 150 mL. The nutrient medium was prepared according to the composition used in Angelidaki et al. [46]. Each bottle was supplemented with 6 g of granular sludge withdrawn from a full-scale UASB reactor (Charalambides Christis Ltd., Limassol, Cyprus) used for the treatment of dairy wastewater at pH 6.8–7.3. The content of sludge in lignin and N-compounds, cellulose, and hemicellulose was measured at 0.35%, 0.30%, and 1.40%, respectively. Granular sludge was washed with distilled water and applied as active inoculum (4% *w/v*), while each bottle was flushed with 100% CO_2 gas to ensure anaerobic conditions. Experiments were performed in triplicate and methane accumulation was determined for 121 days, aiming to evaluate the adaptation of anaerobic digestion to CPW, DCPW, and BR. Following 36 days, a refeed of each material (6 g_{VS} L^{-1}) was applied in each digestion and a subsequent refeed was also conducted at 78 days. The refeed of each material was performed through withdrawal of 50 mL of digested medium, which was replaced with an equal volume of fresh media and substrate. Nevertheless, any remaining substrate and anaerobic sludge could not be removed, while each bottle was subsequently flushed using 100% CO_2 gas. Although the ratio of biomass versus substrate might have been potentially different between each batch, digestion conditions as well as the procedure of the refeed was the same for all materials tested, enabling direct comparison between each flask. Furthermore, in order to test the effect of different initial VS contents of CPW, 3, 6, 12, and 24 g L^{-1} were applied in triplicate anaerobic digestion experiments for 92 days, evaluating the effect of essential oils in the process. The sampling procedure included withdrawing 1 mL of biogas with the use of a gas-tight syringe for determination of biogas composition as described in Section 3.5.4., as well as obtaining biomedium samples for the detection of VFAs. During anaerobic digestion, the volume of biogas, methane, carbon dioxide, oxygen, hydrogen, and nitrogen produced was monitored.

3.5. Analyses

3.5.1. Ethanol Concentration

Gas Chromatography using a flame ionization detector was employed for the determination of ethanol concentration. A Shimadzu GC-2014 (Shimadzu, Milton Keynes, UK) and a 30 m long Zebron ZB-5 capillary column (Phenomenex, Macclesfield, UK) with a 0.25 mm internal diameter was used. The stationary phase of the column was 5% phenyl- and 95% dimethylpolysiloxane, while the mobile phase applied was nitrogen. Samples obtained during ethanol fermentations were centrifuged for 3 min at 13,000 rpm and the supernatant was filtered through 0.2 μm syringe filters. Ethanol was extracted by vortexing 1 mL of the filtered sample with 2 mL of hexane for 1 min. Approximately 1 μL of the extract was injected and the temperature of the column was kept constant at 40 °C for 3 min. Ethanol concentration was calculated, interpolating from a previously established calibration curve and the coefficient of variation for 3 samples was 1.47% at a concentration level of 4 g L^{-1}. Furthermore, the respective LoB, LoD, and LoQ values were calculated at 0, 0.07, and 1.7 g L^{-1}, respectively.

3.5.2. Reducing Sugars

During ethanol fermentations the content of reducing sugars was analyzed by the phenol–sulfuric acid method [47]. The LoB, LoD, and LoQ values were calculated at 0, 1.48, and 11.18 mg L^{-1}, respectively.

3.5.3. Total and Volatile Solids

The total solids (TS) and VS content of granular sludge, CPW, DCPW, and BR was determined according to Standard Methods [48]. The LoB/LoD/LoQ values for the determination of formic acid, acetic acid, propionic acid, butyric acid, and valeric acid concentrations were calculated at 0/0.002/0.3 g L^{-1}, 0/0.01/0.3 g L^{-1}, 0/0.01/0.2 g L^{-1}, 0/0.01/0.2 g L^{-1}, and 0/0.003/0.2 g L^{-1}, respectively.

3.5.4. Biogas Composition

Biogas composition (H_2, O_2, N_2, CH_4, and CO_2) was analyzed using a gas chromatograph (Agilent Technologies, 7820OA, Santa Clara, CA, USA) fitted with a ShinCarbon ST 50/80 (2 m length, 2.2 mm ID) mesh column (Restek Corporation, Bellefonte, PA, USA) and thermal conductivity detector as described by Vardanyan et al. [49]. The LoB/LoD/LoQ values for the determination of hydrogen, oxygen, nitrogen, methane, and carbon dioxide volumes were calculated at 0/0.4/8.4 mL, 0/0.4/3.4 mL, 0/1.2/10.5 mL, 0/0.2/7.0 mL, and 0/0.3/1.1 mL, respectively.

3.5.5. Volatile Fatty Acids (VFAs)

The concentration of VFAs (acetate, formate, butyrate, propionate, and valerate) formed during anaerobic digestion was measured through High Pressure Liquid Chromatography (HPLC). Culture samples were centrifuged at 13,000 rpm for 3 min and filtered using 0.22 mm syringe filters. A Shimadzu LC-20AD liquid chromatograph (Shimadzu, Milton Keynes, UK) equipped with a Shimadzu SPD-20A UV/VIS detector, a Shimadzu SIL-20A HT auto sampler, and a CTO-10AS VP column oven was used. The column was eluted isocratically at a rate of 0.7 mL min^{-1} from an organic analysis column (Rezex ROA-Organic Acid column, Phenomenex, Torrance, CA, USA) with 5 mM H_2SO_4 at 55 °C, while the injection volume was 1 μL.

4. Conclusions

This paper has demonstrated the potential of a novel zero-waste biorefinery for the production of renewable biofuels from CPW through low environmental impact technologies. It was confirmed that the combination of sequential acid and enzyme hydrolysis for CPW pretreatment enhanced the fermentative production of ethanol. Moreover, the thermotolerant *P. kudriavzevii* KVMP10 was a more efficient ethanol producer at elevated fermentation temperatures as compared to industrial yeasts (*S. cerevisiae* and *K. marxianus*). The use of BR in anaerobic digestion for methane generation resulted in increased production of the biofuel. The study presents important operational parameters reducing the energy requirements of the biorefinery.

Author Contributions: Conceptualization, M.K. and I.V.; Methodology, M.K., M.P., and I.V.; Validation, M.P., C.G.S, E.P., and S.S.; Investigation, M.P., C.G.S, E.P., and S.S.; Writing—Original Draft Preparation, M.K., M.P., and I.V.; Writing—Review and Editing, M.K., M.P., and I.V.; Supervision, M.K.

Funding: This research received no external funding.

Acknowledgments: We would like to thank KEAN Soft Drinks Ltd. for the provision of citrus peel waste.

Conflicts of Interest: The authors declare no conflict of interest.

References

1. Kosseva, M.R. Sources, characterization, and composition of food industry wastes. In *Food Industry Wastes*, 1st ed.; Kosseva, M.R., Webb, C., Eds.; Elsevier Inc.: London, UK, 2013; pp. 37–60.

2. FAO. *Citrus Fruit Statistics 2015*; FAO: Rome, Italy, 2016; p. 53.

3. Marin, F.R.; Soler-Rivas, C.; Benavente-Garcia, O.; Castillo, J.; Perez-Alvarez, J.A. By-products from different citrus processes as a source of customized functional fibres. *Food Chem.* **2007**, *100*, 736–741. [CrossRef]

4. Wilkins, M.R. Effect of orange peel oil on ethanol production by *Zymomonas mobilis*. *Biomass Bioenergy* **2009**, *33*, 538–541. [CrossRef]

5. Negro, V.; Bernardo, R.; Fino, D. Recovery of energy from orange peels through anaerobic digestion and pyrolysis processes after D-limonene extraction. *Waste Biomass Valoriz.* **2018**, *9*, 1331–1337. [CrossRef]

6. Mamma, D.; Christakopoulos, P. Biotransformation of citrus by-products into value added products. *Waste Biomass Valoriz.* **2014**, *5*, 529–549. [CrossRef]

7. Ruiz, B.; Flotats, X. Citrus essential oils and their influence on the anaerobic digestion process: An overview. *Waste Manag.* **2014**, *34*, 2063–2079. [CrossRef] [PubMed]

8. Lopez, J.A.S.; Li, Q.; Thompson, I.P. Biorefinery of waste orange peel. *Crit. Rev. Biotechnol.* **2010**, *30*, 63–69. [CrossRef] [PubMed]

9. Li, Q.; Siles, J.A.; Thompson, I.P. Succinic acid production from orange peel and wheat straw by batch fermentations of *Fibrobacter succinogenes* S85. *Appl. Microbiol. Biotechnol.* **2010**, *88*, 671–678. [CrossRef]

10. Martin, M.A.; Siles, J.A.; Chica, A.F.; Martin, A. Biomethanization of orange peel waste. *Bioresour. Technol.* **2010**, *101*, 8993–8999. [CrossRef]

11. Siles, J.A.; Vargas, F.; Gutierrez, M.C.; Chica, A.F.; Martin, M.A. Integral valorisation of waste orange peel using combustion, biomethanisation and co-composting technologies. *Bioresour. Technol.* **2016**, *211*, 173–182. [CrossRef]

12. Liao, J.C.; Mi, L.; Pontrelli, S.; Luo, S. Fuelling the future: Microbial engineering for the production of sustainable biofuels. *Nat. Rev. Microbiol.* **2016**, *14*, 288–304. [CrossRef]

13. Jiang, Y.; Ding, D.; Zhao, S.; Zhu, H.; Kenttamma, H.I.; Abu-Omar, M.M. Renewable thermoset polymers based on lignin and carbohydrate derived monomers. *Green Chem.* **2018**, *20*, 1131–1138. [CrossRef]

14. Mohanty, S.K.; Swain, M.R. Bioethanol production from corn and wheat: Food, fuel and future. In *Bioethanol Production from Food Crops*, 1st ed.; Ray, R.C., Ramachandran, C., Eds.; Elsevier Inc.: London, UK, 2019; pp. 45–49.

15. Oberoi, H.S.; Vadlani, P.V.; Nanjundaswamy, A.; Bansal, S.; Singh, S.; Kaur, S.; Babbar, N. Enhanced ethanol production from Kinnow mandarin (*Citrus reticulata*) waste via a statistically optimized simultaneous saccharification and fermentation process. *Bioresour. Technol.* **2011**, *102*, 1593–1601. [CrossRef] [PubMed]

16. Wilkins, M.R.; Widmer, W.W.; Grohmann, K.; Cameron, R.G. Simultaneous saccharification and fermentation of citrus peel waste by *Saccharomyces cerevisiae* to produce ethanol. *Process Biochem.* **2007**, *42*, 1614–1619. [CrossRef]

17. Forgacs, G.; Pourbafrani, M.; Niklasson, C.; Taherzadeh, M.J.; Hovath, I.S. Methane production from citrus wastes: Process development and cost estimation. *J. Chem. Technol. Biotechnol.* **2011**, *87*, 250–255. [CrossRef]

18. Sandhu, S.K.; Oberoi, H.S.; Dhaliwal, S.S.; Babbar, N.; Kaur, U.; Nanda, D.; Kumar, D. Ethanol production from Kinnow mandarin (*Citrus reticulata*) peels via simultaneous saccharification and fermentation using crude enzyme produced by *Aspergillus oryzae* and the thermotolerant *Pichia kudriavzevii* strain. *Ann. Microbiol.* **2012**, *62*, 655–666. [CrossRef]

19. Choi, I.S.; Kim, J.H.; Wi, S.G.; Kim, K.H.; Bae, H.J. Bioethanol production from mandarin (*Citrus unshiu*) peel waste using popping treatment. *Appl. Energy* **2013**, *102*, 204–210. [CrossRef]

20. Gunaseelan, V.N. Biochemical methane potential of fruits and vegetable solid waste feedstocks. *Biomass Bioenergy* **2004**, *26*, 389–399. [CrossRef]

21. Oberoi, H.S.; Vadlani, P.V.; Madl, R.L.; Saida, L.; Abeykoon, J.P. Ethanol production from orange peels: Two-stage hydrolysis and fermentation studies using optimized parameters through experimental design. *J. Agric. Food Chem.* **2010**, *58*, 3422–3429. [CrossRef] [PubMed]

22. Mohsenzadeh, A.; Zamani, A.; Taherzadeh, M.J. Bioethylene production from ethanol: A review and techno-sconomical evaluation. *ChemBioEng Rev.* **2017**, *2*, 75–91. [CrossRef]

23. Lin, Y.; Tanaka, S. Ethanol fermentation from biomass resources: Current state and prospects. *Appl. Microbiol. Biotechnol.* **2006**, *69*, 627–642. [CrossRef] [PubMed]

24. Taghizadeh-Alisaraei, A.; Hosseini, S.H.; Ghobadian, B.; Motevali, A. Biofuel production from citrus wastes: A feasibility study in Iran. *Renew. Sustain. Energy Rev.* **2017**, *69*, 1100–1112. [CrossRef]

25. Kyriakou, M.; Chatziiona, V.K.; Costa, C.N.; Kallis, M.; Koutsokeras, L.; Constantinides, G.; Koutinas, M. Biowaste-based biochar: A new strategy for fermentative bioethanol overproduction via whole-cell immobilization. *Appl. Energy* **2019**, *242*, 480–491. [CrossRef]

26. Tavares, B.; de Almeida Felipe, M.G.; dos Santos, J.C.; Pereira, F.M.; Gomes, S.D.; Sene, L. An experimental and modeling approach for ethanol production by *Kluyveromyces marxianus* in stirred tank bioreactor using vacuum extraction as a strategy to overcome product inhibition. *Renew. Energy* **2019**, *131*, 261–267. [CrossRef]

27. Kaya, M.; Sousa, A.G.; Crepeau, M.J.; Sorensen, S.O.; Ralet, M.C. Characterization of citrus pectin samples extracted under different conditions: Influence of acid type and pH of extraction. *Ann. Bot.* **2014**, *114*, 1319–1326. [CrossRef] [PubMed]

28. Zhang, M.; Shukla, P.; Ayyachamy, M.; Permaul, K.; Singh, S. Improved bioethanol production through simultaneous saccharification and fermentation of lignocellulosic agricultural wastes by *Kluyveromyces marxianus* 6556. *World J. Microbiol. Biotechnol.* **2010**, *26*, 1041–1046. [CrossRef]

29. Almeida, J.R.M.; Modig, T.; Petersson, A.; Hahn-Hagerdal, B.; Liden, G.; Gorwa-Grauslund, M.F. Increased tolerance and conversion of inhibitors in lignocellulosic hydrolysates by *Saccharomyces cerevisiae*. *J. Chem. Technol. Biotechnol.* **2007**, *82*, 340–349. [CrossRef]

30. Patsalou, M.; Menikea, K.K.; Makri, E.; Vasquez, M.I.; Drouza, C.; Koutinas, M. Development of a citrus peel-based biorefinery strategy for the production of succinic acid. *J. Clean. Prod.* **2017**, *166*, 706–716. [CrossRef]

31. Talebnia, F.; Pourbafrani, M.; Lundin, M.; Taherzadeh, M.J. Optimization study of citrus wastes saccharification by dilute-acid hydrolysis. *Bioresources* **2008**, *3*, 108–122.

32. Pourbafrani, M.; Forgacs, G.; Horvath, I.S.; Niklasson, C.; Taherzadeh, M.J. Production of biofuels, limonene and pectin from citrus wastes. *Bioresour. Technol.* **2010**, *101*, 4246–4250. [CrossRef]

33. Boluda-Aguilar, M.; Garcia-Vidal, L.; Gonzalez-Castaneda, F.D.P.; Lopez-Gomez, A. Mandarin peel wastes pretreatment with steam explosion for bioethanol production. *Bioresour. Technol.* **2010**, *101*, 3506–3513. [CrossRef]

34. Lennartsson, P.R.; Ylitervo, P.; Larsson, C.; Edebo, L.; Taherzadeh, M.J. Growth tolerance of *Zygomycetes Mucor indicus* in orange peel hydrolysate without detoxification. *Process Biochem.* **2012**, *47*, 836–842. [CrossRef]

35. Harun, R.; Danquah, M.K.; Forde, G.M. Microalgal biomass as a fermentation feedstock for bioethanol production. *J. Chem. Technol. Biotechnol.* **2010**, *85*, 199–203. [CrossRef]

36. Torija, M.J.; Beltran, G.; Novo, M.; Poblet, M.; Rozes, N.; Guillamon, J.M. Effect of the nitrogen source on the fatty acid composition of Saccharomyces cerevisiae. *Food Microbiol.* **2003**, *20*, 255–258. [CrossRef]

37. Seong, Y.J.; Lee, H.J.; Lee, J.E.; Kim, S.; Lee, D.Y.; Kim, K.H.; Park, Y.C. Physiological and metabolomic analysis of *Issatchenkia orientalis* MTY1 with multiple tolerance for cellulosic bioethanol production. *Biotechnol. J.* **2017**, *12*, 1–11. [CrossRef] [PubMed]

38. Koppar, A.; Pullammanappallil, P. Anaerobic digestion of peel waste and wastewater for on site energy generation in a citrus processing facility. *Energy* **2013**, *60*, 62–68. [CrossRef]

39. Sanjaya, A.P.; Cahyanto, M.N.; Millati, R. Mesophilic batch anaerobic digestion from fruit fragments. *Renew. Energy* **2016**, *98*, 135–141. [CrossRef]

40. Calabro, P.S.; Pontoni, L.; Porqueddu, I.; Greco, R.; Pirozzi, F.; Malpei, F. Effect of the concentration of essential oil on orange peel waste biomethanization: Preliminary batch results. *Waste Manag.* **2016**, *48*, 440–447. [CrossRef] [PubMed]

41. Lotito, A.M.; Sanctis, M.D.; Pastore, C.; Iaconi, C.D. Biomethanization of citrus waste: Effect of waste characteristics and of storage on treatability and evaluation of limonene degradation. *J. Environ. Manag.* **2018**, *215*, 366–376. [CrossRef]

42. Wilkins, M.R.; Widmer, W.W.; Grohmann, K.; Cameron, R.G. Hydrolysis of grapefruit peel waste with cellulase and pectinase enzymes. *Bioresour. Technol.* **2007**, *98*, 1596–1601. [CrossRef]

43. Faravash, R.S.; Ashtiani, F.Z. The effect of pH, ethanol volume and acid washing time on the yield of pectin extraction from peach pomace. *Int. J. Food Sci. Technol.* **2007**, *42*, 1177–1187. [CrossRef]

44. Wilkins, M.R.; Widmer, W.W.; Cameron, R.G.; Grohmann, K. Effect of seasonal variation on enzymatic hydrolysis of valencia orange peel waste. *Proc. Fla. State Hortic. Soc.* **2005**, *118*, 419–422.

45. Koutinas, M.; Patsalou, M.; Stavrinou, S.; Vyrides, I. High temperature alcoholic fermentation of orange peel by the newly isolated thermotolerant *Pichia kudriavzevii* KVMP10. *Lett. Appl. Microbiol.* **2015**, *62*, 75–83. [CrossRef] [PubMed]

46. Angelidaki, I.; Alves, M.; Bolzonella, D.; Borzacconi, L.; Campos, J.L.; Guwy, A.J.; Kalyuzhnyi, S.; Jenicek, P.; Van Lier, J.B. Defining the biomethane potential (BMP) of solid organic wastes and energy crops: A proposed protocol for batch assays. *Water Sci. Technol.* **2009**, *59*, 927–934. [CrossRef] [PubMed]

47. Dubois, M.; Gilles, K.A.; Hamilton, J.K. Colorimetric method for determination of sugars and related substances. *Anal. Chem.* **1956**, *28*, 350–356. [CrossRef]

48. Rice, E.W.; Baird, R.B.; Eaton, A.D.; Clesceri, L.S. *Standard Methods for the Examination of Water and Wastewater*, 22nd ed.; American Public Health Association, Ameican Water Works Association, Water Environment Federation: Denver, CO, USA, 2012; p. 541.

49. Vardanyan, A.; Kafa, N.; Konstantinidis, V.; Shin, S.G.; Vyrides, I. Phosphorus dissolution from dewatered anaerobic sludge: Effect of pHs, microorganisms, and sequential extraction. *Bioresour. Technol.* **2018**, *249*, 464–472. [CrossRef] [PubMed]

Sample Availability: Samples of the compounds are available from the authors.

Article

Enhanced Enzymatic Hydrolysis of *Pennisetum alopecuroides* by Dilute Acid, Alkaline and Ferric Chloride Pretreatments

Shangyuan Tang [1, †], Chunming Xu [2, †], Linh Tran Khanh Vu [3], Sicheng Liu [1], Peng Ye [1], Lingci Li [1], Yuxuan Wu [1], Mengyu Chen [1], Yao Xiao [1], Yue Wu [1], Yining Wang [1], Qiong Yan [1,*] and Xiyu Cheng [1,*]

[1]   College of Life Sciences and Bioengineering, School of Science, Beijing Jiaotong University, Beijing 100044, China; 16121638@bjtu.edu.cn (S.T.); 16292010@bjtu.edu.cn (S.L.); 17121611@bjtu.edu.cn (P.Y.); 18121614@bjtu.edu.cn (L.L.); 15272020@bjtu.edu.cn (Y.W.); 18272032@bjtu.edu.cn (M.C.); 17271250@bjtu.edu.cn (Y.X.); 18272052@bjtu.edu.cn (Y.W.); 17272018@bjtu.edu.cn (Y.W.)
[2]   Key Laboratory of Cleaner Production and Integrated Resource Utilization of China National Light Industry, Beijing Technology and Business University, Beijing 100048, China; xucm@th.btbu.edu.cn
[3]   Faculty of Chemical and Food Technology, Ho Chi Minh City University of Technology and Education, No. 1 Vo Van Ngan Street, Linh Chieu Ward, Thu Duc District, Ho Chi Minh City 71307, Viet Nam; linhvtk@hcmute.edu.vn
*   Correspondence: qyan@bjtu.edu.cn (Q.Y.); xycheng@bjtu.edu.cn (X.C.); Tel.: +86-10-51684351-209 (X.C.)
†   These authors contributed equally.

Academic Editor: Ivet Ferrer
Received: 8 April 2019; Accepted: 30 April 2019; Published: 2 May 2019

Abstract: In this study, effects of different pretreatment methods on the enzymatic digestibility of *Pennisetum alopecuroides*, a ubiquitous wild grass in China, were investigated to evaluate its potential as a feedstock for biofuel production. The stalk samples were separately pretreated with H_2SO_4, NaOH and $FeCl_3$ solutions of different concentrations at 120 °C for 30 min, after which enzymatic hydrolysis was conducted to measure the digestibility of pretreated samples. Results demonstrated that different pretreatments were effective at removing hemicellulose, among which ferric chloride pretreatment (FCP) gave the highest soluble sugar recovery (200.2 mg/g raw stalk) from the pretreatment stage. In comparison with FCP and dilute acid pretreatment (DAP), dilute alkaline pretreatment (DALP) induced much higher delignification and stronger morphological changes of the biomass, making it more accessible to hydrolysis enzymes. As a result, DALP using 1.2% NaOH showed the highest total soluble sugar yield through the whole process from pretreatment to enzymatic hydrolysis (508.5 mg/g raw stalk). The present work indicates that DALP and FCP have the potential to enhance the effective bioconversion of lignocellulosic biomass like *P. alopecuroides*, hence making this material a valuable and promising energy plant.

Keywords: *Pennisetum alopecuroides*; dilute alkaline pretreatment; ferric chloride pretreatment; enzymatic hydrolysis

---

1. Introduction

With the growth of vehicles and an over-reliance on fossil fuels in industrial development, biofuel production from non-food crop feedstocks (e.g. agricultural residues, forest residues and industrial wastes) has drawn considerable attention [1–8]. In China, anaerobic digestion of organic wastes has been successfully used for biogas production both domestically, as well as at a larger scale [2–6]. Different biomass substrates, such as cornstalk, rice straw, wheat straw and pine foliage, have also been investigated for the production of bioethanol [9–12]. In addition to these lignocellulosic biomass wastes,

Pennisetum alopecuroides is a potential prolific renewable herbaceous plant that is widely distributed in many provinces of China. Some *Pennisetum* species are cultivated as an important feedstock for the animal feed industry in north China, northeast China, southwest China and the middle-lower Yangtze Plain [8]. The annual dry matter yield of *Pennisetum* grass ranges from 40 to 50.2 t/ha, which is much higher than the yields of sugarcane and corn (13~21 t/ha) and comparable to that of miscanthus (27~38 t/ha) [9,10]. While *Miscanthus* species and some other energy plant candidates (e.g., switchgrass) have attracted widespread interest in the bio-energy field, studies on *P. alopecuroides* have been relatively limited. Given its rapid growth, high yield, high adaptability, low cost and environmentally benign production, *P, alopecuroides* was selected as a substrate in the current study to evaluate its application prospects in biofuel production.

The recalcitrant structure of lignocellulosic biomass is the main constraint of its bioconversion [13–18]. Varied pretreatment strategies such as chemical methods (e.g., acid, bases, salts and solvents), physico-chemical processes (e.g., steam explosion, liquid hot water (LHW) and ammonia fiber expansion (AFEX) and biological methods have been developed in attempts to remove hemicellulose and/or lignin from lignocellulosic wastes and reduce the crystallinity of cellulose [13,19–24]. It is widely accepted that efficient pretreatment should avoid the use of expensive chemicals, improve fiber reactivity and maximize the recovery/formation of fermentable sugars, avoid formation of enzyme inhibitory byproducts, preserve cellulose and hemicellulose fractions that are easily accessible to hydrolysis enzymes and minimize energy requirements [11–14]. However, no single strategy could efficiently meet all these criteria due to the variations in material characteristics. The chemical pretreatment of lignocellulosic materials has been widely employed in many pilot and large-scale cellulosic ethanol plants because it is ideal for low-lignin materials and has high reactivity at mild conditions [12–14]. A chemical method is hence a suitable pretreatment strategy for *P. alopecuroides,* a low-lignin material.

Among several chemical methods, dilute acid pretreatment is most commonly used, due to its advantages in cost and process severity [13,14]. One major limitation of acid pretreatment is its requirement of corrosion-resistant reactors [13]. On the other hand, corrosion problems and sugar degradation are less severe in alkaline processes than in acid pretreatment. Alkaline pretreatment is also effective in delignifying the biomass [7,20,22,24]. A mild alkali concentration (<4% *w/w*) favors enzymatic hydrolysis especially for low-lignin materials [13,14]. Previous studies also showed that pretreatment of lignocellulosic biomass using Lewis acids such as $FeCl_3$ enhances digestibility of biomass, producing reusable solubilized hemicellulose [25–27]. Operating at milder temperature conditions (120 °C vs 160~260 °C in steam explosion and LHW pretreatments) in the current study means that dilute acid pretreatment (DAP), dilute alkaline pretreatment (DALP) and ferric chloride pretreatment (FCP) reduce energy consumption and the formation of enzyme inhibitory byproducts [11]. Up to date, DAP, DALP and FCP have been used on a wide range of low-lignin biomasses ranging from wood (cedar, pine, hemlock) to agricultural residues (corn stover, wheat and barley straw, switchgrass, *Miscanthus*) [18–24]. To our best knowledge, the effects of these methods on the biodigestibility of *P. alopecuroides* has not been systematically studied to identify the ideal pretreatment process or to evaluate the potential of *P. alopecuroides* biomass in the bioconversion industry. Moreover, some previous studies have also shown that effective removal of lignin and/or hemicellulose in acid/alkaline pretreatments did not result in a significant increase in reducing sugar yields (only 91.4−92 mg/g) [19,21]. These results indicate that the exact roles of different pretreatments in the improvement of hydrolysis efficiency were complicated, thereby necessitating further research to better understand the mechanism.

In the present study, three chemical pretreatments including DAP, DALP and FCP were systematically investigated to develop an efficient pretreatment strategy for enhancing enzymatic hydrolysis of *P. alopecuroides* biomass. The composition and microstructure of substrates in response to these pretreatments were investigated to better understand the exact roles of each pretreatment in changing biomass recalcitrance and subsequent enzymatic hydrolysis enhancement. The fermentable sugar production from the pretreated *P. alopecuroides* biomass was also studied to evaluate its application prospects for biofuel production.

2. Results and Discussion

2.1. Effect of Different Pretreatments on Biomass Composition

The pretreatment process decreases the recalcitrance of lignocellulosic substrates by removing lignin and hemicellulose components, thereby exposing cellulose to the hydrolysis enzyme [28,29]. The *P. alopecuroides* samples were subjected to different pretreatments, including DAP, DALP and FCP. The solid yield and compositional change of the stalk samples are important indices to evaluate the effectiveness of their pretreatments. As shown in Table 1, the dry matter retained after different pretreatments was about 53.3–58.2%. A *P. alopecuroides* sample pretreated by DALP had a higher solid yield as compared with those pretreated with DAP and FCP. The weight loss of *P. alopecuroides* biomass could be attributed to the solubilization of its components into the aqueous solution. The higher solid yield (or lower weight loss) indicate that less lignocellulosic components were converted into soluble substances. In comparison to DALP, DAP and FCP thereby gave higher soluble sugar concentrations (86.7 mg/g raw stalk (RS) vs 112.2 mg/g RS and 193.4 mg/g RS, respectively).

Table 1. Effect of DAP/DALP/FCP of *Pennisetum alopecuroides* biomass on its chemical composition.

Different Pretreatment	Solid Yield (%)	Hemicellulose Content (%)	Cellulose Content (%)	Lignin Content (%)	Soluble Sugar from Pretreatment (mg/g RS) [1]
Control	–	28.7 ± 0.4	41.8 ± 0.9	17.5 ± 0.6	20.0 ± 0.4
DAP, 3% H_2SO_4	53.3 ± 1.7	15.0 ± 1.7	55.9 ± 2.3	17.1 ± 1.5	112.2 ± 2.0
DALP, 1.0%NaOH	58.2 ± 1.2	15.3 ± 0.1	64.1 ± 1.7	11.7 ± 0.6	86.7 ± 0.2
FCP, 3.2% $FeCl_3$	55.9 ± 1.8	11.9 ± 1.3	60.8 ± 1.3	16.4 ± 0.7	193.4 ± 8.7

[1] The soluble sugar yield in the pretreatment process was calculated based on per g raw stalk (RS). Values are means of triplicate ± standard deviation.

Compositional analysis showed that the weight loss of the stalk samples was mainly due to hemicellulose hydrolysis by both DAP and FCP (Table 1). A previous study also reported that soluble sugars predominantly originated from the hemicellulose fraction of lignocellulosic substrates [23]. The dilute acid pretreatment and $FeCl_3$ pretreatment can markedly solubilize hemicellulose into monomeric sugars and soluble oligomers [23,27]. Results in Table 1 indicate that DALP significantly removed both lignin and hemicellulose, while DAP and FCP preferentially removed hemicellulose. Samples pretreated by DALP had the lowest lignin content of 11.7%. A similar delignification effect was also observed in alkaline pretreatment and microwave-assisted alkaline pretreatment processes for different stalk wastes [6,15,30,31].

2.2. Effect of Different Pretreatments on Cell Structure

Cell structure changes of the stalk samples pretreated by DAP, DALP and FCP were also studied by a scanning electron microscopy (SEM) at the same magnification (× 500) (Figure 1).

As shown in Figure 1a, the untreated *P. alopecuroides* sample had a smooth and intact surface with an unchanged fibrous structure organization. On the other hand, the cell walls of the DAP sample were obviously destroyed. It can be seen in Figure 1b that the rectangular cell wall boundaries became blurry to some extent, indicating that a portion of cellulosic components could have been removed by DAP. The cell surfaces of the FCP sample were also not smooth, and partly broken (Figure 1d). This clearly verifies that the removal of lignocellulosic components such as hemicellulose by FCP resulted in morphological surface modification. Stronger morphological changes were observed from the DALP sample. As can be seen in Figure 1c, the surfaces and cell walls of the DALP sample were significantly destroyed and distorted. Enhanced destruction of plant cell wall was also found in previous studies with microwave-assisted acid/alkaline pretreatments of lignocellulosic wastes [22,23,32]. The present study suggests that the strong delignification effect and partial removal of hemicellulose by DALP remarkably enhances the destruction effect to cell wall structure (Table 1 and Figure 1). The formation of cracks, fragments and the distortion of cell structure induced by DALP could lead to the formation of many reactive sites on the biomass surface. Such changes increasingly benefit the accessibility of stalk samples to enzymes and bacteria, and consequently accelerate the subsequent bioconversion process.

(**a**) Untreated sample

(**b**) DAP sample

(**c**) DALP sample

(**d**) FCP sample

Figure 1. SEM images of *P. alopecuroides* samples with and without pretreatments (500×): (**a**) untreated sample; (**b**) sample with DAP; (**c**) sample with DALP; (**d**) sample with FCP.

2.3. Effect of Different Pretreatments on Enzymatic Hydrolysis

Enzymatic hydrolysis was carried out to evaluate the enhancement of biodegradability of *P. alopecuroides* biomass pretreated with DAP, DALP and FCP. As shown in Figure 2, the untreated sample displayed its lowest reducing sugar yield after 72 h of enzymatic hydrolysis (134.8 mg/g raw stalk). Meanwhile, samples pretreated with 3.0% H_2SO_4 solution and 3.2% $FeCl_3$ solution under 121 °C for 30 min exhibited varied increases in enzymatic hydrolysis efficiency. The reduced sugar yields of the pretreated samples by DAP and FCP reached 258.3 mg/g pretreated stalk (PS) and 293.7 mg/g PS, increasing by 92% and 118% as compared with the untreated samples, respectively. DALP appeared to be the most efficient pretreatment technique by which to obtain the highest reducing sugar yield among all tested pretreatment methods (669.7 mg/g PS) (Figure 2).

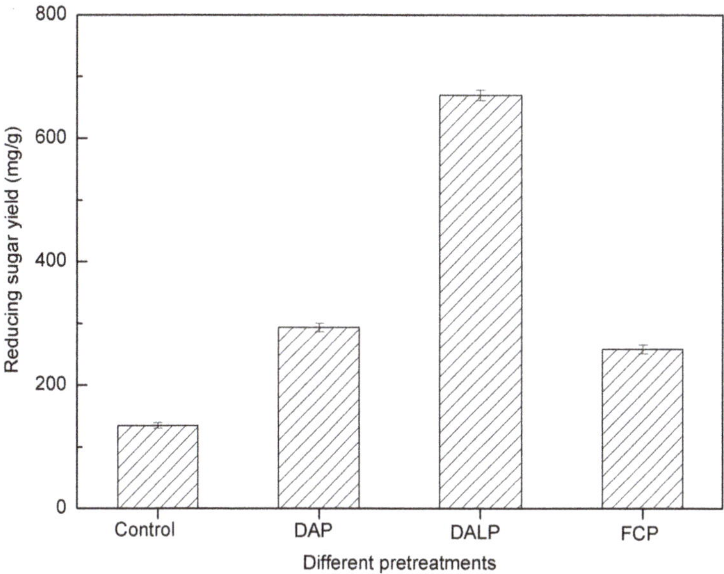

Figure 2. Effect of DAP, DALP and FCP on enzymatic hydrolysis.

The highly ordered crystalline structure of cellulose fibrils, the presence of lignin polymer and the recalcitrant cellulose–hemicellulose–lignin network structure in lignocellulosic biomass severely hindered enzymatic hydrolysis [33,34]. As shown in Table 1 and Figure 1, decreases in hemicellulose content and changes in cell structure, emerging after acid pretreatments (especially FCP), were responsible for the significant increase in reducing sugar yield. It is obvious that the pretreated sample by DALP, in which much more lignin components were removed, showed a much higher reducing sugar yield than those of DAP and FCP samples. The connection between these results strengthens a hypothesis that lignin removal is relatively important for improving enzymatic hydrolysis. This is because delignification not only reduces the adsorption of cellulase onto lignin, but also produces higher cellulose substrate content, consequently making cellulose more accessible to enzymes [30,31,33,34]. A stronger destruction of cell wall and higher increase of cellulose accessibility in DALP samples also produced a positive impact on enzymatic hydrolysis, affirmed by SEM observation (Figure 1).

2.4. Effect of H_2SO_4/NaOH/FeCl$_3$ Concentrations on Biomass Composition

The effects of H_2SO_4/NaOH/FeCl$_3$ concentrations in different pretreatment processes on the chemical characteristics of pretreated samples were further studied. The corresponding acid, alkali and FeCl$_3$ concentrations in DAP, DALP and FCP were selected based on previous studies [7,22,23,27]. In DAP, as the H_2SO_4 concentrations increased from 1.0% to 4.0%, the solid yields decreased from 59.4% to 52.6%. The reduction in solids was mainly ascribed to the degradation of lignocellulosic components. As can be seen in Table 2, the corresponding hemicellulose contents decreased from 16.9% to 12.2%, consistent with the decrease of solid yields. Removed cellulosic components were partially converted into soluble sugars, and the corresponding sugar concentrations in the aqueous phase significantly increased from 40.0 mg/g RS to 119.3 mg/g RS (Table 2). Due to the removal of hemicellulose, cellulose contents increased.

Table 2. The effects of H_2SO_4/NaOH/$FeCl_3$ concentrations during pretreatments of *P. alopecuroides* on its chemical composition.

Different Pretreatment	H_2SO_4/NaOH/$FeCl_3$ Concentrations (%)	Solid Yield (%)	Hemicellulose Content (%)	Cellulose Content (%)	Lignin Content (%)	Soluble Sugar from Pretreatment (mg/g RS) [1]
Control	–	–	28.7 ± 0.4	41.8 ± 0.9	17.5 ± 0.6	20.0 ± 0.4
DAP with H_2SO_4	1%	59.4 ± 1.0	16.9 ± 1.6	51.8 ± 1.9	15.7 ± 1.1	40.0 ± 1.3
	2%	55.0 ± 0.9	15.5 ± 1.3	55.0 ± 0.5	16.8 ± 1.2	69.5 ± 1.2
	3%	53.3 ± 1.7	15.0 ± 1.7	55.9 ± 2.3	17.1 ± 1.5	112.2 ± 2.0
	4%	52.6 ± 2.1	12.2 ± 0.2	56.5 ± 1.6	17.3 ± 1.4	119.3 ± 0.7
DALP with NaOH	0.4%	74.6 ± 2.5	22.0 ± 0.9	53.3 ± 0.3	14.5 ± 0.4	50.5 ± 0.2
	0.6%	64.7 ± 1.7	17.4 ± 0.9	59.0 ± 1.4	14.0 ± 0.1	63.4 ± 2.0
	0.8%	59.4 ± 2.1	16.0 ± 0.2	63.2 ± 0.6	12.6 ± 0.1	83.5 ± 0.7
	1.0%	58.2 ± 1.2	15.3 ± 0.1	64.1 ± 1.7	11.7 ± 0.6	86.7 ± 0.2
	1.2%	53.9 ± 1.7	14.8 ± 0.3	68.0 ± 0.7	10.2 ± 0.8	107.3 ± 0.6
FCP with $FeCl_3$	0.8%	68.8 ± 3.1	15.2 ± 1.7	56.9 ± 1.6	18.0 ± 1.6	90.2 ± 2.2
	1.6%	60.8 ± 1.7	13.5 ± 1.5	60.7 ± 1.1	17.3 ± 0.1	163.6 ± 3.2
	3.2%	55.9 ± 1.8	11.9 ± 1.3	60.8 ± 1.3	16.4 ± 0.7	193.4 ± 8.7
	4.8%	53.1 ± 2.3	9.7 ± 1.7	62.7 ± 1.5	15.3 ± 1.5	200.2 ± 6.7

[1] The soluble sugar yield in the pretreatment process was calculated based on per g raw stalk (RS). Values are means of triplicate ± standard deviation. RS: raw stalk.

In the case of FCP, severe pretreatment conditions (3.2% and 4.8% $FeCl_3$) also effectively removed cellulosic components, resulting in lower solid yields of 53.1~55.9% as compared with those of less severity conditions (0.8~1.6% $FeCl_3$) (Table 2). The significant decrease in hemicellulose content and the slight increase in cellulose content suggest that hemicellulose was more easily degraded than cellulose during FCP pretreatment [7,18]. The corresponding soluble sugar content in the aqueous phase also increased from 90.2 mg/g RS to 200.2 mg/g RS (Table 2). Due to the acidic condition provided by H_2SO_4 and $FeCl_3$ during DAP and FCP, the generated soluble sugars could be dehydrated to furfural and 5-hydroxymethylfurfural (5-HMF) [23,27]. Although the solid yields obtained by DAP and FCP were similar, the soluble sugar concentrations in FCP samples were much higher than those observed in DAP samples (193.4~200.2 mg/g RS vs 112.2~119.3 mg/g RS), indicating that more generated soluble sugars may have been further converted into other byproducts by H_2SO_4 [23,27].

The mass loss observed after acid pretreatments can be mainly attributed to the removal of hemicellulose components [23,27]. Acid pretreatments such as DAP and FCP randomly break glycosidic bonds, removing hemicellulose while improving the cellulose content of lignocellulosic biomass and consequently increasing the accessibility of cellulose to hydrolytic enzymes [28,29]. On the other hand, alkaline pretreatments produce nucleophilic attacks that break the lignin structure, solubilizing lignin fragments or hemicellulose from α-O-4 linkages [35–37]. Results in Table 2 show that relatively higher solid yields of 53.9~74.6% were observed in DALP samples. As the NaOH concentrations increased from 0.4% to 1.2%, the hemicellulose contents decreased from 22.0% to 14.8%, and cellulose contents significantly increased from 53.3% to 68.0%. The corresponding soluble sugar concentrations also increased from 41.8 mg/g RS to 107.3 mg/g RS (Table 2). It should be noted that DALP, with 1.0~1.2% NaOH, more significantly reduced the lignin contents of biomass to 10.2~11.7% as compared with those achieved by DAP and FCP (Table 2).

2.5. Effect of H_2SO_4/NaOH/$FeCl_3$ Concentrations on Enzymatic Hydrolysis

The effects of different H_2SO_4/NaOH/$FeCl_3$ concentrations on the enzymatic hydrolysis of *P. alopecuroides* biomass were also investigated. As shown in Figure 3, samples pretreated with DAP, DALP and FCP had much higher reducing sugar yields after enzyme hydrolysis, 1.4~5.5 times higher than that of the untreated sample (134.8 mg/g). In particular, the sample pretreated with 1.0% H_2SO_4 solution under 121 °C for 30 min exhibited an obvious increase in enzymatic hydrolysis efficiency, and the corresponding reducing sugar yield reached 258.1 mg/g PS. Increased acid concentrations showed a beneficial effect on cellulosic component removal and subsequent enzymatic hydrolysis. The reducing sugar yield reached a maximum value of 336.4 mg/g PS when *P. alopecuroides* biomass was pretreated with 4.0% H_2SO_4. Similar improvements were observed during the FCP process. As can be seen in Figure 3, the reducing sugar yield of the sample pretreated with 0.8% $FeCl_3$ solution was 192.0 mg/g PS. The highest reducing sugar yield of 279.3 mg/g PS was obtained when $FeCl_3$ concentration was increased to 4.8%.

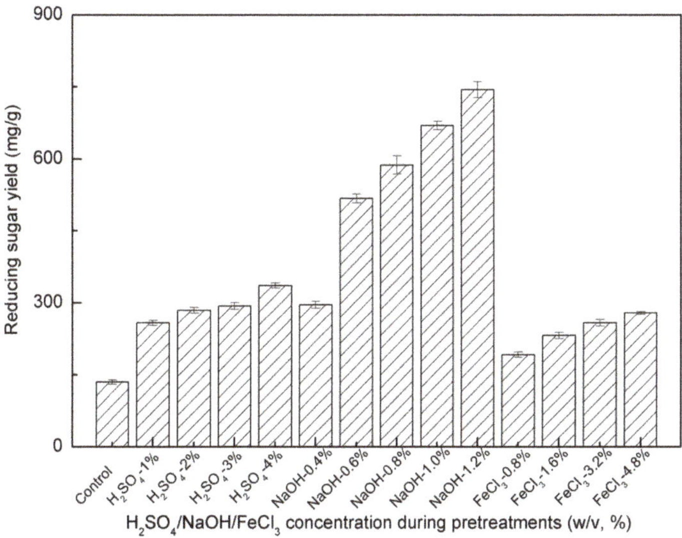

Figure 3. Effects of H_2SO_4/NaOH/$FeCl_3$ concentrations during pretreatments on enzymatic hydrolysis.

As can be seen in Figure 3, the enzymatic hydrolysis efficiency of DALP samples was the highest among the untreated sample and all pretreated samples. At the lowest NaOH concentration of 0.4%, the reducing sugar yield of the DALP sample reached 296.1 mg/g PS, comparable to that obtained in samples pretreated with 3% H_2SO_4 and 4.8% $FeCl_3$. DALP with higher NaOH concentrations (0.6~1.2%) resulted in a significant improvement in hydrolysis performance. Compared to all tested samples, the highest reducing sugar yield was 744.4 mg/g PS using 1.2% NaOH. It should be also noted that some undesirable enzyme-inhibiting byproducts (mainly phenolic compounds, furans and organic acids) might be formed under certain pretreatment processes, especially those conducted at severe conditions (e.g., steam explosion, LHW, chemical treatments, 160~240 °C, 5~45 min, pretreatment severity factor (log R_0) of 2.8~4.8) [11,38–40]. However, tested pretreatment strategies were still effective because cellulase was not inhibited in the subsequent enzymatic hydrolysis of the pretreated biomass, as illustrated by high reducing sugar yields of the pretreated samples (Figure 3). These results also confirmed the potential of these chemical pretreatment methods. The low inhibiting effect could be mainly attributed to the milder operation conditions (low acid/alkaline/$FeCl_3$ concentrations, 120 °C, 30 min, pretreatment severity factor (log R_0) of 2.1), which can lead to lower levels of byproducts [11,20,41–43]. In addition, most byproducts, such as the phenolics formed by the NaOH pretreatment, were retained in the pretreated liquors, resulting in a substantially lower level of phenolics in the enzyme-saccharified hydrolysates [44].

The effects of different pretreatment strategies on removing the physical barrier of lignin, reducing the dense crystalline structure and coating effect of hemicellulose and enhancing the accessibility of the pretreated biomass to hydrolytic enzymes were relatively complicated [11,12,45,46]. Results clearly indicate that both DAP and FCP were effective at removing hemicellulose components and destroying cell wall structure. The hemicellulose removal in DAP and FCP exhibited a positive relationship with the reducing sugar yields during the subsequent enzymatic hydrolysis (Table 2 and Figure 3).

These results further support the observation that under acid/Lewis acid pretreatments, cellulose saccharification is linearly proportional to the amount of hemicellulose (mainly xylan) removed, since hemicellulose removal helps increase cellulose accessibility [27,45]. The reducing sugar yields (279.3~336.4 mg/g PS) and conversion ratios (34.6~44.5%) by DAP and FCP were comparable to those observed in the pretreated bamboo (77 mg/g PS), rice grass (457 mg/g PS) and pine foliage (588 mg/g PS) pretreated with 1~2% H_2SO_4 at 121 °C for 60 min [19–21]. Notably, the simultaneous removals of lignin and hemicellulose were also positively related with the reducing sugar yields of the DALP samples (Table 2 and Figure 3). Despite lower hemicellulose removal, DALP provided a much higher delignification (Table 2) and reducing sugar recovery as compared with DAP and FCP (744.4 mg/g PS vs 279.3~336.4 mg/g PS, respectively). In addition to the obvious impact on delignification, significant enhancements on fermentable sugar releases were also observed in different biomass samples pretreated with NaOH under similar conditions, including bamboo, pine foliage, *Pennisetum purpureum*, wheat straw and *Eucalyptus* (324~629 mg/g PS) [17,20,34,44]. These results indicate that effective lignin removal, as well as cellulose swelling induced by DALP, appeared to be more important factors in decreasing biomass recalcitrance and increasing enzymatic digestibility as compared with hemicellulose removal in DAP and FCP.

2.6. Analysis of Mass Balance and Prospects for P. alopecuroides

A promising pretreatment method not only enables the ability to obtain readily digestible substrates, but also maximizes the total yield of fermentable sugars. Table 3 shows the mass balance of stalk samples under different pretreatments and subsequent enzymatic hydrolysis. The highest reducing sugar yields in the aqueous phases under the optimal DAP, DALP and FCP conditions were 119.3 mg/g RS (DAP-4% H_2SO_4), 107.3 mg/g RS (DALP-1.2% NaOH) and 200.2 mg/g RS (FCP-4.8% $FeCl_3$). The solid yields under the above conditions were 52.6%, 53.9% and 53.1%, respectively. The corresponding reducing sugar yields of enzymatic hydrolysis of the DAP, DALP and FCP samples were 336.4, 744.4 and 279.3 mg/g PS, respectively. Based the above data, the total soluble sugar yields from both pretreatment and enzymatic hydrolysis processes were 296.2, 508.5 and 348.6 mg/g RS, respectively. DAP had the lowest sugar recovery, partly because of its low enzymatic hydrolysis efficiency. Another reason was that cellulosic components, once removed, may be further converted into other byproducts, and thus the sugar yield from the aqueous phase during DAP was not high despite of its high weight loss. Among experimental conditions tested, DALP yielded the highest total sugar recovery through the whole process from pretreatment to enzymatic hydrolysis, while FCP indicated the most efficient method to recover soluble sugars only at pretreatment stage.

Table 3. Mass balance of the untreated and pretreated stalks.

Different Methods	Solid Yield (%)	Soluble Sugar from Pretreatment (mg/g RS)	Soluble Sugar from Enzymatic Hydrolysis (mg/g PS)	Soluble Sugar from Enzymatic Hydrolysis (mg/g RS)	Total Soluble Sugar Yield (mg/g RS) [1]
Control	–	–	–	134.8	134.8
DAP	52.6	119.3	336.4	176.9	296.2
DALP	53.9	107.3	744.4	401.2	508.5
FCP	53.1	200.2	279.3	148.4	348.6

[1] The total soluble sugar yield through the whole process from pretreatment to enzymatic hydrolysis was calculated based on per g raw stalk. RS: raw stalk; PS: pretreated stalk.

The fermentable sugar recovery from the *P. alopecuroides* sample pretreated by DALP was comparable to those obtained in the enzymatic hydrolysis of different lignocellulosic biomasses, as well as *Pennisetum* grass species [9,17,19,20,34,46,47]. As shown in Table 4, the maximum released sugar yields of 362.3~629 mg/g PS were obtained by acid or alkaline pretreatments of various biomass wastes such as wild rice grass, pine foliage, *Eucalyptus* and bamboo [19,20,34,46]. The optimal reducing sugar yield was only 146.9 mg/g pretreated elephant grass (*P. purpureum*) pretreated with 1.5% NaOH at

121 °C for 60 min [17]. Total fermentable sugar yields of 324~537 mg/g Napier grass (*P. purpureum*) were reported for samples pretreated with 2% Ca(OH)$_2$ or 2% NaOH at 121 °C for 60 min [47]. In this study, the yield obtained so far on *P. alopecuroides* (pretreated with 1.2% NaOH at 121 °C for 30 min) was 744.4 mg/g PS (508.5 mg/g RS, conversion ratio of 85.4%), opening a potential avenue for efficient biofuel production from *Pennisetum* grass.

Table 4. Comparison of fermentable sugar recovery from different biomass.

Biomass	Pretreatment Conditions	Sugar Yield (mg/g PS) [1]	Conversion Ratio (%)	Ref.
P. alopecuroides	1.2% NaOH, 121 °C, 30 min	744.4	85.4	This study
Wild rice grass	2% H$_2$SO$_4$, 121 °C, 60 min	457	93.2	[19]
Bamboo	1% NaOH, 3% Tween 80,121 °C, 60 min	629	75.4	[46]
Bamboo	1% H$_2$SO$_4$, 3% Tween 80,121 °C, 60 min	153	24.7	[46]
Pine foliage	1% C-TAB, 1% H$_2$SO$_4$, 121 °C, 60 min	588	98.1	[20]
Pine foliage	1% PEG-6000, 1% NaOH, 121 °C, 60 min	477	88.4	[20]
Eucalyptus	12.5% [TBA][OH], ultrasound irradiation (at a power of 360 W for 60 min)	426.6	51.5	[34]
Eucalyptus	2% NaOH, ultrasound irradiation (at a power of 360 W for 60 min)	362.3	56.6	[34]
P. purpureum Schum	0.5% NaOH, 90 °C, 60 min	(glucose yield: 245 mg/g RS)	NA	[9]
P. purpureum	1.5% NaOH, 121 °C, 60 min	146.9	24.7	[17]
P. purpureum	2% Ca(OH)$_2$ or NaOH, 121 °C, 60 min	324~537[2]	65.5~88.7	[47]

[1] Soluble sugar yields were calculated based on per g pretreated stalk. [2] Sugar yields and conversion ratios were calculated on the basis of the reported data of the cellulose/hemecellulose contents and glucose/xylose/reducing sugar yields in the corresponding references. RS: raw stalk; PS: pretreated stalk; Ref.: references; NA: not available.

Results obtained in this study clearly illustrated that digestibility of *P. alopecuroides* biomass were significantly improved by DALP, hence this pretreated biomass could be used in bioethanol or biogas production [1,12–14]. In fact, the DALP sample could be directly fermented by mixed microorganisms for biogas production, hence omitting the enzymatic hydrolysis step using expensive cellulase [1,3–5]. The techno-economic feasibility of the DALP sample integrated with biogas production, in which energy recovery ranges from 50% to 85%, has already been proven in several industrial biogas plants in China [1,48,49].

The current pretreatment strategies can also be integrated with ethanol fermentation. An energy recovery of 37% was observed in full-scale ethanol production using starch substrates [49]. Lignocellulosic bioethanol production is also energetically sustainable based on heat/electricity production from fermentation residues, while greenhouse gas (GHG) emissions are decreased by 50~93% in this process [50]. It should be noted that the industrial application of lignocellulosic bioethanol production was still limited to some extent due to the high cost of cellulase (about USD 0.50 per gallon ethanol, accounting for 20%–30% of total costs [51]). However, recent techno-economic analysis indicated that the evaluated minimum ethanol selling price (MESP) decreased from USD 4.58 per gallon to USD 1.91~2.46 per gallon, which comes close to the market price of ethanol (USD 2.50~3.10 per gallon) [52–55]. Additionally, many studies have also been conducted to further resolve the bottleneck of enzymatic hydrolysis in lignocellulosic bioethanol production as follows: 1) improving production/activity of cellulase using mutagenesis, co-culturing and heterologous gene expression of cellulases; 2) reusing enzymes by immobilization; and 3) process optimization and integration (e.g., simultaneous saccharification and fermentation processes, cost-effective pretreatment, etc.) for reducing the cost [12–14]. These research avenues are undoubtedly making lignocellulosic bioethanol more economically viable.

It should be noted that due to limited farmland resources in China, planting energy crops on available marginal land, which is estimated to be about 5.5 million ha, is regarded as one of the most promising choices for the production of biofuel feedstocks [56]. If 20% of the marginal land area is used for planting *P. alopecuroides* and the dry biomass yield is about 30 t/ha, theoretically about eight million tons of cellulosic ethanol can be produced annually, assuming a 12% biomass moisture content and an ethanol yield of 264 kg/t of dry biomass [52]. This estimated potential yield would almost reach the 2020 ethanol target (10 million tons per year) in China.

3. Materials and Methods

3.1. Materials

Wild *P. alopecuroides* was manually collected from Fujian, China, and dried in the sun. After that, *P. alopecuroides* samples were dried in an oven at 60 °C for at least 24 h to a constant weight and then milled to pass through a 20-mesh sieve using a plant miller. The main composition of stalks (on a dry weight basis) was as follows: cellulose 41.8%, hemicellulose 28.7% and lignin 17.5%.

3.2. Pretreatment Process

Dried samples were added to glass bottles containing 1.0~4.0% (w/v) sulfuric acid (DAP), 0.4~1.2% (w/v) NaOH (DALP) and 0.8~4.8% (w/v) $FeCl_3$ solutions (FCP), respectively, based on a solid loading rate of 10% [7,22,23,27]. The above samples for all three pretreatments were then autoclaved at 121 °C for 30 min [7,22,23,27]. After the pretreated samples were centrifuged, the supernatants were collected and stored at −20 °C for further analysis. The solid residues were washed with deionized water until the filtrates were neutral. The solids were then dried in an oven at 105 °C to a constant weight. The dried solids were sealed in plastic bags and stored in a desiccator at room temperature until composition analysis and/or enzymatic hydrolysis.

3.3. Enzymatic Hydrolysis

The cellulase used for enzymatic hydrolysis were donated by Hunan Youtell Biochemical Co., Ltd. The protein content of the cellulase was 35 mg/mL. The activities of cellulase, β-glucosidase and endoglucanase were 30 FPU/mL, 6.8 U/mL and 165 U/mL, respectively. The enzymatic hydrolysis was performed in a 250-mL conical flask using 50-mM sodium acetate buffer (pH 5) containing 40 μL tetracycline hydrochloride with 2.5% solid loading at 50 °C and a 150-rpm agitation rate for 72 h. The enzyme was loaded at 15 FPU/g for the untreated and pretreated samples (i.e., 17.5 mg enzyme protein/g solids). One unit of cellulase activity is defined as the amount of the enzyme that releases 1 μmol of glucose per minute in the reaction mixture at 50 °C and pH 5. After solid–liquid separation via centrifugation, reducing sugars in the hydrolysate were analyzed by the 3, 5-dinitrosalicylic acid (DNS) assay [57]. All experiments were carried out in triplicate and all values were the means of triplicate ± SD.

The biomass conversion to fermentable sugar was calculated using the equation [19,47]:

$$\text{Conversion ratio } (\%) = 100 \times Y_{TRS}/(1.111 \times C_C + 1.136 \times C_{HC}) \tag{1}$$

where Y_{TRS} (g/g pretreated sample) is the total reducing sugar yield per gram of pretreated sample in the enzymatic hydrolysate; the constants 1.111 and 1.136 are the conversion factors for cellulose/hemicellulose to the equivalent reducing sugars; and C_C and C_{HC} (g/g pretreated sample) are the cellulose and hemicellulose contents per gram of pretreated sample.

3.4. Scanning Electron Microscopy (SEM) Observation

The stalk samples were used to observe cell destruction before and after different pretreatments via scanning electron microscopy (SEM). The dried samples were fixed on a specimen holder with aluminium tape, then sputtered with gold in a JEOL JEC-1200 sputter-coater (Tokyo, Japan). All specimens were examined with a JEOL JSM-5600 LV scanning electron microscope (Tokyo, Japan) under high vacuum and at an accelerating voltage of 5.0 kV (10 μm, 500× magnification).

3.5. Analytical Methods

The sample mixtures were centrifuged and separated after pretreatments. The supernatants were collected to analyze total soluble sugar content. The solid residues were dried, and solid yields were recorded. Total solid and total soluble sugars were analyzed according to the standard methods [57–59]. The cellulose, hemicellulose and lignin contents were determined according to the standard method of Goering and Van-Soest [60], and calculated on the basis of residual total solids after pretreatments. All experiments were performed in triplicate, and all values were the means of triplicate ± SD.

Pretreatment severity factor ($\log R_0$) was calculated with the following equation [61,62]:

$$R_0 = t^*\exp[(T_r\text{-}100)/14.75] \tag{2}$$

where T_r is the reaction temperature (°C), 100 is the reference temperature (°C) and t is the reaction time (min). The fitted value (14.75) is the arbitrary constant ω.

4. Conclusions

Results in this study show that DAP, DALP and FCP resulted in the obvious structural changes and a high degree of hemicellulose and/or lignin removal from *P. alopecuroides* samples. Among all pretreatment methods tested, FCP produced the highest soluble sugar recovery (200.2 mg/g raw stalk) at the pretreatment stage. In comparison with FCP and DAP, DALP offered a much higher delignification and stronger morphological changes, which could significantly enhance the accessibility of the pretreated stalks to the enzymes, and thereby improve the performance of enzymatic hydrolysis. DALP gave the highest total soluble sugar yield of the pretreatment enzymatic-hydrolysis process (508.5 mg/g raw stalk). These results indicate that *P. alopecuroides*, a popular grass in huge quantities in China, could be used as a promising feedstock for biofuel production.

Author Contributions: Conceptualization, S.T. and C.X.; methodology, S.T., C.X., S.L. and L.L.; software, P.Y.; validation, M.C., S.L. and Y.W. (Yue Wu); formal analysis, C.X., Y.W. (Yining Wang) and X.X.; investigation, C.X., S.T., S.L., Y.W. (Yuxuan Wu), Y.X. and Q.Y.; resources, C.X., Q.Y. and X.C.; data curation, S.T.; writing—original draft preparation, S.T. and X.C.; writing—review and editing, X.C., L.V., and Q.Y.; supervision, X.C.; project administration, X.C.; funding acquisition, X.C.

Funding: This study was financially funded by Fundamental Research Funds for the Central Universities (grant number 2017JBM073), Beijing Jiaotong University Training Program of Research for Undergraduates (grant number 180170142), the Open Research Fund Program of Key Laboratory of Cleaner Production and Integrated Resource Utilization of China National Light Industry (grant number CP-2019-YB6) and the National Natural Science Foundation of China (No. 21306009).

Conflicts of Interest: The authors declare no conflict of interest.

References

1. Caruso, M.C.; Braghieri, A.; Capece, A.; Napolitano, F.; Romano, P.; Galgano, F.; Altieri, G.; Genovese, F. Recent updates on the use of agro-food waste for biogas production. *Appl. Sci.* **2019**, *9*, 1217. [CrossRef]
2. You, Z.; Zhang, S.; Kim, H.; Chiang, P.C.; Sun, Y.; Guo, Z.; Xu, H. Effects of corn stover pretreated with NaOH and CaO on anaerobic co-digestion of swine manure and corn stover. *Appl. Sci.* **2019**, *9*, 123. [CrossRef]
3. Cheng, X.Y.; Zhong, C. Effects of feed to inoculum ratio, co-digestion and pretreatment on biogas production from anaerobic digestion of cotton stalk. *Energy Fuels* **2014**, *28*, 3157–3166. [CrossRef]
4. Cheng, X.Y.; Liu, C.Z. Enhanced coproduction of hydrogen and methane from cornstalks by a three-stage anaerobic fermentation process integrated with alkaline hydrolysis. *Bioresour. Technol.* **2012**, *104*, 373–379. [CrossRef]
5. Cheng, X.Y.; Li, Q.; Liu, C.Z. Coproduction of hydrogen and methane via anaerobic fermentation of cornstalk waste in continuous stirred tank reactor integrated with up-flow anaerobic sludge bed. *Bioresour. Technol.* **2012**, *114*, 327–333. [CrossRef] [PubMed]

6. Kang, X.H.; Sun, Y.M.; Li, L.H.; Kong, X.Y.; Yuan, Z.H. Improving methane production from anaerobic digestion of *Pennisetum Hybrid* by alkaline pretreatment. *Bioresour. Technol.* **2018**, *255*, 205–212. [CrossRef] [PubMed]

7. Camesasca, L.; Ramı'rez, M.B.; Guigou, M.; Ferrari, M.D.; Lareo, C. Evaluation of dilute acid and alkaline pretreatments, enzymatic hydrolysis and fermentation of napiergrass for fuel ethanol production. *Biomass Bioenergy* **2015**, *74*, 193–201. [CrossRef]

8. Wang, S.D.; Chen, J.H.; Yang, G.H.; Gao, W.H.; Chen, K.F. Efficient conversion of Hybrid *Pennisetum* to glucose by oxygen-aqueous alkaline ionic liquid media pretreatment under benign conditions. *Bioresour. Technol.* **2017**, *243*, 335–338. [CrossRef]

9. Tsai, M.H.; Lee, W.C.; Kuan, W.C.; Sirisansaneeyakul, S.; Savarajara, A. Evaluation of different pretreatments of Napier grass for enzymatic saccharification and ethanol production. *Energy Sci. Eng.* **2018**, *6*, 683–692. [CrossRef]

10. Scordia, D.; Cosentino, S.L.; Jeffries, T.W. Effectiveness of dilute oxalic acid pretreatment of *Miscanthus×giganteus* biomass for ethanol production. *Biomass Bioenergy* **2013**, *59*, 540–548. [CrossRef]

11. Kim, D. Physico-chemical conversion of lignocellulose: Inhibitor effects and detoxification strategies: A mini review. *Molecules* **2018**, *23*, 309. [CrossRef]

12. Kumari, D.; Singh, R. Pretreatment of lignocellulosic wastes for biofuel production: A critical review. *Renew. Sustain. Energy. Rev.* **2018**, *90*, 877–891. [CrossRef]

13. Bensah, E.C.; Mensah, M. Chemical pretreatment methods for the production of cellulosic ethanol: Technologies and innovations. *Int. J. Chem. Eng.* **2013**, *2013*, 1–21. [CrossRef]

14. Latika, B.; Sonia, J.; Rumana, A. An economic and ecological perspective of ethanol production from renewable agro waste: A review. *AMB Express* **2012**, *2*, 65. [CrossRef]

15. Kucharska, K.; Rybarczyk, P.; Holowacz, I.; Lukajtis, R.; Glinka, M.; Kamiński, M. Pretreatment of lignocellulosic materials as substrates for fermentation processes. *Molecules* **2018**, *23*, 2937. [CrossRef]

16. Fu, S.F.; Chen, K.Q.; Zhu, R.; Sun, W.X.; Zou, H.; Guo, R.B. Improved anaerobic digestion performance of *Miscanthus floridulus* by different pretreatment methods and preliminary economic analysis. *Energ. Convers. Manag.* **2018**, *159*, 120–128. [CrossRef]

17. Eliana, C.; Jorge, R.; Juan, P.; Luis, R. Effects of the pretreatment method on enzymatic hydrolysis and ethanol fermentability of the cellulosic fraction from elephant grass. *Fuel* **2014**, *118*, 41–47. [CrossRef]

18. Sabanci, K.; Buyukkileci, A.Q. Comparison of liquid hot water, very dilute acid and alkali treatments for enhancing enzymatic digestibility of hazelnut tree pruning residues. *Bioresour. Technol.* **2018**, *261*, 158–165. [CrossRef]

19. Sahoo, D.; Ummalyma, S.B.; Okram, A.K.; Pandey, A.; Sankar, M.; Sukumaran, R.K. Effect of dilute acid pretreatment of wild rice grass (*Zizania latifolia*) from Loktak Lake for enzymatic hydrolysis. *Bioresour. Technol.* **2018**, *253*, 252–255. [CrossRef] [PubMed]

20. Pandey, A.K.; Negi, S. Impact of surfactant assisted acid and alkali pretreatment on lignocellulosic structure of pine foliage and optimization of its saccharification parameters using response surface methodology. *Bioresour. Technol.* **2015**, *192*, 115–125. [CrossRef] [PubMed]

21. Li, K.N.; Wan, J.M.; Wang, X.; Wang, J.F.; Zhang, J.H. Comparison of dilute acid and alkali pretreatments in production of fermentable sugars from bamboo: Effect of Tween 80. *Ind. Crops Products* **2016**, *83*, 414–422. [CrossRef]

22. Cheng, X.Y.; Liu, C.Z. Enhanced biogas production from herbal-extraction process residues by microwave-assisted alkaline pretreatment. *J. Chem. Technol. Biotechnol.* **2010**, *85*, 127–131. [CrossRef]

23. Liu, C.Z.; Cheng, X.Y. Improved hydrogen production via thermophilic fermentation of corn stover by microwave-assisted acid pretreatment. *Int. J. Hydrogen Energy* **2010**, *35*, 8945–8952. [CrossRef]

24. Menegol, D.; Schol, A.L.; Dillon, A.J.; Camassola, M. Influence of different chemical pretreatments of elephant grass (*Pennisetum purpureum, Schum.*) used as a substrate for cellulase and xylanase production in submerged cultivation. *Bioprocess Biosyst. Eng.* **2016**, *39*, 1455–1464. [CrossRef]

25. Kamireddy, S.R.; Li, J.B.; Tucker, M.; Degenstein, J.; Ji, Y. Effects and mechanism of metal chloride salts on pretreatment and enzymatic digestibility of corn stover. *Ind. Eng. Chem. Res.* **2013**, *52*, 1775–1782. [CrossRef]

26. Kang, K.E.; Park, D.H.; Jeong, G.T. Effects of inorganic salts on pretreatment of *Miscanthus* straw. *Bioresour. Technol.* **2013**, *132*, 160–165. [CrossRef]

27. Zhang, H.D.; Lyu, G.J.; Zhang, A.P.; Li, X.; Xie, J. Effects of ferric chloride pretreatment and surfactants on the sugar production from sugarcane bagasse. *Bioresour. Technol.* **2018**, *265*, 93–101. [CrossRef]

28. Meng, X.; Foston, M.; Leisen, J.; Demartini, J.; Wyman, C.E.; Ragauskas, A.J. Determination of porosity of lignocellulosic biomass before and after pretreatment by using Simons'stain and NMR techniques. *Bioresour. Technol.* **2013**, *144*, 467–476. [CrossRef]

29. Brienzo, M.; Fikizolo, S.; Benjamin, Y.; Tyhoda, L.; Görgens, J. Influence of pretreatment severity on structural changes, lignin content and enzymatic hydrolysis of sugarcane bagasse samples. *Renew. Energy* **2017**, *104*, 271–280. [CrossRef]

30. Wang, W.H.; Zhang, C.Y.; Tong, S.S.; Cui, Z.Y.; Liu, P. Enhanced enzymatic hydrolysis and structural features of corn stover by NaOH and ozone combined pretreatment. *Molecules* **2018**, *23*, 1300. [CrossRef]

31. Li, H.; Chen, X.; Wang, C.; Sun, S.; Sun, R. Evaluation of the two-step treatment with ionic liquids and alkali for enhancing enzymatic hydrolysis of Eucalyptus: Chemical and anatomical changes. *Biotechnol. Biofuels* **2016**, *9*, 166. [CrossRef]

32. Yan, Y.H.; Zhang, C.H.; Lin, Q.X.; Wang, X.H.; Cheng, B.G.; Li, H.L.; Ren, J.L. Microwave-assisted oxalic acid pretreatment for the enhancing of enzyme hydrolysis in the production of xylose and arabinose from bagasse. *Molecules* **2018**, *23*, 862. [CrossRef]

33. Jin, S.; Zhang, G.; Zhang, P.; Li, F.; Wang, S.; Fan, S.; Zhou, S. Microwave assisted alkaline pretreatment to enhance enzymatic saccharification of catalpa sawdust. *Bioresour. Technol.* **2016**, *221*, 26–30. [CrossRef]

34. Wang, Z.N.; Hou, X.F.; Sun, J.; Li, M.; Chen, Z.Y.; Gao, Z.Z. Comparison of ultrasound-assisted ionic liquid and alkaline pretreatment of *Eucalyptus* for enhancing enzymatic saccharification. *Bioresour. Technol.* **2018**, *254*, 145–150. [CrossRef]

35. Shimizu, F.L.; Monteiro, P.Q.; Ghiraldi, P.H.C.; Melati, R.B.; Pagnocca, F.C.; Souza, W.; Sant'Anna, C.; Brienzo, M. Acid, alkali and peroxide pretreatments increase the cellulose accessibility and glucose yield of banana pseudostem. *Ind. Crops Prod.* **2018**, *115*, 62–68. [CrossRef]

36. Thomas, H.L.; Seira, J.; Escudi, R.; Carrère, H. Lime pretreatment of miscanthus: Impact on bmp and batch dry co-digestion with cattle manure. *Molecules* **2018**, *23*, 1608. [CrossRef]

37. Duque, A.; Manzanares, P.; González, A.; Ballesteros, M. Study of the application of alkaline extrusion to the pretreatment of *Eucalyptus* biomass as first step in a bioethanol production process. *Energies* **2018**, *11*, 2961. [CrossRef]

38. Ximenes, E.; Kim, Y.; Mosier, N.; Dien, B.; Ladisch, M. Deactivation of cellulases by phenols. *Enzyme. Microb. Technol.* **2011**, *48*, 54–60. [CrossRef]

39. Ximenes, E.; Kim, Y.; Mosier, N.; Dien, B.; Ladisch, M. Inhibition of cellulases by phenols. *Enzyme. Microb. Technol.* **2010**, *46*, 170–176. [CrossRef]

40. Michelin, M.; Ximenes, E.; de Lourdes Teixeira de Moraes Polizeli, M.; Ladisch, M.R. Effect of phenolic compounds from pretreated sugarcane bagasse on cellulolytic and hemicellulolytic activities. *Bioresour. Technol.* **2016**, *199*, 275–278. [CrossRef]

41. Monlau, F.; Sambusiti, C.; Barakat, A.; Quéméneur, M.; Trably, E.; Steyer, J.P.; Carrère, H. Do furanic and phenolic compounds of lignocellulosic and algae biomass hydrolyzate inhibit anaerobic mixed cultures: A comprehensive review. *Biotechnol. Adv.* **2014**, *32*, 934–951. [CrossRef]

42. Imman, S.; Arnthong, J.; Burapatana, V.; Champred, V.; Laosiripojana, N. Effects of acid and alkali promoters on compressed liquid hot water pretreatment of rice straw. *Bioresour. Technol.* **2014**, *171*, 29–36. [CrossRef]

43. Rivera, O.M.P.; Moldes, A.B.; Torrado, A.M.; Domínguez, J.M. Lactic acid and biosurfactants production from hydrolyzed distilled grape marc. *Process Biochem.* **2007**, *42*, 1010–1020. [CrossRef]

44. McIntosh, S.; Vancov, T. Optimisation of dilute alkaline pretreatment for enzymatic saccharification of wheat straw. *Biomass Bioenergy* **2011**, *35*, 3094–3103. [CrossRef]

45. Meng, X.Z.; Wells, T.; Sun, Q.N.; Huang, F.; Ragauskas, A. Insights into the effect of dilute acid, hot water or alkaline pretreatment on the cellulose accessible surface area and the overall porosity of *Populus*. *Green Chem.* **2015**, *17*, 4239–4246. [CrossRef]

46. Obeng, A.K.; Premjet, D.; Premjet, S. Fermentable sugar production from the peels of two durian (*Durio zibethinus Murr.*) cultivars by phosphoric acid pretreatment. *Resources* **2018**, *7*, 60. [CrossRef]

47. Phitsuwan, P.; Sakka, K.; Ratanakhanokchai, K. Structural changes and enzymatic response of Napier grass (*Pennisetum purpureum*) stem induced by alkaline pretreatment. *Bioresour. Technol.* **2016**, *218*, 247–256. [CrossRef]

48. Jiao, X.X.; Jin, H.Y.; Wang, M.M. Research progress of straw pretreatment for anaerobic fermentation producing biogas in china. *China Biogas* **2011**, *29*, 29–39.
49. Yu, Y.; Zheng, P.; Chen, X.G.; Cai, J. Three typical fermentation utilization modes of biomass energy. *Bull. Sci. Technol.* **2009**, *25*, 854–859. (In Chinese) [CrossRef]
50. Moreno, J.; Dufour, J. Life cycle assessment of lignocellusosic bioethanol: Environmental impacts and energy balance. *Renew. Sustain. Energy. Rev.* **2015**, *42*, 1349–1361. [CrossRef]
51. Du, J. Novozymes accelerates cellulosic ethanol commercialized. *China WTO Tribune* **2010**, *10*, 81. (In Chinese) [CrossRef]
52. Duque, S.H.; Cardona, C.A.; Moncada, J. Techno-economic and environmental analysis of ethanol production from agroindustrial residues in Colombia. *Energy Fuels* **2015**, *29*, 775–783. [CrossRef]
53. Chovau, S.; Degrauwe, D.; Van der Bruggen, B. Critical analysis of techno-economic estimates for the production cost of lignocellulosic bio-ethanol. *Renew. Sustain. Energy Rev.* **2013**, *26*, 307–321. [CrossRef]
54. Gubicza, K.; Nieves, I.U.; Sagues, W.J.; Barta, Z.; Shanmugam, K.T.; Ingram, L.O. Techno-economic analysis of ethanol production from sugarcane bagasse using a liquefaction plus simultaneous saccharification and co-fermentation process. *Bioresour. Technol.* **2016**, *208*, 42–48. [CrossRef]
55. Denis, B.; Mark, M.W.; Robert, B. More than ethanol: A techno-economic analysis of a corn stover-ethanol biorefinery integrated with a hydrothermal liquefaction process to convert lignin into biochemicals. *Biofuel. Bioprod. Bior.* **2018**, *12*, 497–509. [CrossRef]
56. Chen, Y.Q.; Guo, X.D.; Liu, J.J.; Zhang, Z.J. Assessment of marginal land potential for energy plants in China. *Land Develop. Eng. Res.* **2017**, *2*, 1–7. (In Chinese)
57. Miller, G.L. Use of dinitrosalicylic acid reagent for determination of reducing sugar. *Anal. Chem.* **1959**, *31*, 426–428. [CrossRef]
58. State Environmental Protection Administration of China (SEPAC). *The Methods for Water and Wastewater Monitoring and Analysis*, 4th ed.; China Environmental Science Press: Beijing, China, 2002.
59. APHA. *Standard Methods for the Examination of Water and Wastewater*, 20th ed.; American Public Health Association: Washington, DC, USA, 1998.
60. Goering, H.K.; Van-Soest, P.J. *Agricultural handbook No. 379. Forage fiber analyses, apparatus, reagents, procedures and some applications*; U.S. Department of Agriculture: Washington, DC, USA, 1970.
61. Overend, R.P.; Chornet, E.; Gascoigne, J.A. Fractionation of lignocellulosics by steam-aqueous pretreatments [and discussion]. *Philos. Trans. R. Soc. Lond. A* **1987**, *321*, 523–536. [CrossRef]
62. Pedersen, M.; Meyer, A.S. Lignocellulose pretreatment severity-relating pH to biomatrix opening. *New Biotechnol.* **2010**, *27*, 739–750. [CrossRef]

Sample Availability: Samples of the chemical compounds used in this study are available from the authors.

Article

The Effect of the Chemical Character of Ionic Liquids on Biomass Pre-Treatment and Posterior Enzymatic Hydrolysis

Joana R. Bernardo, Francisco M. Gírio and Rafał M. Łukasik *

Unidade de Bioenergia, Laboratório Nacional de Energia e Geologia, I.P., Estrada do Paço do Lumiar 22, 1649-038 Lisboa, Portugal; joana.bernardo@lneg.pt (J.R.B.); francisco.girio@lneg.pt (F.M.G.)
* Correspondence: rafal.lukasik@lneg.pt

Academic Editors: Ivet Ferrer, Cigdem Eskicioglu, Georgia Antonopoulou and Audrey Battimelli
Received: 3 February 2019; Accepted: 20 February 2019; Published: 23 February 2019

Abstract: Ionic liquids have been recognised as interesting solvents applicable in efficient lignocellulosic biomass valorisation, especially in biomass fractionation into individual polymeric components or direct hydrolysis of some biomass fractions. Considering the chemical character of ionic liquids, two different approaches paved the way for the fractionation of biomass. The first strategy integrated a pre-treatment, hydrolysis and conversion of biomass through the employment of hydrogen-bond acidic 1-ethyl-3-methyimidazolim hydrogen sulphate ionic liquid. The second strategy relied on the use of a three-step fractionation process with hydrogen-bond basic 1-ethyl-3-methylimidazolium acetate to produce high purity cellulose, hemicellulose and lignin fractions. The proposed approaches were scrutinised for wheat straw and eucalyptus residues. These different biomasses enabled an understanding that enzymatic hydrolysis yields are dependent on the crystallinity of the pre-treated biomass. The use of acetate based ionic liquid allowed crystalline cellulose I to change to cellulose II and consequently enhanced the glucan to glucose yield to 93.1 ± 4.1 mol% and 82.9 ± 1.2 mol% for wheat straw and eucalyptus, respectively. However, for hydrogen sulphate ionic liquid, the same enzymatic hydrolysis yields were 61.6 ± 0.2 mol% for wheat straw and only 7.9 ± 0.3 mol% for eucalyptus residues. These results demonstrate the importance of both ionic liquid character and biomass type for efficient biomass processing.

Keywords: biomass; valorisation; ionic liquid; crystallinity; enzymatic hydrolysis; pre-treatment

---

1. Introduction

Lignocellulosic biomass is a renewable, sustainable, abundant, and CO_2 neutral alternative to fossil feedstock for a portfolio of fuels, chemicals and materials. Composed of crystalline cellulose nanofibrils embedded in an amorphous matrix of cross-linked lignin and hemicelluloses, lignocellulose shows a natural recalcitrance that impedes enzyme and microbial accessibility, resulting in the relatively low digestibility of raw lignocellulosic materials [1]. Thus, an efficient pre-treatment, and consequently a deconstruction of the lignocellulosic biomass, makes these fractions susceptible for more favourable transformation to value-added products [2,3]. However, many pre-treatment methods require harsh conditions, especially temperature and/or pressure that often result in undesired by-products, which decrease the sugar yields and inhibit enzymatic hydrolysis and further bioconversion [2].

In recent years, ionic liquids (ILs) have gained increasing interest for biomass processing due to their capacity to dissolve lignocellulosic biomass by an effective disruption of the complex network of noncovalent interactions between carbohydrates and lignin [4–6]. A main function of IL in lignocellulosic biomass pre-treatment is the modification the fibrillary structure of cell walls in order to: (i) decrease cellulose crystallinity, (ii) increase cellulose surface accessibility by the removal of lignin and/or hemicellulose, and (iii) promotion of a swelling effect of the biomass [7]. Imidazolium-based ILs are among the most extensively studied ILs and have demonstrated that either a cation or an anion is of considerable importance in biomass processing [8,9]. Swatloski et al. showed that a high concentration of chloride anions is effective in breaking down the hydrogen-bond network of cellulose [10]. A similar effect was reported for the acetate ([OAc]) anion, which was demonstrated to be efficient in the dissolution of lignocellulosic biomass [11]. It was reported that a key reason for this was the high hydrogen bond acceptor capacity (β) of the [OAc] anion ($\beta = 1.201$) in comparison to previously mentioned chloride anion ($\beta = 0.83$) [12]. Due to this, 1-ethyl-3-methylimidazolium acetate IL is confirmed to be one of the best and is one of the most commonly used [13] ILs, able to dissolve a large variety of lignocellulosic biomass and to fractionate it into cellulose- and hemicellulose-rich fractions, as well as to produce high pure lignin [9,14,15].

An alternative approach to biomass processing with ILs is the employment of ILs as both solvent and catalyst. In these processes, ILs hydrolyse mainly polysaccharide fractions without the presence of other catalysts [5,16,17]. Therefore, ILs with an acidic character have demonstrated an ability to selectively hydrolyse hemicellulose [18,19], both cellulose and hemicellulose [20], or lignin [21]. Some of the most commonly used ILs in such approaches are those based on hydrogen sulphate ([HSO$_4$]). They are able to catalyse a selective hemicellulose hydrolysis [17–19]. Furthermore, [HSO$_4$]-based ILs have been increasingly used because of their acidic properties and due to their low cost when compared to other ILs [22]. The use of acidic IL, e.g., 1-butyl-3-methylimidazolium hydrogen sulphate, [bmim][HSO$_4$], allowed achievement of up to 90% fermentable glucose after enzymatic saccharification of the cellulose-rich *Miscanthus* pulp [18]. The same IL was also reported as being able to hydrolyse and to convert the hemicellulose fraction of wheat straw with no additional catalyst used [17]. The pre-treatment with this IL produced a liquor enriched in hemicellulosic sugars, furans and organic acids, and a solid fraction constituted mainly of cellulose and lignin. Furthermore, water was confirmed to have an influence on the equilibrium of the hemicellulose hydrolysis. The increase of the water content close to 10% (w/w) in the reaction system disfavoured xylose dehydration, and thus allowed the production of hemicellulose-derived monosaccharides to increase significantly [19].

Building on previous works about the processing of biomass with hydrogen-bond basic ([emim][OAc]) [11] and hydrogen-bond acidic ([emim][HSO$_4$]) [17,19] ILs, this work aimed to demonstrate the importance of biomass type on the efficiency of the biomass pre-treatment, as well as on the efficiency of subsequent enzymatic hydrolysis. This was examined using two very distinct types of biomasses, namely, herbaceous (wheat straw) and hardwood *Eucalyptus globulus*, allowing the elucidation of changes observed in the chemical structure of the biomass, cellulose crystallinity and consequently the effects of these on the cellulose-rich pulp hydrolysability.

2. Results and Discussion

2.1. Hydrogen-Bond Acidic IL

2.1.1. Biomass Pre-Treatment with [emim][HSO$_4$]

The first methodology used focused on the biomass pre-treatment with [emim][HSO$_4$]. As stated above, this approach allowed the integration of biomass pre-treatment, hydrolysis and conversion in a single-step process. The acidic character of the [HSO$_4$] anion of IL, promoted a selective hydrolysis of the hemicellulose fraction, and the resulting products (mainly pentoses and furfural) were kept in the liquid phase. Both biomasses, wheat straw and eucalyptus residues, were subject to processing with [emim][HSO$_4$] IL at 140 °C for 90 min at varied [emim][HSO$_4$]/H$_2$O mass ratio and with fixed 10 wt.% of dry biomass in the reaction mixture.

The pentose and furfural yields obtained in these trials are depicted in Figure 1.

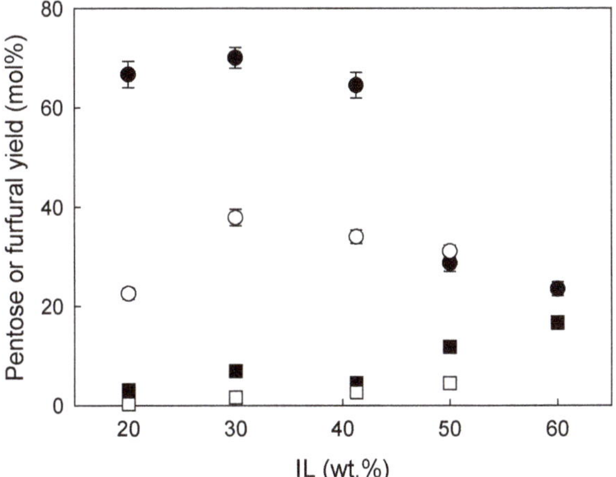

Figure 1. The yields of (●○) pentose (sum of arabinose and xylose) or (■□) furfural as a function of ionic liquid (IL) concentration (wt.%) obtained from pre-treatment of wheat straw (filled symbols) and eucalyptus residues (open symbols) performed at 140 °C for 90 min.

As can be seen, at IL concentration of 30 wt.%, the pentose yields peaked for both biomasses. For eucalyptus residues or wheat straw, a further increase in the IL concentration was demonstrated to have a negative effect on the pentose yield as it was counterbalanced by predominant production of furfural. These results are not surprising since a high performance of acidic ILs towards furfural is often reported in the literature [23–25]. One of the reasons for this is that pentose conversion to furfural is a dehydration reaction; therefore, a higher concentration of IL, or in other words a lower concentration of water, may disturb the equilibrium existing in the system promoting the dehydration of pentoses towards furfural production. On the other hand, the presence of water in the system allowed protection of pentoses from dehydration, and consequently, it was possible to obtain a high pentose yield for IL concentration below 30 wt.%. Although this observation is valid for either wheat straw or eucalyptus residues, the yield of hemicellulose hydrolysis, as well as pentose and furfural yields, seems to be strongly dependent on the nature of the biomass. As can be seen in Figure 1, in the best conditions, i.e., 30 wt.% IL, pre-treatment of the eucalyptus residues with [emim][HSO$_4$] allowed achievement of only 37.9 ± 1.7 mol% pentoses, while for wheat straw the maximum pentose concentration was almost double, i.e., 70.1 ± 0.5 mol%. This difference is also reflected in the composition of pre-treated solids and the solid yields. The latter was very high for eucalyptus residues and varied between 81 and 85 wt.%, while for pre-treated wheat straw it oscillated between 68 and 75 wt.%. For both biomasses, the lowest solid yields were observed for solids obtained following pre-treatment with an IL concentration of 30 wt.%. This is a direct consequence of the most pronounced hydrolysis of hemicellulose. Further increase in the IL concentration increased the solid yield for wheat straw up to 74.9 ± 3.0 wt.%. This unexpected increase of the solid yield with an increase in the pre-treatment intensity might be justified by the formation of pseudo-lignins, also called humins [26]. As pseudo-lignins are quantified gravimetrically [27], they contribute to an increase in the lignin content detected, which is clearly visible in Figure 2. For two the highest IL concentrations tested, the lignins recovered exceeded 100 wt.% of the lignin present in the native biomass. This, in turn may confirm this hypothesis.

Similar effects were not observed for eucalyptus residues, for which either solid yield appeared to be constant or the lignin content in the solids pre-treated at various IL concentrations did not change, as demonstrated in Figure 3.

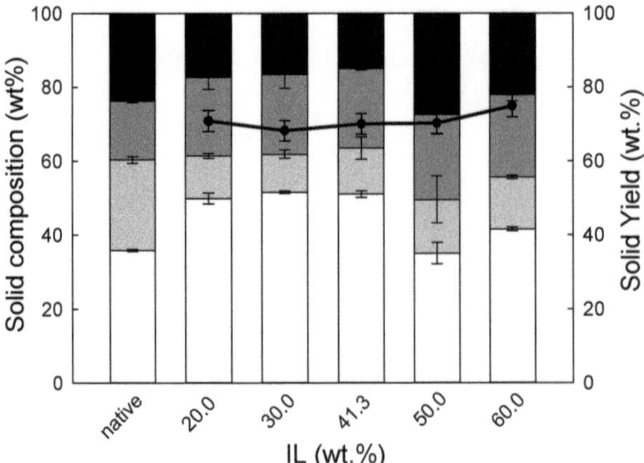

Figure 2. The solid phase composition (white bar–cellulose; light grey bar–hemicellulose; dark grey bar–Klason lignin; black bar–others) of wheat straw pre-treatment at 140 °C and 90 min with [emim][HSO$_4$] at various % IL. The solid line represents the solid yield (wt.%) of pre-treated solids recovered after pre-treatment.

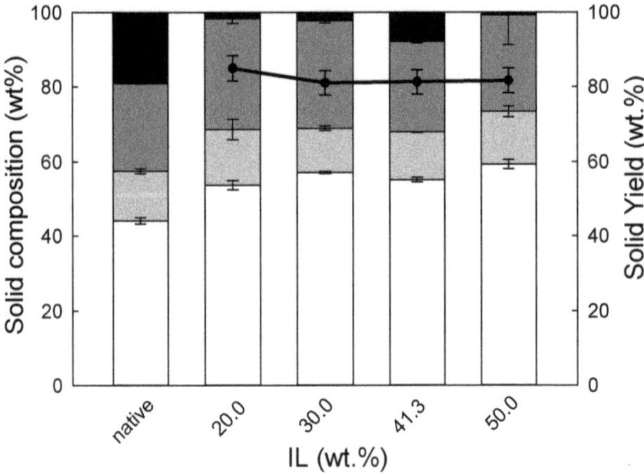

Figure 3. The solid phase composition (white bar–cellulose; light grey bar–hemicellulose; dark grey bar–Klason lignin; black bar–others) of eucalyptus residues pre-treatment at 140 °C and 90 min with [emim][HSO$_4$] at various % IL. The solid line represents the solid yield (wt.%) of pre-treated solids recovered after pre-treatment.

The aforementioned analysis of the hemicellulose hydrolysis also reflects the macromolecular composition of the pre-treated leftovers of both biomasses. It is clear that the IL pre-treatment induced the reduction of the hemicellulose fraction in comparison to native biomasses, as xylan was found in amounts not exceeding 15 wt.% of the pre-treated biomasses. Consequently, the major components of the solids produced were glucan, followed by lignin, as already discussed. This indicates that contrary to a great performance by aqueous [emim][HSO$_4$] solution in processing hemicellulose, a cellulose hydrolysis in this catalytic system seems to be very inefficient. For IL concentrations up to the studied 41.3 wt.%, the cellulose yield was kept constant, but in case of wheat straw, for higher IL concentrations (>50 wt.%), the cellulose content started to diminish. This demonstrates that, only under these conditions, the cellulose fraction of wheat straw was also susceptible to hydrolysis, and for pre-treatment with IL concentration of 60 wt.% as much as 20 wt.% of cellulose, in comparison to native biomass, was removed. In the case of eucalyptus residues, in the range of IL concentrations studied, 20 wt.% of IL already allowed the removal of about 1/3 of hemicellulose and further increases in the IL concentration in the reaction mixture resulted in no further significant changes in the composition of the pre-treated solids, as can be seen in Figure 3. These results confirm again that although ILs are capable of processing various types of herbaceous and woody biomass as a single feedstock, the conditions for efficient pre-treatment are dependent upon the biomass type, with softwoods or hardwoods recognised as the most difficult to process when compared to herbaceous biomass [6]. Indeed, Xu et al. found that the eucalyptus residues treatment catalysed by 0.5% (v·v^{-1}) [bmim][HSO$_4$] yielded only $\approx$ 25 mol% of pentose at 190 °C [28]. Other reports in the literature depict the use of the acidic ILs in the pre-treatment of biomass as well. For example, Li et al. demonstrated acidic IL to be an efficient system for the hydrolysis of lignocellulosic materials. They obtained total reducing sugars (TRS) yields of 23% and 15% from corn stalk after only 5 min of the reaction at 100 °C in [bmim][HSO$_4$] and [C$_4$SO$_3$Hmim][HSO$_4$], respectively. On the other hand, longer reaction times provided lower TRS yields, suggesting that these strongly acidic ILs resulted in the promotion of more advanced degradation of polysaccharide fraction at the pre-treatment step [20]. Other work showed that in case of *Miscanthus*, pre-treatment at a higher temperature (120 °C) and for 4 h with [bmim][HSO$_4$] and 20 vol.% (17 wt.%) of H$_2$O content, close to 16% of hemicellulose sugar monomers were obtained [18]. In a different study, Carvalho et al. showed that the same IL with 9.22 wt.% H$_2$O content in the pre-treatment of wheat straw (125 °C and 82.1 min) allowed a 40.1 mol% pentose yield to be obtained [19]. Thus, comparing the data in the literature to the data presented in this work, it can be stated that a higher yield of pentoses was selectively achieved without excessive amounts of IL being present in the system.

2.1.2. Enzymatic Hydrolysis of Pre-Treated Solids

The efficiency of pre-treatment of wheat straw and eucalyptus residues by [emim][HSO$_4$] was evaluated by enzymatic hydrolysis. The enzymatic hydrolysis yield of pre-treated wheat straw is presented in Figure 4.

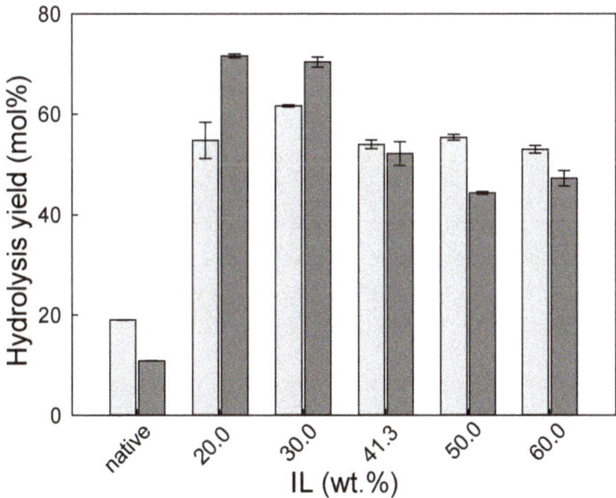

Figure 4. The enzymatic hydrolysis yield (glucan to glucose–light grey bar; and xylan to xylose–dark grey bar) of wheat straw pre-treated solids as a function of IL concentration used in the pre-treatment. The enzymatic hydrolysis yield for native wheat straw is presented for comparison.

As can be observed in Figure 4, the pre-treatment of wheat straw with [emim][HSO$_4$] allowed the enzymatic hydrolysis of cellulose to glucose to increase by 3-fold in comparison to native biomass. Interestingly, the enzymatic hydrolysis yield of cellulose (light grey bars) changed only slightly with the IL concentration used in pre-treatment and varied from 53.0 ± 0.7 mol.% to 61.6 ± 0.2 mol% for 60 and 30 wt.% of IL, respectively. On the other hand, the xylan hydrolysis yield decrease was more pronounced, with an increase of IL concentration from 71.7 ± 0.4 mol% to 47.3 ± 1.5 mol% for 20 and 60 wt.% of IL concentration, respectively. The reason for this might be that with an increase in the reaction intensity, more hemicellulose was extracted from the biomass, and consequently less hemicellulose accessible for enzymatic attack was present in the pre-treated solid. In addition, the formation of pseudo-humins may interfere with the accessibility of enzymes to hemicellulose, and lignin presence affects an enzymatic hydrolysis because of unproductive and irreversible cellulase enzyme adsorption on lignin [29].

Analogous to wheat straw, pre-treatment of eucalyptus residues with [emim][HSO$_4$] improved the enzymatic hydrolysability in comparison to native biomass as can be seen in Figure 5.

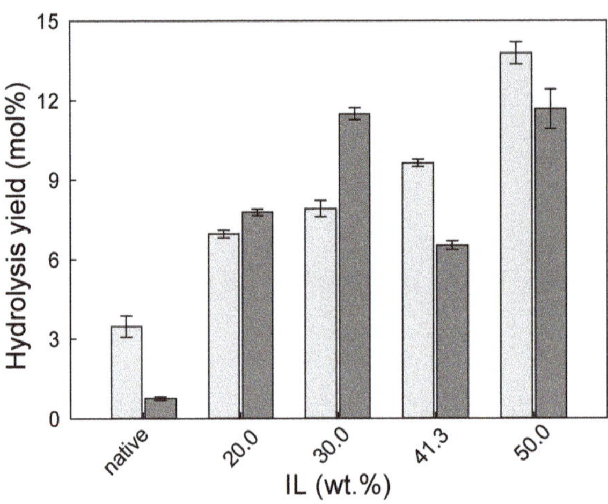

Figure 5. The enzymatic hydrolysis yield (glucan to glucose–light grey bar; and xylan to xylose–dark grey bar) of eucalyptus residues pre-treated solids as a function of IL concentration used in the pre-treatment. The enzymatic hydrolysis yield for native eucalyptus residues is presented for comparison.

Furthermore, for eucalyptus residues pre-treated solids, an increase of IL concentration had a generally positive effect on the yield of enzymatic hydrolysis. For example, an enhancement of glucan to glucose yield was found with an increase of IL concentration; however, the hydrolysis yields detected were very low and did not exceed 14 mol%, i.e., 4.5-fold lower than observed for wheat straw. In previous work, Xu et al. verified that the cellulose-rich solids from eucalyptus residues after [bmim][HSO$_4$]-catalysed hydrothermal microwave pre-treatment, allowed the achievement of greater glucose conversion yield (89.2%). However, this was only possible when temperatures as high as 200 °C were used in pre-treatment and enzymatic hydrolysis was performed at 2 w/v% of substrate concentrations and 20 FPU/g substrate after 72 h [28]. Brand et al. demonstrated that the pre-treatment time has a significant effect on the enzymatic saccharification [18]. The pre-treatment of *Miscanthus* with [bmim][HSO$_4$] at 120 °C for 8 h resulted in a solid which produced ~80% glucose and ~30% hemicellulose release. It is noteworthy that enzymatic saccharification was performed according to very favourable NREL conditions for enzymatic hydrolysis [30].

These very different results for wheat straw and eucalyptus residues drove us to employ the approach with hydrogen-bond basic IL, namely [emim][OAc].

2.2. Hydrogen-Bond Basic IL

2.2.1. Biomass Pre-Treatment with [emim][OAc]

The biomass pre-treatment with hydrogen-bond basic [emim][OAc] IL allowed biomass dissolution and fractionation into cellulose-, hemicellulose- and lignin-rich fractions. For this purpose, pre-treatment of wheat straw or eucalyptus residues in [emim][OAc] at 120 and 140 °C, for 2 h at a biomass/IL ratio of 1/20 (*w/w*) was examined on the basis of the methodology presented in the literature [11]. The composition of the obtained solids is given in Figures 6 and 7.

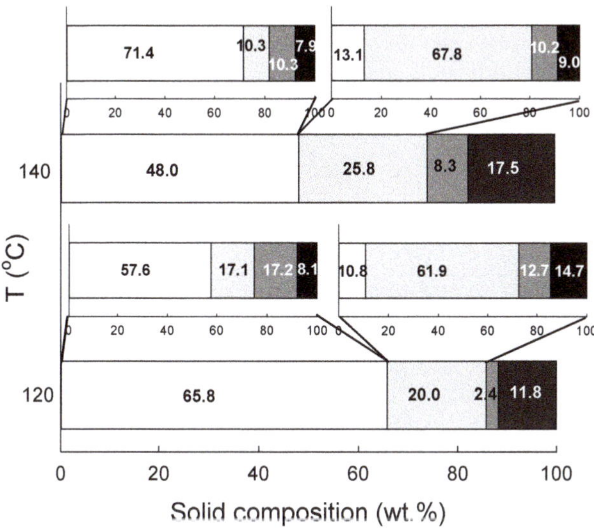

Figure 6. The cellulose- (white bar), hemicellulose- (light grey bar), and Klason lignin-rich (dark grey bar) fractions obtained from wheat straw pre-treated with [emim][OAc] at 120 °C and 140 °C and 2 h. Inserts present the composition of cellulose- and hemicellulose-rich fractions (the same colours were used to differentiate each individual component). Numbers presented in the figure indicate the composition (expressed in wt.%). For exact values, refer to Table S1 in the Electronic Supplementary Information (ESI). Black bars correspond to other non-identified components and were calculated as the difference between biomass used for process or characterised fraction and the cellulose, hemicellulose and lignin determined.

The results obtained demonstrate that temperature has an effect on the selectivity of the fractionation because, although a significantly lower amount of cellulose-rich fraction was obtained at 140 °C in comparison to 120 °C (48.0 wt.% vs. 65.8 wt.% for 140 and 120 °C, respectively), it contained more cellulose, i.e., 57.6 ± 0.3 wt.% vs. 71.4 ± 0.6 wt.% for 120 and 140 °C, respectively. However, as can be seen in Figure 6, a major reason for this was insufficient fractionation of hemicellulose and lignin because the cellulose-rich sample obtained at 120 °C still contained 17.1 ± 3.4 and 17.2 ± 1.0 wt.% of hemicellulose and lignin, respectively. The same sample obtained at 140 °C had much lower hemicellulose and lignin content with only 10.3 ± 1.8 wt.% of each. These relevant differences in more selective fractionation of biomass achieved at higher temperatures also found confirmation in the hemicellulose-rich sample. Contrary to the cellulose-rich sample, the hemicellulose-rich fraction obtained at 140 °C contained more solid, i.e., 25.8 wt.% in comparison to 20.0 wt.% for 120 °C, and this solid was enriched in hemicellulose as it encompassed 67.8 ± 1.6 wt.% of hemicellulose, while at 120 °C it was only 61.9 ± 1.5 wt.%. Consequently, it was predominantly counterbalanced by a difference in lignin and other component contents found in both hemicellulose-rich solids. Interestingly, the lignin-rich fraction obtained at 120 °C was negligible (2.4 wt.%), while that produced at 140 °C was more than three times higher and was equal to 8.3 wt.%. This again clearly indicates that a higher temperature was more effective in the fractionation of wheat straw and allowed the production of cellulose-, hemicellulose- and lignin-rich fractions characterised by higher purity.

Labbé et al. also concluded that at high temperatures, [emim][OAc] is able to cleave the acetyl groups covalently attached, mostly to the hemicellulose component of yellow poplar [31]. Therefore, at a higher temperature this IL can effectively disrupt the carbohydrate–lignin linkages favouring hemicellulose release.

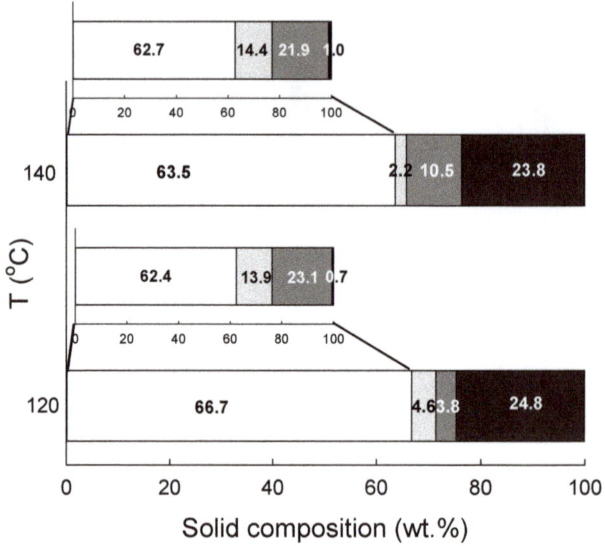

Figure 7. The cellulose- (white bar), hemicellulose- (light grey bar), and Klason lignin-rich (dark grey bar) fractions obtained from eucalyptus residues pre-treated with [emim][OAc] at 120 °C and 140 °C and 2 h. Inserts present the composition of cellulose- and hemicellulose-rich fractions (the same colours were used to differentiate each individual component). Numbers presented in the figure indicate the composition (expressed in wt.%). For exact values, refer to Table S2 in the ESI. Black bars correspond to other non-identified components and were calculated as the difference between biomass used for process or characterised fraction and cellulose, hemicellulose and lignin.

As demonstrated in Figure 7, eucalyptus residues processing with [emim][OAc] at different temperature has almost no effect on the selectivity of biomass fractionation. An increase of temperature by 20 °C from 120 to 140 °C, similar to wheat straw, reduced the amount of cellulose-rich sample recovered from 66.7 wt.% to 63.5 wt.% and enhanced its purity by less than 1 wt.%. Although the trends are the same as those observed for wheat straw, the changes are negligible when compared to those presented in Figure 6. Hemicellulose-rich fractions were recovered in very small quantities, which made their characterisation impossible. Consequently, it can be concluded that although biomass fractionation to cellulose-, hemicellulose- and lignin-rich fraction occurred the eucalyptus residues make the process with hydrogen-bond basic IL less efficient than is the case for herbaceous biomass and others presented in the literature [9,11,32,33].

2.2.2. Enzymatic Hydrolysis of Pre-Treated Solids

Regardless the efficiency of both biomass fractionation processes, cellulose-rich samples were subject to the enzymatic hydrolysis according to the method presented in the experimental section. The obtained results are presented in Figure 8 and are compared to those achieved for hydrogen-bond acidic IL.

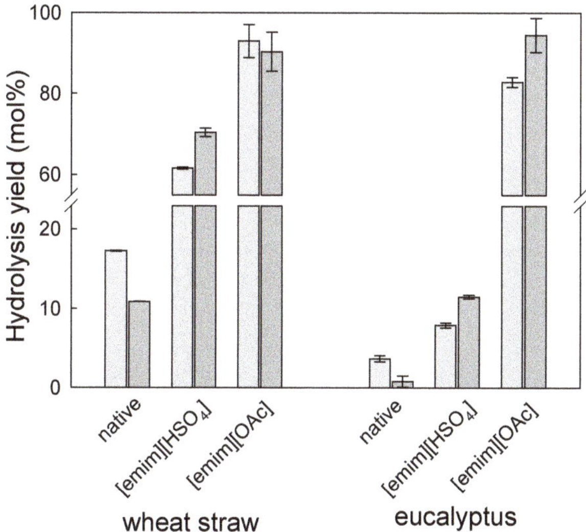

Figure 8. The enzymatic hydrolysis yield (glucan to glucose–light grey bar; and xylan to xylose–dark grey bar) of pre-treated wheat straw and eucalyptus residues solids produced in processes at 140 °C with aqueous [emim][HSO$_4$] (30 wt.%, 90 min) and [emim][OAc] (2 h). The enzymatic hydrolysis yield for native biomasses is presented for comparison.

The results obtained demonstrate clearly that pre-treatment of either wheat straw or eucalyptus residues with [emim][OAc] dramatically enhanced the enzymatic hydrolysis yields. For example, for the cellulose-rich sample of wheat straw obtained from pre-treatment with [emim][OAc], the glucan to glucose yield was as high as 93.1 ± 1.1 mol%, while for the same biomass pre treated with [emim][HSO$_4$] a maximum hydrolysis yield of 61.6 ± 0.2 mol% was achieved. Although this 50% enhancement is remarkable, the same process for eucalyptus residues demonstrated an even more astonishing improvement of glucan to glucose hydrolysis yield. As was shown above, the pre-treated solids obtained after reaction with [emim][HSO$_4$] allowed achievement of a very modest enzymatic hydrolysis yield of 7.9 ± 0.3 mol%, but the enzymatic hydrolysis yield of cellulose-rich solid obtained from the pre-treatment of eucalyptus residues with [emim][OAc] was as high as 82.9 ± 1.2 mol%. In terms of potential valorisation of cellulose present in the native biomass, a switch from [emim][HSO$_4$] to [emim][OAc] was also demonstrated to be a better choice for both biomasses. The yield of glucose released after both steps (pre-treatment with [emim][OAc] and posterior enzymatic hydrolysis) was as high as 91.4 mol% and 74.9 mol% for wheat straw and eucalyptus residues, respectively. The same yield for both biomasses pre-treated with [emim][HSO$_4$] with 30 wt.% of IL was only 62.3 mol% and 7.9 mol% for wheat straw and eucalyptus, respectively.

These results indicate that a change of the IL used, from hydrogen-bond acidic to hydrogen-bond basic IL, promoted a significant change in the pre-treated solids, as the enzymatic hydrolysis yield increased by more than 1000%. The results in the literature also confirm similar, although not such pronounced, differences. For example, Bian et al. studied the effect of IL pre-treatment on enzymatic hydrolysis of cellulose as a function of chemical and physical structure changes [34]. In a case of cellulose isolated from sugarcane bagasse subjected to IL ([emim][OAc]), dissolution at a mild temperature (90 °C) followed by a solid regeneration in water, resulted in an increase in glucose content from 80.0–83.3 wt.% to 91.6–92.8 wt.%, a decrease in the degree of polymerisation from 974–1039 units to 511–521 units, a transformation from cellulose I to cellulose II, and an increase of surface area during the pre-treatment. Such cellulose was subsequently hydrolysed by commercial cellulases with 2 w/v% cellulose substrate and enzyme loadings of 35 FPU/g (cellulase) and 40 CbU (β-glucosidase) in relation to the dry weight of cellulose substrates, and allowed achievement of a high glucose conversion yield of 95.2 mol%. These results suggest that pre-treatment led to an effective disruption of cellulose favouring enzyme hydrolysis. Torr et al. also observed an improvement of the glucan to glucose yield after saccharification for 72 h performed at biomass loading of 1.5% (*w/v*) and Celluclast 1.5 L and Novozymes 188 with 40 FPUs/g glucan and β-glucosidades of 50 IU/g glucan, from 5 mol% in the untreated pine wood to 84 mol% in wood pre-treated with [emim][OAc]. The analysis of the substrates revealed that the most important change brought by the pre-treatment was an increase in the accessible surface area. In this case, the delignification was not observed, and loss of cellulose crystallinity only occurred for the highest intensity pre-treatments [35]. Therefore, to verify this, changes in the morphology of pre-treated solid materials were studied using X-ray diffraction.

2.3. Morphological Analysis of Pre-Treated Solids

Crystalline cellulose is the most organised form of cellulose in the biomass [36]. Also, crystallinity of cellulose has been reported as one of the most relevant factors influencing the efficiency of enzymatic hydrolysis [37]. X-ray diffraction permits measurement of the crystallinity of the material as a whole, because it demonstrates either crystalline (organised) or disordered components of the biomass, namely, amorphous cellulose, hemicellulose and lignin [38]. ILs have been shown to affect cellulose crystallinity during pre-treatment and consequently enhancing the enzymatic hydrolysis [39]. Therefore, the effect of different pre-treatment approaches on crystallinity, which could justify the efficiency of enzymatic hydrolysis, was also tested in this work. The results obtained for wheat straw and eucalyptus residues are depicted in Figures 9 and 10, respectively.

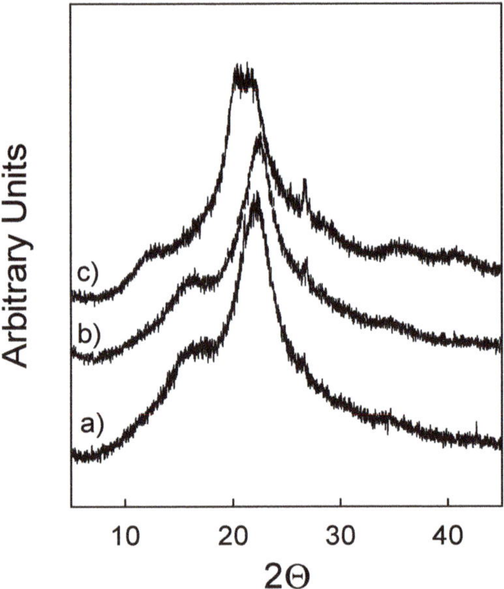

Figure 9. Powder XRD patterns of: (**a**) native wheat straw, (**b**) pre-treated with [emim][HSO₄], and (**c**) pre-treated with [emim][OAc] IL.

Figures 9a and 10a show the diffraction patterns of untreated wheat straw and eucalyptus residues. The main signal can be observed at 22.3° for wheat straw and at 22.5° for eucalyptus residue samples. This signal indicates the distance between hydrogen-bonded sheets in cellulose I and corresponds to the (200) lattice plane. For both biomasses, the second main band observed is a broad signal registered at $2\theta = 16.7°$ and corresponds to overlapping signals of (101) and (10-1) planes. The "valley" at 18.1° is associated with an amorphous region in the biomass and includes disordered cellulose, hemicellulose and lignin. The third, barely noticeable, signal at 34.5° corresponds to one-quarter of the length of one cellobiose unit and arises from ordering along the fibre direction.

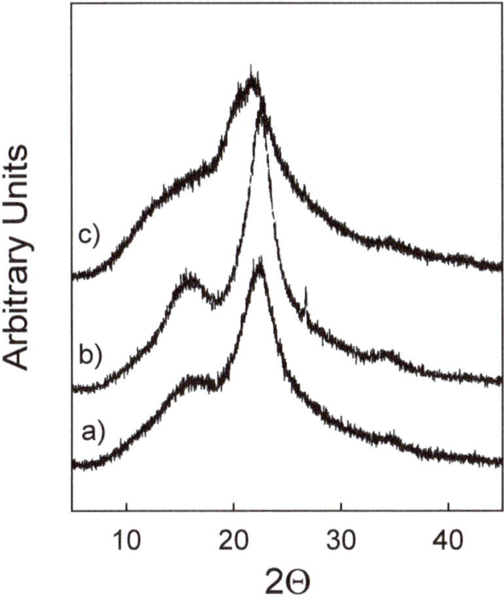

Figure 10. Powder XRD patterns of: (**a**) native eucalyptus residues, (**b**) pre-treated with [emim][HSO$_4$] and (**c**) pre-treated with [emim][OAc] IL.

The XRD patterns of wheat straw and eucalyptus residues pre-treated with [emim][HSO$_4$] indicated in Figures 9b and 10b as diffractograms, respectively, mirror the diffractograms observed for native biomasses (Figures 9a and 10a). This confirms that hydrogen-bond acidic ILs, such as [emim][HSO$_4$] tested in this work, do not induce any qualitative changes in the pre-treated samples. The same cannot be said about the cellulose-rich solids produced in the [emim][OAc] pre-treatment. The most dominant change is a disappearance of signal at 22.5°, and the presence of a broad asymmetric signal consisting of a doublet at 20° and 21.7°, as demonstrated in Figure 9c. Similarly, the broad signal at 16° disappeared and was substituted with a new signal that emerged at 12.1°, as can be seen in the same figure. These changes are characteristic of a transformation of cellulose I to cellulose II and they are the most visible for wheat straw pre-treated samples. For the cellulose-rich sample of eucalyptus residues, similar changes in the diffractogram are visible although they are less notable. For example, as shown in Figure 10c, a broad signal at 20–22° can be found. Additionally, a signal at 16° became very wide and was transformed into the arm of the main signal. These changes in signals confirm the alteration in the cellulose organisation similar to what was observed for wheat straw. At the same time, as these signals are still not complete defined, contrary to what was observed for the wheat straw cellulose-rich sample, it may indicate that in case of eucalyptus residues, the transformation of crystalline cellulose is less effective and may require longer pre-treatment time. Regardless of the reasons, the observed changes in the XRD patterns justify the fact that even partial alteration of biomass crystallinity is sufficient to promote more efficient enzymatic hydrolysis, as demonstrated in this work. Similar results were presented in the literature, where it was confirmed that cellulose II is more readily digested than cellulose I. It has been argued that the van der Waals interactions between hydrogen-bonded sheets in cellulose I are stronger than in cellulose II and that they act as the main factor in resisting cellulose hydrolysis [40]. Li et al. compared the pre-treatment of switchgrass with

[emim][OAc] and with a 1.2% (*w/w*) dilute sulphuric acid [41]. For both untreated and dilute acid pre-treated switchgrass samples, little or no change in cellulose crystallinity was observed, but for the sample obtained after pre-treatment with [emim][OAc] the crystallinity was altered significantly, with a structural transformation from cellulose I to cellulose II observed. This, in turn, promoted better performance of enzymatic hydrolysis.

3. Materials and Methods

3.1. Materials

The wheat straw sample was delivered by ECN (Energy Research Centre of the Netherlands,), from Petten, the Netherlands. The eucalyptus residues were kindly provided by The Navigator Company from their paper mill in Cacia, Portugal. The wheat straw and eucalyptus residues moisture content was found to be 9.8 and 8.4% (*w/w*), respectively and was determined using an AMB-50 moisture analyser.

Both feedstocks were ground with a knife mill IKA® WERKE, MF 10 basic (Staufen, Germany) to particles smaller than 0.5 mm, homogenised in a defined lot, and stored in plastic containers at room temperature prior to further use.

The [emim][HSO$_4$] IL (99 wt.% of purity) was purchased from Iolitec GmBH—Heilbronn, Germany and was used in reactions without any previous purification. The water content in IL was determined by a volumetric Karl-Fischer titration and was 3796 ppm. The [emim][OAc] with stated purity >95% was purchased from Iolitec GmbH—Heilbronn, Germany. Prior to use in the pre-treatment, this IL was subject to drying under vacuum (0.1 Pa) at room temperature for at least 24 h. The water content in this IL was 2800 ppm, determined by a volumetric Karl-Fischer titration as for the other IL.

In pre-treatment experiments, the following reagents were used: 0.1 M and 3% (*w/w*) NaOH aqueous solutions prepared from NaOH pellets (99% purity) supplied by Eka chemicals/ Akzonobel–Bohus, Sweden. The aqueous solutions of 1 M and 4 M HCl were prepared from fuming HCl 37% (*w/w*) with a purity grade for analysis (Merck—Darmstadt, Germany). Ethanol 96% (*v/v*) and acetonitrile of HPLC-gradient purity for analysis (Carlo Erba Group—Arese, Italy) and acetone (98% purity) were supplied by Valente & Ribeiro, Lda.—Belas, Portugal. For the preparation of NaOH and HCl solutions, distilled water (17 M Ω cm^{-1}) and ultrapure water (18.2 MΩ cm^{-1}) both produced by the PURELAB Classic of Elga system were used.

For filtration, paper and glass microfibre filters (Whatman GE Healthcare Bio-Sciences Corp.—Piscataway, NJ, USA) and nylon filters, 0.45 lm HNPW (Merck Millipore—Billerica, MA, USA) were used.

Glucose ($\geq$98 wt.%, Merck—Darmstadt, Germany), xylose ($\geq$98 wt.%, Merck, Germany), arabinose ($\geq$98 wt.%, Merck, Germany), furfural (furan-2-carbaldehyde) (99 wt.%, Sigma-Aldrich—Steinheim, Germany), 5-hydroxymethylfurfural (5-(hydroxymethyl)-2-furaldehyde) (99 wt.%, Sigma-Aldrich, Germany) and acetic acid (glacial, 99.8 wt.% Merck—Darmstadt, Germany) were used for the qualitative and quantitative HPLC analyses of the obtained liquids and solids. Sulphuric acid (96 wt.%, Panreac—Castellar del Vallès, Spain) was used to prepare mobile phase for HPLC analyses (5 mM sulphuric acid).

For the enzymatic hydrolysis assays, 0.1 M sodium citrate buffer (pH 4.8) prepared from citric acid monohydrate (99.7% purity) and tris-sodium citrate (>99% purity) both from VWR International Ltd.—Leicester, England and a 2 wt.% sodium azide solution were used. Celli®CTec2 enzyme solution kindly provided by Novozymes A/S Europe—Bagsvaerd, Denmark was employed in the enzymatic reaction.

3.2. Biomass and Pre-Treated Solid Characterisation

Both biomasses and pre-treated solids were characterised to determine the total moisture [42], total lignin and polysaccharide contents [27]. Acid-insoluble lignin was determined gravimetrically, while acid-soluble lignin was established using UV spectrophotometry. The content of glucan and hemicelluloses (xylan, arabinan, and acetyl groups) was determined using high performance liquid chromatography (HPLC) equipment. Furthermore, for native biomasses, total extractives, ash and protein contents were determined according to standard methods, namely: NREL/TP-510-42619 [43], NREL/TP-510-42622 [44] and ISO 8968-1:2014 [45], respectively.

The composition of both biomasses is presented in Table 1.

Table 1. Wheat straw and eucalyptus residues macromolecular composition.

Components (Dry Weight %)	Wheat Straw	Eucalyptus Residues
Glucan	35.9 ± 0.3	44.1 ± 0.9
Hemicellulose	26.7	19.6
Xylan	22.1 ± 0.6	15.7 ± 0.2
Arabinosyl group	2.0 ± 0.7	0.5 ± 0.1
Acetyl group	2.6 ± 0.9	3.4 ± 0.9
Lignin	16.7	33.8
Acid-insoluble	15.5 ± 0.4	26.4 ± 0.1
Acid-soluble	1.2 ± 0.1	7.4 ± 0.1
Ash	11.4 ± 0.1	1.0 ± 0.1
Extractives		
Water	9.4 ± 1.3	3.3 ± 0.4
Water (not ash)	5.1 ± 0.5	0.2 ± 0.0
Ethanol	1.4 ± 0.1	1.5 ± 0.1

3.3. Biomass Processing

3.3.1. Pre-Treatment of Biomass with [emim][HSO$_4$]

All reactions were performed with a 10 wt.% of dry biomass in the reaction mixture. For this purpose, 0.5 g of dry biomass and 4.5 g of aqueous IL solution with different [emim][HSO$_4$]/H$_2$O ratios, were placed into a 15 mL glass vial (Supelco/Sigma-Aldrich, Bellefonte, PA, USA). Next, a vial was placed into the oil bath pre-heated to the desired temperature (140 °C), and reactions were carried out for 90 min, under continuous magnetic stirring. The reaction condition was selected on the basis of previously published work [17]. After reaction, the mixture was cooled to room temperature and approximately 5.0 mL of ultrapure H$_2$O was added to precipitate non-hydrolysed fractions. The resulting mixture was filtered under vacuum using 0.45 µm nylon membrane filters. The liquid phase was collected and stored in a freezer for posterior analysis by HPLC, while recovered solid biomass was washed with 100 mL of ultrapure H$_2$O (in 10 mL portions) to guarantee removal of IL from the precipitated solid. Next, the obtained solid was placed in the oven at 60 °C for 24 h and afterwards was stored at room temperature for 1 h to analyse the dry mass content. The composition of the solid was analysed as presented in Section 2.2.

3.3.2. Pre-Treatment of Biomass with [emim][OAc]

The pre-treatment was performed according to a method presented in the literature [11]. Reactions were performed at two different temperatures (120 or 140 °C) for 2 h with a 5% (*w/w*) biomass/IL ratio in a 15 mL vial. Following the aforementioned procedure presented in the literature, three solid fractions, rich in cellulose, hemicellulose and lignin were obtained. All of them were characterised according to methods presented in Section 2.2.

3.4. Enzymatic Hydrolysis of Solids

The digestibility of pre-treated solids obtained with [emim][HSO$_4$] or [emim][OAc] was evaluated by enzymatic hydrolysis. The assays were performed using 5% (w/v) solids (dry weight basis) concentration in 50 mL vials with 5 mL of 0.05 M sodium citrate buffer (pH 5), prepared from citric acid monohydrate and tris-sodium citrate and 100 μL of a 2 wt.% sodium azide solution to prevent undesired growth of microorganisms. Distilled water was added to reach 5.0 mL taking into account the volume of enzyme added last. The enzyme dosage used was 10% (w/w cellulose) of Celli$^{®}$CTec2 (199.9 FPU·mL^{-1}). The enzymatic hydrolyses were performed in a shaking incubator (Optic Ivymen system—Madrid, Spain) at 180 rpm and 50 °C for 72 h. After hydrolysis, enzymes were inactivated by freezing the samples. To measure monosaccharide content, the hydrolysates were filtered under vacuum using nylon filters (pore size of 0.45 μm) and analysed by HPLC. The glucose and xylose yields were calculated considering the glucan and xylan contents and factors of (162/180) and (132/150) for dehydration, respectively.

3.5. Chemical Analysis

3.5.1. HPLC Analysis

The liquid phases obtained from the pre-treatment of biomass with [emim][HSO$_4$] and enzymatic hydrolyses, as well as from native or pre-treated biomass characterisation were analysed using Agilent 110 series equipment with a Bio-Rad Aminex HPX-87H column (Hercules, CA, USA). Analyses were performed at 65 °C with 5 mmol·L^{-1} H$_2$SO$_4$ used as the mobile phase at a flow rate of 0.6 mL·min^{-1}. The detection was performed using RID (refractive index detector) for monosaccharides (glucose, xylose and arabinose) and acetic acid and DAD (diode array detector) at 280 nm wavelength for furans (furfural and 5-HMF ≡ 5-hydroxymethylfurfural). The quantitative analyses were performed by external calibration using standard solutions.

3.5.2. XRD Measurements

The crystallinity analyses of untreated biomasses and pre-treated materials were performed by X-ray powder diffraction (XRD). For this purpose, a Rigaku Geigerflex D/MAX-III C X-ray powder diffractometer with vertical goniometer, Bragg-Brentano geometry and graphite monochromator was used. The CuKα radiation (λ = 1.5418 Å) at 45 kV and 20 mA was used. The samples were analysed in the range of 2θ from 5° to 50° with increments of 0.02°.

3.6. Experimental Uncertainty

Each weighing was made with a standard uncertainty (u) u(m) of 0.1 mg. All pretreatment experiments were performed with a u(T) of 2 °C. All enzymatic hydrolysis experiments were performed with a u(T) of 0.1 °C. All other experimental errors related to measurements depended on the calibration technique used to quantify the concentrations of products. All reactions (pre-treatments and enzymatic hydrolyses) and analyses were performed in duplicate and results are given as mean values with the corresponding standard deviation.

4. Conclusions

The present work takes a major step towards providing a comparative framework between two IL-type pre-treatments coupled with enzymatic saccharification, in terms of their performance on converting wheat straw and eucalyptus residues to fermentable sugars. The first strategy relied on the processing of biomass with hydrogen-bond acidic IL. This approach allowed integration of biomass pre-treatment, hydrolysis and conversion of biomass in a single-step process. The acidic character of the [HSO$_4$] anion of IL promoted a selective processing of hemicellulose fraction and the resulting products (mainly pentoses and furfural) were kept in a liquid phase. The solids produced